New Security Challenges

Series Editor
George Christou
University of Warwick
Coventry, UK

The last decade has demonstrated that threats to security vary greatly in their causes and manifestations and that they invite interest and demand responses from the social sciences, civil society, and a very broad policy community. In the past, the avoidance of war was the primary objective, but with the end of the Cold War the retention of military defence as the centrepiece of international security agenda became untenable. There has been, therefore, a significant shift in emphasis away from traditional approaches to security to a new agenda that talks of the softer side of security, in terms of human security, economic security, and environmental security. The topical New Security Challenges series reflects this pressing political and research agenda.

More information about this series at
http://www.palgrave.com/gp/series/14732

Mirva Salminen · Gerald Zojer ·
Kamrul Hossain
Editors

Digitalisation and Human Security

A Multi-Disciplinary Approach to Cybersecurity in the European High North

palgrave
macmillan

Editors
Mirva Salminen
Northern Institute for Environmental and Minority Law
Arctic Centre, University of Lapland
Rovaniemi, Finland

Gerald Zojer
Northern Institute for Environmental and Minority Law
Arctic Centre, University of Lapland
Rovaniemi, Finland

Kamrul Hossain
Northern Institute for Environmental and Minority Law
Arctic Centre, University of Lapland
Rovaniemi, Finland

New Security Challenges
ISBN 978-3-030-48069-1 ISBN 978-3-030-48070-7 (eBook)
https://doi.org/10.1007/978-3-030-48070-7

This Palgrave Macmillan imprint is published by the registered company Springer Nature Switzerland AG
The registered company address is: Gewerbestrasse 11, 6330 Cham, Switzerland

Preface

This book is a result of the *Enablement Besides Constraints: Human Security and a Cyber Multi-Disciplinary Framework in the European High North* (ECoHuCy) research project. The project was conducted as part of the "society, integrity and cyber-security" theme within the Nordic Societal Security Programme run by NordForsk—an organisation run under the auspices of the Nordic Council of Ministers that promotes Nordic cooperation on research and research infrastructure. The Northern Institute for Environment and Minority Law (NIEM) at the Arctic Centre of the University of Lapland hosted this three-year project from 2017 to 2019 alongside partners from three other countries, namely, Norway, Sweden and the United Kingdom. Under the leadership of the NIEM, the project participants included UiT—the Arctic University of Norway, the Institute for Security and Development Policy (ISDP) and the University of Swansea.

In the ECoHuCy project, we explored the connection between the notions of human security and cybersecurity. Thematically, this is a relatively under-explored domain within the field of cybersecurity research. Cybersecurity generally refers to the technical aspects of security, highlighting the safe and secure functioning of computer networks, and to external threats to the systems that control cyberspace operations. Although cybersecurity is often viewed as a significant component of states' national security, we found that the human dimensions of it have largely been ignored in the prior academic research. Hence, our aim with the ECoHuCy project was to consider various aspects of human security, in a somewhat redefined form, in light of the developments taking place in relation to digitalisation and the cybersecurity framework. We offered a wider approach to the issues by integrating, among other aspects, human rights and human development frameworks in order to promote a solid understanding of how, in a specific context, regional characteristics and regional developments reinforce both constraints and enablement as they relate to the broader cybersecurity agenda. While we considered the Arctic to be the general point of reference, our main focus was on the European High North, where the harsh climatic conditions, the relatively poor physical infrastructures and the sparse population—of whom the Sámi and other indigenous peoples form groups with distinct identities that still carry out their traditional livelihood activities—coincide and interact with ongoing regional digital developments.

During the early stages of the project, while the participants from all the partner institutions were meeting in a number of workshops and presenting their research within the project, a call for papers was simultaneously announced to attract external scholars working on related studies. Our aim was to prompt the project participants to engage in dialogue with external scholars within the field. We received significant interest from scholars working on various issues concerning cybersecurity, and a good number of papers were submitted to be presented at the project's mid-term conference, which was held in Rovaniemi in May 2018. At the end of the conference, over a dozen high-quality chapters were submitted for potential inclusion in the book stemming from the project, from which rigorous external and internal review processes resulted in the selection of eleven chapters to be included in the present volume. These chapters cover themes such as the interaction between the overarching frameworks of human security and human rights and the multidisciplinary cyber framework, the relationship between digital infrastructure and human security, the influence of digitalisation and the cyber framework in the regional societal and environmental context, as well as the role of electronic governance in the framing of digital citizenship. The authors of the individual chapters have addressed important and relevant issues with a focus on the European High North, which we believe enhances the value of this book. Both the dialogue fostered through the workshops and the blind external review process have helped to heighten the quality of the research. That said, however, the opinions expressed in the individual chapters belong to the respective authors.

The preparation of the final manuscript took almost a year and a half, and it required a lot of effort and patience on the part of all those involved in the process, including the individual authors, workshop participants and external and internal reviewers. We are grateful to all of them. Our special thanks go to Marcin Dymet, who has coordinated the language review process. In addition, we are also grateful to NordForsk and to the Economic and Social Research Council of the United Kingdom for the generous research funds they provided, without which we could not have accomplished this exploratory research project.

Rovaniemi, Finland
December 2019

Mirva Salminen
Gerald Zojer
Kamrul Hossain

Contents

Notes on Contributors

Alexandra L. Carleton has worked in Sydney and London as a lawyer and has written on Polar law since living in the High Arctic in 2013. She has a Bachelor of Science and Laws, a Master of Laws and three years of doctoral research at Cambridge University. She was an advisor to Ashurst in 2017 on professional conflicts of interest and has been a presenter at the Polar Law Symposium since 2016. She has published in a number of global peer-reviewed journals on international law. She is currently undertaking a professional veterinary medical degree with a focus on emerging pathogens in the High Arctic and Antarctica.

Alan Collins is a Professor of International Relations at Swansea University. He secured his Ph.D. from the University of Wales, Aberystwyth where he also held a British Academy Post-Doctoral Fellowship. His research interests include critical security studies, and he is currently conducting research on access to justice in Southeast Asia, and cyber vulnerabilities in nuclear command and control systems.

Igor Datyev is a senior researcher at the Institute for Informatics and Mathematical Modelling of KSC RAS. Candidate of Technical Sciences (2011). He graduated from the Petrozavodsky University with a degree in information systems and technologies (2004). His research interests include multihop wireless networks, simulation modelling and social networking services.

Andrey Fedorov is a senior researcher at the Institute for Informatics and Mathematical Modelling of KSC RAS. Candidate of Technical Sciences (2005). He graduated from the Petrozavodsky University with a degree in information systems and technologies (2000). His research interests include simulation modelling, social networking services, geo-information systems.

Nadezhda Filimonova is a Ph.D. Candidate at the University of Massachusetts Boston. Her research addresses the topics of Arctic governance, Sino-Russian relations in the Arctic, Russian Arctic policy, and Arctic urban climate governance. She holds two Master's Degrees: one in Political Science and International Studies from Uppsala University and the other in International Relations from St. Petersburg State University. Previously, she has worked in Sweden and Russia at both universities and research institutions.

Kristin Smette Gulbrandsen is a Ph.D. student in the Department of Human Geography at Lund University. She is currently researching digitalisation and regional development in northern Norway. Kristin was previously a Research Assistant at Swansea University where she conducted research on digital service provision in sparsely populated areas from a social exclusion perspective. She holds a M.A. in Security studies from Aberystwyth University and a B.A. in International Relations from the University of Leeds.

Liisa Hokkanen is a Ph.D. researcher and a university lecturer on social work at the University of Lapland in Rovaniemi. Her research interests include the empowerment-advocacy approach in social work practice, social work with adults, disability, volunteering, peer support, welfare mix and participatory research.

Kamrul Hossain is a Research Professor and the Director of the Northern Institute for Environmental and Minority Law (NIEM) at the Arctic Centre in the University of Lapland. He has previously served the Faculty of Law of the University of Lapland as the (acting) Professor of Public International Law. He is also an Adjunct Professor of International Law at the University of Lapland. He leads a number of international and national research projects at the Arctic Centre funded, among others, by the Academy of Finland, NordForsk, Nordic Council of Ministers etc. Prof. Hossain has extensively published scholarly articles in the field of international and environmental law as well as in human rights discipline that broadly apply to the Arctic. He had held visiting fellowships at a number of foreign universities including the University of Toronto, Scott Polar Research Institute of the University of Cambridge and the Muroran Institute of Technology in Japan.

Stefan Kirchner is Associate Professor of Arctic Law at the Arctic Centre of the University of Lapland. Prior to joining the Arctic Centre, he practised public and international law in private and government practice in Germany. He has taught human rights and international law at universities in Germany, Lithuania, Finland, Italy, Ukraine and Greenland.

Ville Kivivirta works as postdoctoral researcher of administrative science at Faculty of Social Sciences, University of Lapland. His current research interests relate to the role of ICT in organising and northern change.

Laura Olsén-Ljetoff is a Ph.D. candidate at the University of Lapland. Her research interests focus on indigenous peoples, traditional Sámi livelihoods, indigenous knowledge, human security and human rights. Previously she has conducted research on several matters connected to the rights and situation of Sámi people living in Finland, for example on the rights and multiple discrimination faced by Sámi persons with disabilities and Sámi identifying as sexual and gender minorities.

Mario Portugal-Ramirez is a Ph.D. Candidate in Global Governance and Human Security at the University of Massachusetts Boston. He has studied sociology at Universidad Autonoma Gabriel René Moreno (Bolivia) and holds a Master's degree in sociology for Facultad Latinoamericana de Ciencias Sociales (Flacso) Ecuador. His current research interests are global digital health, digital sociology, and indigenous people's political movements.

Mirva Salminen conducts her doctoral study on digitalisation and cybersecurity in the European High North at the Arctic Centre, University of Lapland. Previously, she has carried out research for several organisations in both public and private sectors, including universities, research institutes and corporations, publishing primarily on cybersecurity and security commercialisation. Mirva holds M.Soc.Sc. degree in

International Politics and B.Soc.Sc. in Political Science, both from the University of Tampere. Parts of her degrees she has completed in the US and in the UK.

Juha Saunavaara is an Assistant Professor at the Hokkaido University Arctic Research Center (Japan). He is also affiliated with the Global Station for Arctic Research, Global Institution for Collaborative Research and Education in Hokkaido University. Saunavaara earned his Ph.D. from the University of Oulu (Finland) and holds a title of Docent at the University of Turku (Finland). His recent research has concentrated on the Arctic submarine communications cable projects, commercial utilisation of the Northern Sea Route, regional development in Hokkaido and northern Scandinavia, and cooperation between Arctic and non-Arctic states and regions.

Arto Selkälä is a University Lecturer in Quantitative Methods in the University of Lapland. His research interests include survey methodology, web surveys, new interactive measurement types and functions in web surveys, survey quality issues and research methodology in psychology and sociology.

Michael Sheehan is Professor of International Relations at Swansea University. He is a specialist on security studies and is the author or editor of twelve books on international security, including, *Arms Control: Theory and Practice* (Basil Blackwell, 1986), *Balance of Power: History and Theory* (Routledge, 1996), *National and International Security* (Ashgate, 2000), *Security: An Analytical Survey* (Lynne Rienner, 2005), and (with Natalie Bormann), *Securing Outer Space* (Routledge, 2009). He is currently engaged in research into security problems facing arctic indigenous peoples.

Maxim Shishaev is a Chief Researcher at Institute for Informatics and Mathematical Modelling of KSC RAS, professor of Murmansk Arctic State University. Doctor of Technical Sciences (2010). His research interests include distributed intelligent information systems, cognitive human-machine interfaces, P2P systems and networking, information support for regional development and others.

Daria Shvets is a Ph.D. Candidate at the Universitat Pompeu Fabra in Barcelona. Born in Siberia, knowing the severe conditions of the north first-hand, she focuses on legal issues of the Arctic with the emphasis on submarine cables, law of the sea and environmental law matters. Daria holds a law degree from the Lomonosov Moscow State University and currently is a visiting researcher at the Arctic Research Center of Universitat Autònoma de Barcelona.

Leena Viinamäki is a principal lecturer (D.Soc.Sc., social policy) in the Lapland University of Applied Sciences. Her research interests relate to educational and labour market citizenship and welfare citizenship from the welfare policy perspective in northern Finland.

Gerald Zojer is a researcher in the Northern Institute for Environmental and Minority Law at the Arctic Centre, and Ph.D. candidate in Political Sciences at the Faculty of Social Sciences, both of the University of Lapland. His research interests are on hegemony theories, sociology of technology studies, digitalisation and political ecology, with a particular focus on the Arctic region. His most recent work relates to the interplays between socioeconomic development in the Arctic with global economic trends and the adaptation to new technologies in the region.

List of Figures

List of Tables

PART I

Introduction

CHAPTER 1

A Human Security Perspective on Cybersecurity in the European High North

Mirva Salminen and Gerald Zojer

Digitalisation is rapidly changing both societies and human activities worldwide. This trend is also observable in the European High North, that is, in the northernmost parts of Finland, Sweden and Norway, as well as in north-western Russia. National and regional policies, as well as commercial endeavours, have been designed to advance digital development and/or to mitigate its harmful effects, although only limited attention has been paid to the interests, needs and fears of the people and communities experiencing this development. These efforts seek to sustain overall economic growth, to create cost-efficiency gains and/or profits, and to enhance national and/or organisational security. Yet, they mostly treat the people and communities living in the region as objects

M. Salminen (✉) · G. Zojer
Northern Institute for Environmental and Minority Law, Arctic Centre, University of Lapland, Rovaniemi, Finland
e-mail: mirva.salminen@ulapland.fi

G. Zojer
e-mail: gerald.zojer@ulapland.fi

M. Salminen et al. (eds.), *Digitalisation and Human Security*, New Security Challenges,
https://doi.org/10.1007/978-3-030-48070-7_1

of development whose lives will be improved by digitalisation (Salminen & Hossain, 2018). By contrast, the primary aim of digitalisation and cybersecurity efforts should be the advancement of human wellbeing. This book addresses this recognised gap in both research and policy making. To justify the authors' claim, the chapters focus on scrutinising digitalisation and security within the cyber-physical environment from a human security perspective. The concept of human security allows for a broader and deeper understanding of security by extending it to topics beyond mere state security through the inclusion of aspects such as environmental, economic, food, health, personal, community and political security, as well as by including other actors than states as both the objects and subjects of security production (Buzan & Hansen, 2009; Davi, 2009; Sheehan, 2005).

Following the end of the Cold War, the inter-, supra- and transnational security discourses have increasingly acknowledged the human being to be in need of protection and, hence, questioned the primacy of the state as the referent object of security. According to such a widened understanding of security, it is culture, human progress, identity and the physical integrity of individuals that need to be safeguarded (e.g. Commission on Human Security, 2003; Gasper, 2014; Martin & Owen, 2014). The 1994 United Nations Human Development Report contributed to the popularisation of the human security concept. The report, its successors and the work based on them acknowledge the interconnectedness as well as the potentially conflicting aspects of security with regard to human wellbeing and development (United Nations Development Programme [UNDP], 1994). The referent object of security is individuals and/or communities, who are suppressed not only due to the risks stemming from socio-cultural, environmental and economic stagnation, but also due to the inter-linked threats resulting from the practices of technological innovation. At the same time, the concept of human security offers a positive approach to understanding security, for example, through the utilisation of a wide range of opportunities to tackle security threats and to improve human wellbeing in an integrated manner.

While in a state-centred security approach powerful state actors, such as governments or policy-makers, define what is considered a security threat, the human security approach allows individuals and communities to define what they perceive to be threats and/or opportunities in relation to societal wellbeing (e.g. Salminen & Hossain, 2018). As such, a human security approach enables the affected people to voice their concerns, and it can also be used as a tool for empowerment (Hoogensen Gjørv, 2012). The concept of human security requires the endorsement of the linkages and interdependencies of issues, such as technological development and accessibility, societal inclusion, environmental threats, globalisation and human rights, as well as their connectedness to national security. It is inclusive of traditional security concerns, although it promotes a more comprehensive understanding of security (Hossain, Zojer, Greaves, Roncero, & Sheehan, 2017; Zojer & Hossain, 2017). The concept excels in assessing or studying human wellbeing in context-specific approaches in order to raise awareness of an issue or to motivate appropriate responses to it (Gómez & Gasper, n.d.). It provides added value when analysing the impacts of digitalisation on societies at the sub-state or interregional levels, such as the European High North and its particularities (Salminen & Hossain, 2018; Zojer, 2019). Yet, the main criticism levelled at the usefulness of the concept as an analytical tool is that human security is too wide to say anything substantial about security (Davi, 2009; Paris, 2001) and/or fails to challenge the hegemonic tendencies inherent in contemporary governance (Sahle, 2010). Regardless of such criticism, the authors of this book find a human security framework to be useful when analysing contextualised everyday (in) securities in the European High North, which often bring forth the effects of high-level decision-making in a very concrete manner.

While digitalisation can certainly be associated with societal impacts that are similar worldwide, regional peculiarities give rise to distinct sets of opportunities and challenges. The European High North is under scrutiny in this book because it has particular characteristics, such as socio-demographic differences when compared to the more southern parts of these countries, vast distances, a delicate balance between nature and human practices, limited resources and infrastructures, a harsh climate and vulnerability in the face of environmental threats. For instance, the population density in the north is significantly lower than the national average: in both Lapland and Norrbotten the population density is only about 10% of the nationwide population density.[1] Due to this sparsity, there is less infrastructure in the north, meaning that some areas have poor mobile network connectivity, poor public transport and/

[1] In 2017, the population density (persons per square kilometre) was 18.1 in Finland compared to 1.9 in Lapland; 24.7 in Sweden compared to 2.6 in Norrbotten; and 17.1 in Norway compared to 1.6 in Finnmark and 6.6 in Troms (Eurostat, 2019b).

or less health services. The distances to the nearest facilities can be vast, for example, both of the hospitals in Finnish Lapland are located in the southern part of the region (Rovaniemi and Kemi), which is several hundred kilometres and several hours drive away from the northernmost villages of the region. The population in the Nordic countries is generally decreasing, although northern urban areas are experiencing a population increase due to both immigration from outside and urbanisation processes within the region. Of those from within the region, women are more likely to move to urban areas than men, seeking better educational opportunities and more suitable employment. As a result, the rural areas in the north are dominated by a male population that is gradually decreasing (Heleniak & Bogoyavlenskiy, 2014; Rasmussen, Hovelsrud, & Gearheard, 2014). In general, the Nordic populations are ageing (Gassen & Heleniak, 2016), although the European High North has an even higher share of elderly people,[2] a trend that is expected to intensify (Emelyanova & Rautio, 2012). In the northernmost areas, the gross domestic products are lower than the national averages.[3] While in rural areas in particular non-marketed economic activities and subsistence remain important, the unemployment rates in the north are often higher than the national averages.[4] Despite having lower household incomes and less labour market opportunities, many inhabitants of the circumpolar north rate their "satisfaction with life" as being high (Rautio, Poppel, & Young, 2014). Since parts of the European High North extend beyond the Arctic Circle, the winters are long and dark, and temperatures below −40 °C set the environmental conditions apart from those of

[2] In 2018, the median age of the population (in years) was 42.8 in Finland compared to 47 in Lapland; 40.6 in Sweden compared to 44.8 in Norrbotten; and 39.5 in Norway compared to 40.7 in Finnmark and 40 in Troms (Eurostat, 2019c).

[3] In 2016, the gross domestic product at current market prices (in 1000 euros per inhabitant) was 39.3 in Finland compared to 36.2 in Lapland; 46.7 in Sweden compared to 44.5 in Norrbotten; and 64.1 in Norway compared to 51.1 in Finnmark and 53.2 in Troms (Eurostat, 2019a).

[4] In 2017, a number of the municipalities with the highest unemployment rates were located in the northern parts of Sweden and Finland. For instance, in the final quarter of 2017, most of the municipalities in Lapland county had higher unemployment rates than the Finnish average (11.2%), with Salla (20%) and Enontekiö (17.7) having the highest (Statistics Finland, 2018). Further, the municipalities of Haparanda (6.1%) and Pajala (5.4%) had unemployment rates that were significantly above the Swedish average of 4.0% (Regionfakta, 2018).

the southern parts of the countries. Further, the European High North is the homeland of the Sámi, an indigenous people with several small linguistic groups.

These characteristics have only partially been recognised in the national digital frameworks of Finland, Sweden and Norway, or Russia (see Salminen, 2018). Aside from scrutinising the effects of top-down policies in terms of facilitating and securing digitalisation, it is important to investigate how the people and communities in the region interact among themselves. Such interactions will help to illuminate the associated needs and interests, as well as the fears, which people have with regard to intensifying digitalisation. Moreover, they will assist with generating ways and means to convey these needs, interests and fears in a bottom-up fashion, that is, to the decision-making processes through which the outlook for the digital European High North is framed.

The chapters included in this book depict both positive (enabling) and negative (threatening) potentials in relation to the regional digital development of the European High North. They focus on identifying ways to reinforce the positive potentials while diluting the negative ones. Moreover, in contrast to the dominant understanding of cybersecurity, as defined by governmental agencies or cybersecurity corporations, they allow room for local security perspectives. They highlight the societal cleavages that digitalisation brings forth or intensifies, while suggesting ways and means to overcome them. In the process, the chapters treat the people and communities living in the European High North as both the objects and subjects of digitalisation. People ought to have a say in, as well as an ability to influence, the transformations of their everyday living environment. Shifting the focus to the daily experiences of people and communities changes the definition of cyber threats and the resultant security measures. Rather than concentrating on technical and/or national security threats, threats to human wellbeing—as experienced by both individuals and communities—in the digitalising European High North are highlighted (Salminen, 2019; Zojer, 2019). The range of available security measures thus broadens to include, for example, active digital inclusion policies, cross-border infrastructural development and the innovative use of digital technology in anticipating food trends or animal disease outbreaks. This book does not, therefore, discuss mainstream cybersecurity topics such as cyberwar, digital espionage, recently detected advanced persistent threats or the technical specificities of a particular cyberattack.

The book presents some of the results of a three-year research project—Enablement Besides Constraints: Human Security and a Cyber Multi-disciplinary Framework in the European High North (ECoHuCy). The key theoretical and empirical interventions carried out during the project are threefold. First, to examine national digitalisation and cybersecurity policies together, rather than as two separate entities. The prevailing practice of principally associating digitalisation with economic benefits and opportunities, while framing cybersecurity in terms of threats to national security, renders the discussion regarding digital societal trajectories sporadic and spotty. Many important, perhaps even defining, phenomena may thus go unnoticed. Second, to position the human being at the heart of digitalisation, as discussed above. Third, to scrutinise the ongoing transformations in the context of the European High North. Both digitalisation and cybersecurity have thus far been only marginal research areas, for instance, in the field of Arctic studies—a state of affairs that this book aims to help change.

This edited volume is organised into five parts. It begins with this introduction and ends with a concluding chapter summarising the main findings of the research. The second part of the book, **A Multi-Disciplinary Cybersecurity Approach**, develops a human-centred cybersecurity framework of analysis. The chapters in this part discuss how the concepts of human security, human rights and cybersecurity relate and connect to each other. They scrutinise how multifaceted human security, with a focus on individuals and communities, can offer a promising alternative security approach to the traditional—state-centred and business-oriented—notion of security, as well as how human security is interconnected with digitalisation and cybersecurity in the European High North.

This part begins with a chapter entitled **Comprehensive Cybersecurity and Human Rights in the Digitalising European High North** authored by Mirva Salminen, Gerald Zojer and Kamrul Hossain. The chapter presents the interconnections between ever-advancing digitalisation, cybersecurity, human security and human rights in the European High North. While digitalisation provides multiple opportunities for individuals and communities in the area, it also generates concerns for them. The mainstream cybersecurity approaches neglect many of the everyday concerns of people in the area. Therefore, the authors claim that a human security approach to cybersecurity allows for a more inclusive understanding of both the enabling and constraining effects of digitalisation by deepening and widening the prevailing conceptualisations. They add force to their argument by highlighting the interconnections between human security and human rights. Such an approach places individuals and communities at the heart of cybersecurity and examines the changes that digitalisation brings to their everyday life.

In the second chapter, Nadezhda Filimonova and Mario Portugal-Ramirez examine **The New Frontier for Human Cybersecurity: Russia's Cybersecurity Policies in the Arctic**. This chapter investigates the development of Russia's official cybersecurity policies with regard to the Arctic since the early 2000s, which is when the first Strategy for the Development of an Information Society was adopted. The principal focus is on examining how the notion of cybersecurity is framed at the official level and determining who its major target groups are. More specifically, the chapter analyses whether the issues associated with human security are included in Russia's official cybersecurity discourse. The subject of Russia's official discourse concerning cybersecurity in the Arctic is important in terms of revealing the critical actors that will enforce and benefit from cybersecurity during a time characterised by the increasing digitalisation of the Arctic.

Third, Alan Collins, in his chapter entitled **Critical Human Security and Cyberspace: Enablement Besides Constraints**, applies a critical human security perspective to cyberspace. This chapter contends that emancipation provides human security with its core purpose, as it frees people from both the constraints imposed by despotic regimes (freedom from fear) and an exploitative economic system (freedom from want). The idea that human emancipation can challenge the pre-existing order is reflected in the literature that foresees the digital revolution ushering in an age of transparency and empowerment. Conceiving security as emancipation, this chapter utilises the conceptual tool of the "three faces" of power to examine how human agency is enabled by the digital revolution (freedom from), while simultaneously being constrained by it (oppressed by).

The third part of this book is entitled **Human Rights and Digital Infrastructure**. It focuses on the interconnections between adequate and functional information infrastructure and the realisation of human rights. The included chapters touch upon submarine fibre-optic cables as a highway for internet traffic and the international law that governs the development of this cable network. They also consider the question of just how good internet access is good enough for the realisation of human rights and the mitigation of the problems associated with societal exclusion in the European High North. In addition, the chapters scrutinise the inclusion and exclusion that the digitalisation of public services generates or reinforces among different societal groups.

First, in **Social Exclusion as Human Insecurity: A Human-Cybersecurity Framework Applied to the European High North**, Kristin Smette Gulbrandsen and Michael Sheehan explore the dynamics of digital exclusion. Digitalisation is changing the way in which governments provide public services. This has particular significance for geographically peripheral and sparsely populated regions such as the European High North, which is also challenged by an ageing population and labour shortages. Achieving success in the shift from traditional to digital services partly hinges on the availability of adequate digital competence and infrastructure. Without these components, rural villages and especially elderly people are at risk of suffering disadvantage. To address this issue, the chapter presents a cybersecurity framework that borrows from the human security and social exclusion theory. Drawing on key informant interviews conducted in Norway, Sweden and Finland, it finds that while digitalisation provides benefits to remote areas, it can also cause digital exclusion. Therefore, significant attention must be paid to whose voices and interests shape the current digitalisation agenda.

The second chapter in this part is authored by Stefan Kirchner and it is entitled **Mobile Internet Access as a Human Right: A View from the European High North**. Even in developed countries, in many rural areas the mobile phone coverage, to say nothing of the mobile internet coverage, remains limited. Based on the evolution of both mobile internet standards and state actions in this regard, this chapter considers mobile internet access from the perspective of human rights. Tracing the different approaches to internet access that can be found within the framework of human rights, the existence of a right to sufficiently fast mobile internet access without location-specific discrimination is examined. Particular attention is paid to the sparsely populated areas of the European High North, as well as to Finland's contribution to the development of customary international law, wherein such a right is currently emerging. In addition, this chapter investigates the importance of internet access in relation to the realisation of other human rights as well as the enforcement of the emerging social human right to mobile internet access.

The third chapter included in this part, **The Legal Regime of Submarine Telecommunications Cables in the Arctic: Present State and Challenges**, is authored by Daria Shvets. According to her, submarine telecommunications cables play a crucial role in the daily data transmission between continents and states. The number of submarine cable systems is increasing worldwide, which underlines their significance in relation to modern communications. Cable networks have recently reached the Arctic. Several cables have already been laid on the seabed of the Arctic Ocean, while more projects are forthcoming. This chapter examines the general legal regime governing submarine cables under international law, the legislation of the Arctic states regarding submarine cables and the challenges facing the present legal regime applicable to the Arctic in light of both current and upcoming submarine cable projects.

Finally, Juha Saunavaara expands on the topic of submarine fibre-optic cable development in **Connecting the Arctic While Installing Submarine Data Cables Between East Asia, North America and Europe**. The Arctic Connect Project aims to connect East Asia, North America and Europe by running a submarine fibre-optic cable through the Northeast Passage and the European High North. Along with the Quintillion Submarine Cable System, which originates in Alaska, it is one of the most visible aspects of the lively discussion concerning Arctic connectivity that has taken place over the last few years. This chapter clarifies the role of the European High North and other Arctic regions in such initiatives. The human security framework is then utilised as a flexible conceptual tool for analysing the roles of local people and local communities as well as exploring the different meanings, hopes and fears associated with these international market-driven projects. The European High North serves as the spatial focus of the research, although references are also made to other geographical locations.

The fourth part of this book concerns **Society and Environment**. It discusses experienced cybersecurity, identity formation on digital platforms and the opportunities that information and communication technologies provide for digital empowerment and inclusion in the European High North. To understand ongoing societal transformations and the related security questions, it is necessary to recognise how people use widespread digital technologies such as social media and digitalised public services. The abundance of data available via social media, for example, opens up novel research opportunities, although it also raises the questions regarding suitable research methods, research ethics and privacy concerns. Moreover, the availability of data may assist with recognising ongoing environmental changes, and such data could be utilised, for example, in combatting the emergence of pathogeneses, which contribute to food insecurity in the European High North.

This part begins with a chapter on **Cybersecurity of the Digital Citizens in the Remote Areas of the European High North** authored by Ville Kivivirta, Leena Viinamäki and Arto Selkälä. This chapter focuses on how digital citizens in the most remote areas of the European High North experience cybersecurity. The empirical research was conducted in the Finnish regions of Lapland, Ostrobothnia and North Karelia. The research instruments comprised a postal survey of citizens ($n=1063$) living in the three regions, a web-based questionnaire designed for those public sector professionals responsible for e-services, two workshops for experts, as well as additional interviews conducted with Sámi people and third-sector representatives. The gathered data were examined by means of a three-phase gap analysis. Based on the results, citizens in the three regions are slightly more critical than officials with regard to cybersecurity, and they experience cybersecurity in a broader sense than simply as customers. Further, active digital citizens in remote areas experience cybersecurity problems that may eventually decrease the social acceptability of e-services.

In the second chapter in this part, Maxim Shishaev, Andrey Fedorov and Igor Datyev provide an **Analysis of Social Networking in Studying the Identity of Local Communities**. According to them, societal security represents a crucial aspect of human security, which is especially important for local communities in the European High North. As identity is the ultimate criterion for societal security, it is an essential object for studying and monitoring. The traditional means of studying social phenomena such as identity involve conducting time-consuming and expensive sociological investigations. However, as a result of digitalisation, social entities are now widely represented in the virtual space of social networking services, which opens up new avenues for analysing identity on the basis of a social networking analysis. In this chapter, the authors focus on the possible ways of analysing the identities of local communities in the context of societal security issues by using social networking data. Further, they highlight the problems that must be overcome in relation to the implementation of these techniques.

Next, Laura Olsén-Ljetoff and Liisa Hokkanen scrutinise **The Interconnection Between Digitalisation and Human Security in the Lives of Sámi with Disabilities**. This chapter adopts an intersectional approach to examining the interconnections that exist between digitalisation and both the positive and negative aspects of personal and community security in the everyday lives of Sámi persons with disabilities who live in Finland. Based on interviews conducted with Sámi persons with disabilities, the chapter first provides an analysis of the concrete and complex ways in which digitalisation and personal security interact and intersect in the everyday lives of the interviewees. Then, the discussion turns to the intersection of digitalisation and community security, particularly with regard to its cultural aspects, such as languages, traditional livelihoods and connections to Sámi communities. This chapter emphasises the importance of considering the heterogeneity and multiple identities of minority groups—which give rise to special needs—when designing future digital solutions.

Finally, **Emerging Pathogeneses and Satellite Telemetry: Containing Contagion in the European High North** is a chapter authored by Alexandra L. Carleton. According to her, the creation of a digital framework for reporting pathogen incidents across the Arctic and Boreal regions is vital in terms of information-sharing, more so even than filing reports in either specific academic journals or government archives. Regular intervention in relation to species that are susceptible at higher latitudes is difficult, both because such species may be migratory and wild and because the ethics involved in interventionist studies are appropriately strict. Examining the science associated with emerging pathogenic discoveries in the European High North, this chapter highlights the ethical and practical considerations concerning the opening up of northern latitudes. Greater access to endemic sub-Arctic areas may facilitate pathogenic transfer, the survival of various species and, in turn, the livelihoods of traditional peoples. Further, disease mapping is of considerable value to human security when it comes to building both predictive capabilities and outcome resilience.

As the chapters included in this volume make clear, digitalisation and cybersecurity are multifaceted phenomena in the European High North. At best, digitalisation advances human security in the region by allowing for additional information collection, analysis and dissemination, as well as protective or mitigative actions based on such information;

novel solutions to service provision and hence a lessened need to travel; new and/or transformed economic opportunities; improved responsiveness in terms of governance; as well as improved social connectivity and enhanced digital human rights. However, successful digitalisation must build on robust cybersecurity so that digital operations can be relied upon and trusted. In addition to safeguarding information and/ or infrastructure, this requires the protection of human beings amid digital developments in the European High North. Digitalisation transforms the ways in which both individuals and communities construct and (strive to) maintain their identities and cohesion, which may be perceived threatening development locally. Dys- or malfunctioning of digitalised infrastructure may cause significant environmental damage or the temporal isolation of villages or households. In fact, the mere expansion of information infrastructure and computational capacity results in environmental challenges, for example, in the form of enhanced energy consumption (e.g. Hodgson, 2015; Laakso & Terävä, 2019). Human abuse of digital technologies can endanger individuals' psychological integrity, wellbeing and sense of belonging. The failure to recognise digital human rights may result in violations of privacy, digital divides and even extensive information collection without a justifiable purpose. Variations in digital know-how levels position people differently in a digitalising society in which physical service provision is being withdrawn, especially from sparsely populated rural areas. Finally, differentiating between genuine and/or accurate information and tampered with and/or disinformation is increasingly difficult in the European High North, as it is elsewhere.

This book represents an opening for studying both digitalisation and cybersecurity from a human security perspective and in the context of the European High North. The proper contextualisation of such an examination highlights human security questions related to digitalisation that may otherwise go unnoticed. Therefore, the aim of this work is to inspire further inquiries into what cybersecurity may mean in the digitalising European High North as well as how it can be best addressed in decision-making at all levels.

References

Buzan, B., & Hansen, L. (2009). *The Evolution of International Security Studies*. Cambridge: Cambridge University Press.

Commission on Human Security. (2003). *Human security now*. New York, NY: United Nations.

Davi, M. (2009). *Human security as the one size fits all policy approach?* Paper presented at the European Security and Defence Forum WS2, Chatham House. Retrieved from https://www.chathamhouse.org/sites/files/chathamhouse/public/Research/International%20Security/1109esdf_davi.pdf.

Emelyanova, A., & Rautio, A. (2012). Aging population of the Barents Euro-Arctic region. *European Geriatric Medicine, 3*(3), 167–173. https://doi.org/10.1016/j.eurger.2012.03.005.

Eurostat. (2019a). *Gross domestic product (GDP) at current market prices by NUTS 3 regions* (nama_10r_3gdp) [data file]. Retrieved from https://appsso.eurostat.ec.europa.eu/nui/show.do.

Eurostat. (2019b). *Population density by NUTS 3 region* (demo_r_d3dens) [data file]. Retrieved from https://appsso.eurostat.ec.europa.eu/nui/show.do.

Eurostat. (2019c). *Population: Structure indicators by NUTS 3 region* (demo_r_pjanind3) [data file]. Retrieved from https://appsso.eurostat.ec.europa.eu/nui/show.do.

Gasper, D. (2014). Human security: From definitions to investigating a discourse. In M. Martin & T. Owen (Eds.), *Routledge handbook of human security* (pp. 28–42). London and New York, NY: Routledge/Taylor Francis Group.

Gassen, N. S., & Heleniak, T. (2016). *The impact of migration on projected population trends in Denmark, Finland, Iceland, Norway and Sweden: 2015–2080* (Nordregio Working Paper No. 2016:5). Stockholm: Nordregio.

Gómez, O. A., & Gasper, D. (n.d.). *Human security guidance note. A thematic guidance note for regional and national human development report teams*. United Nations Development Programme, Human Development Office. Retrieved from http://hdr.undp.org/sites/default/files/human_security_guidance_note_r-nhdrs.pdf.

Heleniak, T., & Bogoyavlenskiy, D. (2014). Arctic populations and migration. In J. Nymand Larsen & G. Fondahl (Eds.), *Arctic human development report: Regional processes and global linkages* (pp. 53–104). Copenhagen: Nordic Council of Ministers.

Hodgson, C. (2015, November 13). Can the digital revolution be environmentally sustainable? *The Guardian*. Retrieved from https://www.theguardian.com/global/blog/2015/nov/13/digital-revolution-environmental-sustainable.

Hoogensen Gjørv, G. (2012). Security by any other name: Negative security, positive security, and a multi-actor security approach. *Review of International Studies, 38*(4), 835–859.

Hossain, K., Zojer, G., Greaves, W., Roncero, J. M., & Sheehan, M. (2017). Constructing Arctic security: An inter-disciplinary approach to understanding security in the Barents region. *Polar Record, 53*(1), 52–66.

Laakso, V., & Terävä, H. (2019, June 26). Netti syö kasvavalla tahdilla sähköä ja suurin syyllinen ovat nettivideot – "Epämiellyttävä totuus, josta ei haluta puhua" [Internet consumes electricity at an increasing rate and videos are the guilty party—"An unpleasant truth nobody wants to talk about"]. *Yle.* Retrieved from https://yle.fi/uutiset/3-10832413.

Martin, M., & Owen, T. (2014). Introduction. In M. Martin & T. Owen (Eds.), *Routledge handbook of human security* (pp. 1–14). London and New York, NY: Routledge/Taylor Francis Group.

Paris, R. (2001). Human security: Paradigm shift or hot air? *International Security, 26*(2), 87–102. Retrieved from http://aix1.uottawa.ca/~rparis/Paris.2001.IS.Human%20Security.pdf.

Rasmussen, R. O., Hovelsrud, G. K., & Gearheard, S. (2014). Community viability and adaptation. In J. Nymand Larsen & G. Fondahl (Eds.), *Arctic human development report: Regional processes and global linkages* (pp. 423–473). Copenhagen: Nordic Council of Ministers.

Rautio, A., Poppel, B., & Young, K. (2014). Human health and well-being. In J. Nymand Larsen & G. Fondahl (Eds.), *Arctic human development report: Regional processes and global linkages* (pp. 297–346). Copenhagen: Nordic Council of Ministers.

Regionfakta. (2018, January 15). *Unemployed.* Retrieved from http://www.regionfakta.com/Norrbottens-lan/IN-ENGLISH-/Workxx/Unemployed.

Sahle, E. N. (2010). *World orders, development and transformation.* London: Palgrave Macmillan.

Salminen, M. (2018). Digital security in the Barents region. In K. Hossain & D. Cambou (Eds.), *Society, environment and human security in the Arctic Barents region* (pp. 187–204). Abingdon and New York, NY: Routledge.

Salminen, M. (2019). Refocusing and redefining cybersecurity: Individual security in the digitalising European High North. *The Yearbook of Polar Law, 10,* 321–356.

Salminen, M., & Hossain, K. (2018). Digitalisation and human security dimensions in cybersecurity: An appraisal for the European High North. *Polar Record, 54*(2), 1–11.

Sheehan, M. (2005). *International security: An analytical survey.* Boulder, CO: Lynne Rienner.

Statistics Finland. (2018, February 21). *Unemployed jobseekers, persons in services and vacant jobs at the end of the month* (tyonv_001_201801) [data file]. Retrieved from http://stat.fi/index_en.html.

United Nations Development Programme. (1994). *Human development report 1994.* New York and Oxford: Oxford University Press.

Zojer, G. (2019). The interconnectedness of digitalisation and human security in the European High North: Cybersecurity conceptualised through the human security lens. *The Yearbook of Polar Law, 10,* 297–320.

Zojer, G., & Hossain, K. (2017). *Rethinking multifaceted human security threats in the Barents region: A multilevel approach to societal security.* Juridica Lapponica 42. Rovaniemi: University of Lapland Printing Centre.

PART II

A Multi-Disciplinary Cybersecurity Approach

CHAPTER 2

Comprehensive Cybersecurity and Human Rights in the Digitalising European High North

Mirva Salminen, Gerald Zojer and Kamrul Hossain

2.1 Introduction

Finland, Sweden and Norway are generally considered to be highly digitalised societies. For example, according to the European Commission's Digital Economy and Society Index (DESI)[1] for 2019,

[1] The DESI is a composite index that summarises the relevant indicators into five dimensions: connectivity, human capital, use of internet services, integration of digital technology and digital public services. It is intended to observe Europe's digital performance and to track the evolution of EU member states in terms of digital competitiveness (DESI, 2019). More information about the methodology used is available on the website.

M. Salminen (✉) · G. Zojer · K. Hossain
Northern Institute for Environmental and Minority Law,
Arctic Centre, University of Lapland, Rovaniemi, Finland
e-mail: mirva.salminen@ulapland.fi

G. Zojer
e-mail: gerald.zojer@ulapland.fi

K. Hossain
e-mail: khossain@ulapland.fi

M. Salminen et al. (eds.),
Digitalisation and Human Security, New Security Challenges,
https://doi.org/10.1007/978-3-030-48070-7_2

Finland (overall score of 69.9) and Sweden (overall score of 69.5) ranked the highest of all the European Union (EU) member states (average overall score of 52.5). Moreover, they are said to be "among the global leaders in digitalisation" (DESI, 2019, n.p.). Despite not being an EU member state, Norway's DESI country report is also published annually. In 2019, Norway would have been ranked fifth (overall score of 66.0) if counted among the EU member states. The European Commission's Digital Transformation Monitor tracks the adoption of digital technologies by businesses across the EU. According to its Digital Transformation Scoreboard for 2018,[2] "Scandinavian and western European economies [...] dominate the top positions" (p. 4). Further, "Finland performs strongly in digital transformation and stands out in e-leadership, investments and access to finance, and digital infrastructure, yet challenges remain in the fields of entrepreneurial culture and ICT start-ups" (p. 130), while Sweden "performs particularly well in six out of the seven dimensions, its strongest asset being the supply and demand of digital skills. Nevertheless, [the country] has some room for improvement [...] in areas such as digital transformation and infrastructure" (p. 132). Moreover, the northern European countries have acknowledged the vulnerabilities embedded in their approach to digitalisation and, therefore, developed national frameworks for information security and cybersecurity—sometimes combining the two (e.g. Government Offices of Sweden, Ministry of Justice, 2016; Ministry of the Interior, 2019; Ministry of Transport and Communications, 2016; Norwegian Ministries, 2012; Norwegian Ministry of Foreign Affairs, 2017; Secretariat of the Security Committee, 2013). The main objectives set out in these frameworks include the advancement of national security, the protection of human rights, as well as the provision of support for the national economy. Internet access is considered to be an everyday necessity and national programmes for enhancing the development of the knowledge society and improving the connectivity and accessibility of the online environment have been in place since the 1990s.

[2]The Digital Transformation Scoreboard for 2018 "surveyed EU companies [...], screened national policies and analysed in a macro-perspective approach a wide range of national data sourced from Eurostat, national statistics offices and international organisations" and, further, it analysed "information collected online from open sources including online press, blogs, corporate web sites, social media, forums, broadcast television" (Digital Transformation Scoreboard, 2018, n.p.).

While the general statistics and descriptions of digitalisation in northern Europe appear to be extremely good, they actually tell us little about the regional variation within these countries. For instance, access to digital networks, the availability of digital services, access to (relevant) information, the state of digital inclusion and the realisation of human rights, as well as the levels of cybercrime and digital abuse can vary regionally (see Salminen, 2018). In this chapter, we are interested in the interconnections between digitalisation, cybersecurity, human security and human rights, in the northernmost areas of Finland, Sweden and Norway. For analytical purposes, we refer to these areas (Lapland in Finland, Norrbotten in Sweden, as well as Troms, Nordland and Finnmark in Norway[3]) as the European High North. No fixed reference point for this term exists, as the chapters included in this volume make clear. There are three main reasons why we chose to focus on the Arctic and the borderline Arctic areas of mainland Europe. First, these areas are similar enough to be discussed together. That is, their political, economic, social, cultural, geographical and climatological profiles are highly similar. Second, digitalisation in these areas has proceeded along similar lines. Third, when compared to the population and power centres of southern Finland, Sweden and Norway, these areas are generally considered to be peripheral and, thus, of only limited interest. Digitalisation is arguably making everyday life easier and more comfortable throughout the European High North, although it is simultaneously generating problems that did not previously exist.

In sum, and as verified by prior research (e.g. Dymet, 2019; Hossain, 2019; Kilpeläinen, 2016; Kurtti & Berg, 2018; Salminen, 2019; Zojer, 2019a; see also, chapters in this volume), people's everyday life is changing in the European High North due to the development and ever-expanding application of information and communication technologies (ICT) in both the public and private sectors. Regardless of the common presumption that ICT are neutral and so treat people equally, the implications of digitalisation are, in fact, region specific (e.g. Carr, 2012, p. 175; Schou & Hjelholt, 2018, pp. vi, 6, 10; Webster, 2006, pp. 11–12). Similarly, individuals' and communities' interests, needs and challenges with regard to digitalisation vary, for example, between

[3]Troms and Finnmark will merge into one county on 1st January 2020. At the time of writing, however, they are two independent counties.

different socioeconomic and/or cultural groupings (Lindell, 2018; Mendez Garcia, 2011; Mesch, 2012; Nam & Park, 2016; Subramony, 2007; Zambianchi et al., 2019). Thus far, little research has been conducted concerning digitalisation and the related security and rights questions in the European High North. Our concern is, therefore, that the mainstream policy approaches to both digitalisation and cybersecurity in northern Europe are somewhat blind to the related interests, needs and fears of people in different areas of these countries. In this case, individuals and communities become receivers of ICT, as well as the services that such technologies enable, without being able to influence their development (cf. Salminen & Hossain, 2018, p. 113). In addition, the more digital that societies become, the greater divider fast and reliable access to related services may become between the included and the excluded (e.g. Mesch, 2012; Nam & Park, 2016). The aim of this chapter is to provide a theoretical framework for a more nuanced investigation of the issues. We believe that a well-contextualised human security approach based on human rights represents a comprehensive means of scrutinising the effects (both positive and negative) of digitalisation on society. At best, it can enhance inclusion, empowerment and, hence, security in digitalising societies, and it can also contribute to meaningful and targeted policies in area-specific contexts.

In this chapter, we first examine the connections between digitalisation and cybersecurity. That is, we consider how the inherent vulnerabilities of the ever-advancing digitalisation of societies finally prompted a reaction from states in the form of cybersecurity. This section summarises the mainstream cybersecurity agenda, which we suggest is too narrow to address the everyday uncertainties that people face in their digitalising living environments. Second, we investigate the concept of human security as well as the opportunities it provides for generating a more comprehensive agenda that encompasses both digitalisation and cybersecurity. Third, we add some rigour to the elusive concept of human security by basing it on human rights. The basic requirement is that the rights that apply in the physical environment also apply in the digital environment. The discussion regarding emerging digital human rights, such as the right to internet access (see Kirchner, Chapter 6), which, for instance, Finland has declared to be a basic right, is also taken into consideration. We end the chapter by offering some highlights from the proposed framework.

2.2 Interconnections Between Digitalisation and Cybersecurity

Commercial interest in the digitalisation of the European High North remains limited, especially outside its urban centres. The sparse population, regional lack of a skilled workforce, relatively long distances and harsh climate render the construction and maintenance of digital networks expensive. Finland, Sweden and Norway have tried to solve the problems associated with the extension of fast and reliable digital networks to their northernmost areas through different forms of public intervention (see Salminen, 2018, pp. 190–192). This policy can be considered to have been somewhat successful because, for example, the DESI 2019 country reports state that fixed broadband covers 94% of households in Finland and 4G covers 99% (DESI 2019 for Finland, p. 5), fixed broadband and 4G cover 96% of households in Sweden (DESI 2019 for Sweden, p. 5), while fixed broadband covers 97% and 4G 100% of households in Norway (DESI 2019 for Norway, p. 4). However, it is recognised that extending coverage to the remaining households will be both expensive and difficult (e.g. the "last mile" discussion in Adrian & Brown, 2012). For instance, Lapland in Finland has been one of the main receivers of national and EU network construction funding for years. Yet, people there still report problems in relation to coverage, the reliability of networks and redundancy (Salminen & Hossain, 2018).

Similarly, the digitalisation of health and social security services, schooling and education, administration and a great number of other services is currently taking place, with or without national and EU-wide funding schemes (e.g. Kilpeläinen, 2016; Kurtti & Berg, 2018; Salminen, 2019). The justification for this transformation is that it is the only financially viable means of continuing service provision in areas with diminishing populations. People's need to travel decreases, although digital service provision simultaneously enables the withdrawal of physical services from rural areas (Salminen & Hossain, 2018). In addition, some novel applications intended to meet the particular needs of people living in the northernmost areas have been developed, while people have been creative in adapting the available technology to serve their interests, for instance, in relation to reindeer husbandry (e.g. Kotituomi, Huju, Kulmala, & Lahtela, 2019; McGwin, 2019; Rätti & Wallén, 2017; Zojer, 2019a, 2019b). Nationwide accessibility programmes designed to improve the inclusion of people with disabilities in the digital sphere have been running for years (e.g. in Finland from 2005 onwards [see the publications of the Ministry of Transport and Communications]). Different language versions of public information and service platforms are becoming increasingly available (cf. Dymet [2019] on the online viability of minority languages from the European High North). The infrastructures upon which modern societies run are also becoming digitised in the European High North. For instance, the supply of heat, electricity and water is now digitally controlled and steered—not to mention monetary transactions, logistics or traffic in general (e.g. Ministry of the Interior, 2019; Norwegian Ministry of Foreign Affairs, 2017, p. 7). Digitalisation is, therefore, transforming everyday life in profound ways that affect people's ways of life, socio-economic positions and the related opportunities and concerns, livelihoods, as well as what they need to spend time on during the day. It may also generate tensions within or between cultural groupings and enable or hinder the discussion of those tensions (Hossain, 2019; Olsén-Ljetoff & Hokkanen, Chapter 11).

As a rule, the further that digitalisation penetrates a society, the more dependent that society becomes on the functioning of ICT and the related services. Parallel manual structures either decay or are dismantled. The dependency on digital tools and services comes up frequently in discussions with people living in the European High North. These people admit to their own addictions and dependencies, but they also tell vivid stories about situations in which the digitalised infrastructure failed without them being able to do much about it.[4] ICT and digital networks are inherently vulnerable, for example, due to potential technical failures and interoperability problems, the complexity of connections and human error in relation to coding or connecting. Further vulnerabilities emerge as a result of individuals' limited awareness of digital security questions and limited skills when using and operating technologies. The recognition of vulnerabilities, possible technical malfunctions as well as some people's and organisations' willingness to abuse such vulnerabilities has led to the adoption of information-cybersecurity frameworks. Once societies' ICT dependency has become severe, the question of security is lifted from the level of individuals and organisations (technical and administrative information security) to the level of the entire society (strategic

[4] Data collected in November 2018 in Enontekiö by Mirva Salminen for another article (case study).

cybersecurity). This novel framework is referred to as cybersecurity and it largely encompasses information security. However, the biggest difference is that, in the case of cybersecurity, the referent object of security is extended from information to those digitalised infrastructures deemed critical for the functioning of society. Alongside skilled individuals, organisations and technologies, the state carries responsibility for the provision of security (e.g. Salminen, 2019). Yet, it remains unclear who holds the main responsibility, what the means of cybersecurity provision are, and how successful such efforts have been thus far.

No unanimous or univocal conceptualisation of cybersecurity currently exists. What the concept actually entails and how it differs from conceptualisations of information security, network security, information hygiene and the like have been topics of discussion for years (e.g. de Oliveira Albuquerque, Garcia Villalba, Sandoval Orozco, de Sousa Jr., & Kim, 2016; Stevens, 2018; von Solms & van Niekerk, 2013; von Solms & von Solms, 2018). For the sake of clarity and consistency, in this chapter, we lean on the conceptualisations provided in the Finnish, Swedish and Norwegian strategies. Finland's Cybersecurity Strategy 2019 is sparse in terms of definitions, although the earlier strategy from 2013 states that "[c]yber security means the desired end state in which the cyber domain is reliable and in which its functioning is ensured" (p. 1). Building on this definition, the 2019 strategy sets three pivotal goals for developing the cyber domain and safeguarding the functions that depend on it that are vital to society, as well as for supporting the availability of reliable digital services and business development (p. 4). These three pivotal goals include the development of international cooperation for the borderless protection of the cyber domain; the improved coordination of cybersecurity leadership and management, planning and preparation; and the development of cybersecurity know-how. A future cybersecurity development programme will indicate the measures through which these goals are to be achieved.

The Swedish National Cyber Security Strategy (2016) begins with the need to "manage the risks inherent in digital transformation" (p. 3),[5] which impacts both prosperity and security. Similar to the Finnish

[5] Digital transformation is, according to the strategy, said to mean that "an increasing proportion of all activities in society is to varying degrees dependent on networks and information systems, and thereby on cyber security" (Swedish National Cyber Security Strategy, 2016, p. 4).

approach, cybersecurity "concerns the whole society" and "[e]veryone needs to take responsibility for cyber security issues" (p. 3). Collaboration and the need for a common direction within society are highlighted, as are international cooperation and the promotion of knowledge and expertise. The objectives of cybersecurity are aligned with those of national security, namely "protecting the lives and health of the population, the functioning of society, and […] capacity to uphold fundamental values such as democracy, the rule of law and human rights and freedoms", as well as fostering national growth and competitiveness (pp. 3–4, 7). However, the definition of cybersecurity is simply "a set of security measures to preserve the confidentiality, integrity and availability of information" (p. 4), which entails the protection of both information itself and the systems used to store and transfer it (p. 5). This is important for maintaining the quality and effectiveness of societal functions, guaranteeing individuals' ability to exercise their rights and freedoms based on accurate and available information, protecting sensitive information, as well as safeguarding "digital information and control systems that continuously handle large quantities of sensitive information in order to control, e.g. electricity distribution, water supply, transportation, transport infrastructure or hospital equipment" or industrial activities (p. 5).

Finally, the Cyber Security Strategy for Norway (2012) defines cybersecurity as the "[p]rotection of data and systems connected to the Internet" (p. 28). It also provides a definition of information security as the "[p]rotection of the confidentiality, integrity and availability of information" (p. 28), which comes close to the definition of cybersecurity given in the Swedish strategy. The Norwegian strategy points to the significant changes that ICT have generated in society, which are generally seen as gains. ICT are "integrated into all aspects of society" and so as "critical for society to function normally", while "[i]n many cases, the population is dependent on ICT in order to receive a service" (p. 8). This ubiquitous nature of ICT has rendered them a strategic security challenge, the response to which is the responsibility of all stakeholders. It is pivotal to ensure confidence in the reliability of digital services (p. 8). The Norwegian strategy sets four goals for information security, including better coordination and common situational understanding, robust and secure ICT infrastructure for everyone, the ability to handle adverse ICT events and a high level of competence and security awareness (p. 17). It is noteworthy in all these strategies that only a limited role is allocated to human beings and their everyday experiences. People and their wellbeing are dependent on ICT, which may cause problems through negligence or malevolence, and they play a role in the provision of cybersecurity. However, they are not the main referent object of cybersecurity, as the cyber domain, information, data and systems are.

Digitalisation is favoured and advanced because it provides opportunities. It generates new ways of doing things. Further, it enables the improvement of efficiency and creates cost savings by enhancing the old ways of doing things. While doing so, it also creates friction and dissatisfaction, since it does not only ease everyday life but also generates novel vulnerabilities, problems and exclusions (e.g. Gulbrandsen & Sheehan, Chapter 5; Kurtti & Berg, 2018), as well as an experienced lack of human interaction. For instance, it can help to enhance economic globalisation and to enable local businesses to advertise, sell or customise their products for global markets, which may contribute to employment and/or strengthen economic diversification (Keskitalo & Southcott, 2014). However, data from the United States indicate that e-commerce may concentrate in metropolitan areas rather than create new jobs in rural places (Gebeloff & Russell, 2017). E-commerce also challenges domestic businesses. In the Nordic countries, for example, roughly a quarter of all online purchases come from abroad, while Finnish online shoppers make half of their purchases from abroad (PostNord, 2017). The increasing popularity of e-commerce hence poses challenges for existing retailers. In rural regions, where only a few retail businesses exist, their disappearance would prove particularly challenging for community members who lack digital skills. States' actions may also create new risks at the individual level, as seen from a mainstream cybersecurity perspective. As Dunn Cavelty (2014) points out, the measures through which states aim to secure their information infrastructure may hamper human security. For instance, surveillance challenges the freedom of speech and the right to privacy. Similarly, certain cyber-defence measures, such as states' incentives to maintain backdoors[6] or to buy and utilise zero-day exploits[7] for popular software, may even counteract mainstream cybersecurity itself, which Dunn Cavelty (2014) refers to

[6] A backdoor is a software aspect that allows access to the device on which the software is running by bypassing its security means.

[7] A zero-day exploit refers to exploiting a software security vulnerability on the first day it is discovered/before a security patch has been developed and disseminated, or before it has been updated for devices running that software.

as the cybersecurity dilemma. This is illustrated by the recent case of EternalBlue, an exploit the development of which has been attributed to the United States' National Security Agency (NSA)[8] (CBS News, 2016; Shane, Perlroth, & Sanger, 2017). After the exploit was leaked in 2017, it was abused to create a malware attack that infected the computer systems of over 200,000 users worldwide, including crippling the digital infrastructure of numerous hospitals in the United Kingdom (BBC, 2017), which led to new insecurities concerning health. In 2019, the same exploit was used to launch cyberattacks against the administrations of several US cities (Perlroth & Shane, 2019).

The aim of cybersecurity is to safeguard the opportunities provided by digitalisation. However, it addresses only relatively poorly the disadvantages that people experience as a result of digitalisation. Cybersecurity should ensure the functioning of digitalised infrastructures in all situations so as to keep individuals' and organisations' (sensitive) information protected, to prevent cybercrime and enhance the investigation of it, to counteract digital espionage and to ensure that states thrive in digitalised battlefields—among other things. Nonetheless, it falls short in terms of addressing the concerns and difficulties that people face when they, for example, can no longer discuss financial matters with a familiar client manager at a nearby bank, cannot repair their own cars or snow mobiles, are unable to call for help when the nearest radio mast has lost power or drive hundreds of kilometres for nothing when the automated appointment reservation system has failed to recognise an ad hoc holiday.[9] The technical information security and strategic cybersecurity frameworks are good at protecting certain aspects of digitalisation, but we believe that they are not enough. To instil trust in the digitalising living environment, the digitalisation-related everyday security concerns that individuals and communities have must also be included in the agenda. Moreover, people need to have a say in the processes by which digital security and cybersecurity are defined and the related agendas negotiated.

[8]The NSA's website describes it as leading "the U.S. Government in cryptology that encompasses both signals intelligence (SIGINT) and information assurance (now referred to as cybersecurity) products and services, and enable[ing] computer network operations (CNO) in order to gain a decision advantage for the Nation and [its] allies under all circumstances" (https://www.nsa.gov/about/mission-values/).

[9]Data collected in November 2018 in Enontekiö by Mirva Salminen for another article (case study).

In the following section, we investigate the potentiality of the concept of human security to link the different aspects of digitalisation and security together better. The aim is to provide a more comprehensive cybersecurity agenda that scrutinises the security of both individuals and communities in their digitalising everyday environments so that people can participate in the definition of both security and security-related challenges and solutions. The goal is to develop a security approach that addresses both the opportunities and challenges generated by digitalisation. Moreover, the new approach should bring bottom-up responses to the side of top-down solutions.

2.3 A Human Security Approach to Cybersecurity

Security studies generally aim to analyse threats or risks to a certain referent object in order to address crises or conflicts. In the field of international relations, security approaches traditionally focus on military threats to states' sovereignty (e.g. Mearsheimer, 2014; Waltz, 2010). Yet, towards the end of the twentieth century, non-military challenges to states and their inhabitants gained increasing attention in international security debates. During the 1980s and along with the end of the Cold War, the scope of security studies broadened and increasingly included new sectors, such as the environment or societies on a sub-state level (e.g. Buzan, Wæver, & de Wilde, 1998; Heininen, 2013a; Hoogensen Gjørv et al., 2014; McSweeney, 1999). Such broad security approaches focus on human wellbeing. Prominently, the United Nations' Human Development Reports (HDRs) have raised attention to shift the focus of international politics to assure the prosperity of human life. As part of this work, the 1994 HDR contributed to popularising the concept of human security.

Human development and human security are closely interlinked: while human development constitutes a "broad concept, aiming at enlarging people's choices and freedoms" (Gómez & Gasper, n.d., p. 3), human security is concerned with the absence of insecurities in people's everyday life. Such a new security paradigm ought to be better suited to the complex realities and challenges of the twenty-first century (Commission on Human Security, 2003). In particular, when analysing insecurities in relation to the population of the European High North, such a focal shift seems legitimate. Many scholars consider the Arctic to be a region of peace and stability, where conflicts occur between different societal or economic actors within states rather than between states (e.g. Heininen, 2013b; Hossain, Zojer, Greaves, Roncero, & Sheehan, 2017; Nicol & Heininen, 2014; Tamnes & Offerdal, 2014). By shifting attention towards the individual as the referent object of security, human security "sits on the interstices of human rights, human development, and security discourses" (Martin & Owen, 2014, p. 1). It expands security concerns from mere physical integrity to aspects such as culture, identity and human progress. Therefore, it extends "beyond physical violence as the only relevant threat/vector; and beyond physical harm as the only relevant damage" (Gasper, 2014, p. 32). The aim of human security is "to protect the vital core of all human lives in ways that enhance human freedoms and human fulfilment" and, thus, to protect "freedoms that are the essence of life" (Commission on Human Security, 2003, p. 4). Such a comprehensive approach to security includes the "freedom from fear", that is, the protection of individuals from physical violence, which overlaps with traditional security approaches. Yet, it expands the understanding of security to include "freedom from want", that is, the absence of threats to material or ideational freedoms.

As no universally accepted definition of human security exists, many scholars build their definitions on the seven key areas defined in the 1994 HDR, including economic, food, health, environmental, personal, political and community security. While these key areas are all individually important, they are also interconnected or may even conflict with one another. Digital security or cybersecurity has not been considered part of the human security approach thus far, although subsequent HDRs have increasingly included digitalisation-related issues in their reporting, despite never integrating them into security contexts. Due to the ongoing digital changes worldwide, our view is that the lack of discussion of human security in relation to cybersecurity results in further human vulnerabilities that need to be seriously considered and addressed. The lack of a universal or satisfactory definition of human security, as well as the complex relationships between the key areas of human security, however, leave the concept vulnerable to criticism. Krause (2004) notes that in order to be useful, human security must avoid becoming "a loose synonym for 'bad things that can happen'" (p. 367), while Paris (2001) argues that if "human security means almost anything, then it effectively means nothing" (p. 91). Yet, when the concept of human security is not confined to a narrow definition or a strict list of issues and is instead used in a flexible manner, it allows for a deeper understanding of the root causes of insecurity in specific contexts. Additionally, it generates capacities to address those root causes (Tadjbakhsh, 2014). Moreover, human security is not only an academic theory, as it can also be used as a tool for identifying challenges and threats related to specific issues or communities (e.g. Hossain et al., 2017) and, hence, it can be utilised in relation to the policy-making agenda (Floyd, 2007). Therefore, "the very lack of a general theory allows an openness that is important for this kind of work" (Sen, 2014, p. 22). Consequently, utilising or applying human security in a flexible and open manner renders it a meaningful tool for developing responsive policies in thematic or area-specific contexts, and it also serves as a multidimensional analytical framework for assessing threats and challenges to human development (United Nations Trust Fund for Human Security, 2009, 2016).

In the mainstream understanding of security, governmental actors are considered to be the securitising actors, that is, the actors that define threats to societal integrity. However, as discussed above, mainstream cybersecurity policies follow a traditional understanding of security and fall short when it comes to including societal challenges and threats on a sub-state level. When governmental actors consider an issue to be an existential threat or insecurity, it may justify the use of emergency measures (Buzan et al., 1998) or even drive "states to devote disproportionate levels of resources towards a militarised response" (Hoogensen Gjørv, 2012, p. 846). Yet, "[r]isks and insecurities are case- and person-specific, and partly subjective, so human security analysis requires listening to people's 'voices', their fears, perceptions, including the 'voices of the poor' but also of the rich" (Gasper, 2014, p. 34). By addressing the opportunities, risks and challenges associated with digitalisation at a sub-state level, the human security approach can give a voice to individuals and communities with regard to cybersecurity. Moreover, as some challenges that occur in the daily life of people might originate in states' actions or policies, this approach empowers people vis-à-vis states (Hoogensen Gjørv, 2012). Both its flexibility and empowering ability promote the viability of the human security approach as an analytical tool for studying the complex interrelations that exist between digitalisation and human wellbeing in the European High North (see also Zojer, 2019a).

In relation to the intensely digitalising everyday life of the European High North, the application of the human security approach allows for the better realisation of human rights in response to the changing environment. As a policy tool, human security is able to identify urgent issues, thereby allowing for the adoption of measures necessary to repel threats and promote security in a given situation. According to Alkire (2002), the human security approach identifies gaps in the existing security infrastructure, detects ways to mitigate the impact of existing security deficits, strengthens the resilience of individuals and communities against the conditions of insecurity, encourages participatory policy processes and addresses threats both within and across borders. The promotion of human security reinforces both human rights and human development. Therefore, individuals and communities are able to enjoy the "fruits of human development in a safe environment" (Dorn, 2001, para. 12), as interpreted within the normative framework of third-generation human rights, that is, the right to environment or the right to development (Hossain, 2013, p. 496).

Given that digitalisation is changing people's daily life in the European High North and that the human security approach can address the everyday opportunities and challenges brought about by this transformation, the approach ought to be tied to the human rights framework in order to avoid "over-securitisation" or "the loss of focus". Although some scholars perceive tensions between human security and human rights and so claim, for example, that the advancement of human security waters down the firmer legal framework of human rights (Howard-Hassmann, 2012), the frameworks have gained international ground and are compatible. While human rights provide the theoretical and legal rigidity necessary to advance human security, the human security approach provides additional and necessary political flexibility that can serve to advance human rights. The goal of the protection and promotion of human rights is to offer dignity for all individuals as human beings. This dignity lies in the fulfilment of basic needs in a safe environment—very much what human security stands for. However, the problems of human security, unlike those of human rights, are not inherent in the human condition and so are not universally (or equally) applicable to all human beings. Instead, they relate to the circumstances and regional particularities in which people live.

Due to modern societies' increasing dependence on ICT, it is vital to provide and maintain functioning infrastructures and to secure information in all its formats. However, digitalisation significantly affects people's everyday life and the expanding use of ICT has the potential to violate human rights. Therefore, we argue that, similar to the broadening and deepening of the traditional security discourse, a comprehensive cybersecurity approach needs to include the opportunities, risks and challenges that digitalisation brings for individuals and communities in their region-specific settings, as well as to assure that human rights are not violated. Thus, in the following section, we scrutinise how the human rights regime may bring additional firmness to the suggested comprehensive cybersecurity framework by specifying the rights and freedoms of individuals and groups that generate certain obligations on the part of other actors, primarily states.

2.4 Human Rights in the Digitalising European High North

2.4.1 The Human Rights Framework

Human rights are a set of internationally acknowledged rights and interests attributed to individuals and, to a limited extent, to specific groups, such as women, children, minorities or indigenous peoples. In the European High North, such group rights concern the Sámi people, but also other minorities living in the region. These rights and interests are inherent to all human beings, irrespective of their race, sex, ethnicity, religion and so forth. That is, they are universal. Human rights are, on the one hand, inalienable given that they can never be taken away, while on the other hand, they are indivisible, interdependent and interrelated. All human rights—whether political and civil rights or economic, social and cultural rights—enjoy equal importance and status. The denial of one right invariably impedes the enjoyment of other rights. Similarly, the enjoyment of one right cannot be fully realised without the enjoyment of others (United Nations Population Fund [UNFPA], 2005). The human rights regime has been developed in a rather formalised fashion, as it creates an international standard acknowledged by states that is known as international human rights law. The legal development is clear and comprises both customary and treaty law as well as non-binding instruments. In addition, numerous other treaty-based systems, at both the universal and regional levels, include mechanisms for the implementation of human rights and, thus, the completion of these bodies of law.

In general, the human rights framework offers a guarantee for individual rights, which means that individuals are designated as rights holders, while states incur the responsibility of duty bearers. Therefore, a breach committed by a state is expected to prima facie mean that an individual is aggrieved. However, certain violations may affect an entire group in a given context, for example, a group of individuals at a sub-national level or an indigenous community as a whole. Both national and international legal frameworks may offer safeguards in such situations. In the former context, public interest litigation represents a common approach adopted in certain national legal systems whereby an individual asks for remedies on behalf of a group of people. In the latter context, an individual affected "in community with other members" of a group may raise a complaint with human rights monitoring bodies if a state has disregarded its commitment to observe an international standard. For example, the Human Rights Committee (HRC), the treaty-monitoring body under the International Covenant on Civil and Political Rights (ICCPR), can be addressed when remedies at the national level are exhausted.

The ground rule regarding the application of human rights in digitalising daily life is that the rights that apply offline also apply online (e.g. Brown, Esterhuysen, & Knodel, n.d., p. 4). However, as digitalisation does not only refer to the increasing amount of time that human beings spend dealing with ICT, the question stretches further into the foundations of contemporary society. The extent to which, for instance, the infrastructure deemed critical to the functioning of society has been or is being digitalised raises human rights questions that go beyond the freedom of speech and assembly or the right to privacy. If an individual cannot call for help in the case of a serious road accident because there is a simultaneous local blackout in cellular services due to the poor maintenance of the infrastructure in rural areas, the very fundamental right to life comes into question (see also Kirchner, Chapter 6). Therefore, and unsurprisingly, one of the issues currently receiving significant attention in the international debate concerns what human rights entail in the context of the digital transformation and how their realisation is to be organised.

The mainstream human rights instruments, such as the Universal Declaration of Human Rights (UDHR), the ICCPR and the International Covenant on Economic, Social and Cultural Rights (ICESCR), which were drafted between 1948 and 1966, long before the advent of the internet, contain language that supports the recognition of access to the internet as a human right (Handelsman & Kalantar, 2014, p. 33). The most basic rights of all human beings include the freedom of expression, freedom of speech, freedom of opinion, freedom of association and the right to privacy, which are today heavily connected with the use of the internet. Article 19 of the UDHR reads:

> Everyone has the right to freedom of opinion and expression; this right includes freedom to hold opinions without interference and to seek, receive and impart information and ideas through **any media** and regardless of frontiers. (emphasis added)

The right itself applies to an individual's freedom of expression, where the internet can be used as a tool. Article 19(2) of the ICCPR includes similar references to the freedom of expression. Article 12 of the UDHR and Article 17 of the ICCPR guarantee individual freedom from arbitrary interference with one's privacy, family, home or correspondence as well as from any "attacks upon his honour and reputation", which may well occur online. Moreover, the exercise of a number of other rights is increasingly practiced using ICT as a vehicle. This development has prompted a far-reaching discussion on digital rights.

Digital rights, or so-called human rights in the digital era (Hutt, 2015), are a varyingly defined group of rights that relate to digital technologies and that are said to be universal. A number of associations aim to promote digital rights, although they define them only sparsely. For example, the Electronic Frontier Foundation (EFF) lists key "issues" on its website, including free speech (the use of ICT to publish ideas; criticise those in power; gather and report news; make, adapt and share creative works; and allow vulnerable communities to meet, grow and make themselves heard safely), privacy ("[r]espect for individuals' autonomy, anonymous speech, and the right to free association [...] balanced against legitimate concerns like law enforcement"), creativity and innovation ("to protect and strengthen fair use, innovation, open access, net neutrality, and freedom to tinker" so that intellectual property laws "serve the public in a number of ways"), transparency (government use of ICT such as "cell phone location tracking, the use of surveillance drones, secret interpretations of electronic surveillance law, and the expanding use of biometrics"), international advocacy and "security in deep ways including privacy and anonymity, DRM [digital rights management], censorship, and network neutrality". The European Digital Rights (EDRi) organisation, similarly, defines its "central objective [as being] to promote, protect and uphold fundamental human rights and freedoms in the digital environment". Finally, Access Now also lists key "issues" on its website, including privacy, freedom of expression ("which is critical for demonstrating dissent, guaranteeing a free press, and defending human rights"), digital security ("to ensure that [an individual's] online activities are private, safe, and secure"), business and human rights ("companies to make their practices more transparent, accountable, and rights-respecting") and net discrimination ("internet access should be offered to everyone on a nondiscriminatory basis, without favoring certain websites, applications, or services"). As the internet and the technologies that have developed around it are still relatively new and constantly evolving, the discussion of digital rights, their content and their applicability in different situations is likely to continue. What the emerging rights actually entail, as well as how they relate to the formalised human rights regime, remain open questions.

The International Bill of Human Rights (including the aforementioned mainstream human rights instruments) refers to the rights to education, to access the highest attainable standard of physical and mental health, to maintain privacy, to access employment opportunities, to freely participate in the cultural life of a community, to enjoy the arts and to share in scientific advancement and its benefits. The exercise of these rights is becoming increasingly dependent on technology due to the rapid advancement of ICT use. Moreover, the most fundamental of all human rights—the right to life—may be, at least indirectly, threatened by digital activities. For example, a technical disruption or an attack to an information system may jeopardise the right to healthcare offered online. Such a situation has the potential to threaten not only the right to health under Article 12 of the ICESCR, but also the right to life embodied in Article 6 of the ICCPR. In addition, the exercise of such rights as "the right to free speech, taking on the form of hate speech, inciting racial or religious hatred" (Kulesza, 2014, pp. 14–15) may result in societal tensions affecting, for example, the right to life unless states take appropriate steps to secure the online environment from both the dissemination of "hate speech" and the disclosure of individuals' online identities. Similarly, arbitrary interference with an individual's freedom of expression may cause harassment affecting that individual's reputation and honour or imprisonment affecting their personal liberty. Further, the internet has become crucial in relation to exercising procedural human rights, such as the right to participation and inclusion in decision-making processes. These rights are mostly applicable in terms of the promotion of group rights. For example, with regard to indigenous peoples' rights, the International Labour Organization (ILO) Convention No. 169 refers to effective "participation" and "consultation" (Articles 6 and 7. The online environment increasingly offers avenues for such participation and consultation.

The national security focus of the mainstream cybersecurity approach generates further complications, as some of the actions taken to protect the state and/or society may hinder the realisation of human rights and, therefore, prompt a serious human security concern (Cavelty, 2014). This issue has generally been addressed as a necessary balancing act between (national) security and (human) rights (e.g. Hildebrandt, 2013; Kuner, Cate, Millard, & Svantesson, 2013; Naarttijärvi, 2018; Taddeo, 2013). However, the more comprehensive human security approach to cybersecurity enables more nuanced scrutiny of the interconnections between security and rights, and it does not need to position these variables as two ends of a continuum. The freedoms from fear, want and indignity go hand in hand with the realisation of human rights. The enjoyment of human rights, as well as the ability to benefit from using online platforms to exercise them, lie at the core of any democracy. However, the exercise of certain specific human rights can be postponed, in accordance with due process, when in conflict with a state's national security interests. For example, Article 19 of the ICCPR (concerning the freedom of expression) refers to the conditions in which the right can be restricted. While a complete examination of human rights in the context of digitalisation falls outside the scope of the present chapter, the following paragraphs focus on the most relevant digitalisation-related human rights questions in relation to the European High North.

2.4.2 *Right to Internet Access, Human Rights Online and Human Digital Development*

The spirit of human rights in terms of combating inequality or accelerating development and human progress is now dependent on widely available, accessible, affordable and secure connectivity for all segments of the population (Smith, 2012, p. 470). Limiting or denying access to the internet restricts or denies access to information and the ability to freely share one's views. The right to internet access has, therefore, become an important aspect of the discussion concerning the need to protect the substantial integrity of human rights (Kulesza, 2014, pp. 10–11). Access to the internet is a primary pre-requisite for the realisation of human rights online, which the UN Human Rights Council (UNHRC) reaffirmed in 2016 and 2018 (United Nations General Assembly (UNGA), 2016; UNHRC, 2018). The human rights implications of internet access connect the principle of non-discrimination to the issue of network access. It refers to "network equality", which ensures inclusion and equal opportunities for all, not only through political or legal means, but also through the technical functioning of the internet (Kulesza, 2014, pp. 14–15). However, as referred to elsewhere in this chapter, despite the presumption of neutrality found in the use of the term "network equality", the implications of digitalisation are area specific. Thus, they offer distinct challenges to the enjoyment of a number of human rights for a segment of population in the Nordic countries. The empowering aspect of human rights lies in states' obligation to not only prevent a violation of human rights, but also to protect and promote such rights. The protection and promotion of human rights requires the implementation of special measures through *affirmative actions* while taking into consideration the structural realities pertaining to differences in multicultural societies (Romany & Chu, 2004, p. 833). In 2018, the UNHRC reiterated that the intentional disruption or prevention of access to, or the dissemination of, information online represents a violation of international human rights law.

However, the freedom of dissemination of information is not unlimited, given states' obligation to adopt measures intended to restrict the dissemination of information deemed to be criminal or the disclosure of which is deemed criminal. The Convention on Cybercrime (the so-called Budapest Convention, ETS No. 185, 2001), for example, requires the harmonisation and modernisation of criminal law among states in order to prevent computer systems from being used to facilitate the sexual abuse and exploitation of children. Similarly, the EU General Data Protection Regulation (GDPR, EU 2016/679) contains provisions concerning the processing of personal data and its transfer outside the EU/EEA region. All malevolent actions, whether conducted by non-state or state actors, may in one way or another threaten human rights. Additionally, inaction on the part of states may also result in a violation of human rights. For instance, in the case of identity theft or hate speech, states incur responsibility for modifying their national regulations as a measure for promoting human rights. Finland, for example, has relatively recently defined identity theft as a criminal offence (The Criminal Code of Finland 38:9a, amendment 368/2015). Hate speech has not yet been defined in Finnish legislation, although it is punishable as defamation (The Criminal Code of Finland 24:9; 24:10, amendment 879/2013) or incitement to hatred (The Criminal Code of Finland 11:10; 11:10a, amendment 511/2011). Therefore, while the protection of human rights requires the protection of the cyber domain from threats associated with interdependent networks and information infrastructure (Rossini & Green, 2015, p. 10), it also requires the protection of people from one another online. That is, it requires the finding of effective ways to prevent, restrict and sanction such detrimental activities as hate speech, identity theft, defacement, scams, (sexual) exploitation and grooming. Placing human beings at the heart of the cybersecurity discourse and the actions it incentivises represents one way of resisting the prevailing techno-administrative expert understanding of cybersecurity and accommodating it within a more comprehensive framework generated around human rights, wellbeing and democracy (Franklin, 2019, p. 19).

While the internet can be used to promote human rights, it can just as easily be used to violate them. For example, it can be utilised to spread illegal material or to organise criminal activities (Smith, 2012, p. 474). The information infrastructure provides surveillance opportunities, including monitoring, censorship and the exclusion of certain groups or individuals from matters sensitive to the government, thereby violating their rights (Maréchal, 2015, p. 3441). The fast pace of ICT innovation makes it easier for both state and non-state actors to collect, analyse and share data points regarding a person or a group (Maréchal, 2015, p. 3441). As discussed earlier, states incur the legal obligation to prevent human rights violations online, as well as to protect and promote human rights online. The content of such an obligation and the limits of individual privacy permissible under international human rights law represent the core of the challenges posed by online communication (Kulesza, 2014, p. 17). Any unlawful surveillance activity or intervention in with individual's personal liberty is clearly a violation of that individual's human rights. The UNHRC provides clear guidance in its General Comment No. 16 on Article 17 (adopted in 1988) that existing human rights standards should guarantee "the integrity and confidentiality of correspondence ... de jure and de facto" (UN Human Rights Committee, 1988). However, a lawful and non-arbitrary intervention in privacy may be justified, so long as it is in the interests of national security, territorial integrity or public safety. While such an intervention can easily be used to violate human rights, especially in the case of an autocratic regime, the justification for the intervention ought to be the prevention of disorder or crime, the protection of health or morals and/or the protection of reputation, which eventually promotes human rights more generally (e.g. Article 10(2) of the ECHR).

Given the specificity of the European High North, including the pristine environment, the sparsity of the population, the remoteness and distance between settlements, the relatively sparse physical infrastructure, the increasing human activities and the presence of indigenous Sámi and local communities with their own ways of life, the realisation of human rights is increasingly contingent on the availability of both stable and secure digital infrastructure and digital services. Physical services in remote areas are being withdrawn as they are replaced by digital services. However, unless the stability and security of public services run on information infrastructure, as well as the smooth functioning of the infrastructure itself, are ensured, the integrity of human rights may be jeopardised. In the European High North, as people increasingly rely on the internet, insecurities in terms of the integrity of human rights lie in threats to privacy and personal data, the freedom of expression, the freedom of association and so on. However, as discussed earlier, the right to freedom of expression may at times breach other people's enjoyment of the right to maintain personal integrity, honour and reputation. For example, hate speech on social media or via other digital communication channels, particularly against specific groups (such as the Sámi), generates tensions in terms of the human dignity attached to the group. Yet, digitalisation significantly contributes to social progress in favour of the promotion of human rights related to, for example, access to education, healthcare and adequate services, the enjoyment of culture and cultural identity, as well as access to basic needs, such as food, water and energy. In the European High North, such social progress may be seen to be shaped by a constructive understanding of technology, which recognises its cultural meaning(s) to be formed by intrinsic social values attached to its use (see Gulbrandsen & Sheehan, Chapter 5), which eventually contributes to the realisation of the so-called third-generation human rights.

E-health services are of significant importance throughout the European High North, as they indicate the states' adoption of special measures intended to meet the requirement of Article 12 of the ICESCR and Article 25 of the Convention on the Rights of Persons with Disabilities, which both promote the enjoyment of the "highest attainable standard of physical and mental health". Empirical evidence suggests that savings of hundreds of kilometres of travelling to see a doctor for five minutes is a major source of satisfaction among residents (see Gulbrandsen & Sheehan, Chapter 5). In terms of education, the increasing use of e-materials and courses offered online, which is a common practice on the part of educational institutions in the area, represent measures adopted by the states to promote the realisation of the right to education. Finland and Sweden, for instance, have instituted a number of projects to implement education, teaching and learning strategies with digital means (Hammine, 2016; Organisation for Economic Co-operation and Development [OECD], 2008). In addition, the maintenance of ethnic and community cultures includes the right to practice one's own language, given its importance in forming a community identity. Sámi education institutions offer online courses on and in the Sámi languages using interactive digital media such as video conferencing, with the eventual aim being to meet the requirements set out in the framework of the European Charter of Minority Languages (Hammine, 2016, p. 12; Sami Language Education, 2001). In terms of the right to enjoy and take part in the cultural life of a community, as reflected in Article 27 of the ICCPR and Article 15(1) of the ICESCR, it is suggested that culture is an "integrated whole" embracing, among other aspects, all sorts of technological developments (Hossain, 2019). The Sámi, as an indigenous community, enjoy certain distinct rights that are regarded as fundamental to maintain and develop their culture. These rights include, for example, the right to practice traditional livelihoods, such as reindeer herding, and the right to communicate traditional knowledge. Such practices are increasingly being transformed by technological innovation. For instance, in relation to herding practices, Global Positioning System (GPS) tracking methods make it easier to locate the freely roaming reindeer. GPS data are also used by herders to learn about pasture circulation (Hossain, 2019, p. 293). Digitalised narratives concerning Sámi culture offer the means to share their unique culture, traditional knowledge and customary norms and values—if they so choose—thereby complementing the governance of the natural environment on which the enjoyment of human rights is contingent (Mathisen, 2010, p. 64).

The European High North, as part of the broader Arctic region, is governed by a number of high-level institutions of inter-governmental cooperation, including the Arctic Council. The central aims of such institutions include the promotion of environmental protection and sustainable regional development. This entails the adoption of measures intended to foster conditions for the promotion of the right to enjoy a healthy and sound environment, which represents, according to Heinämäki (2010), the right to be part of nature. The right to environment in this context goes hand in hand with the right to development, that is, the promotion of social and economic wellbeing. The right to development, according to Article 1 of the Declaration on the Right to Development (DRD, 1986), is "an inalienable human right by virtue of which every human person and all peoples are entitled to participate in, contribute to, and enjoy economic, social, cultural and political development, in which all human rights and fundamental freedoms can be fully realized". Today, both economic and social development are integrally linked to the digital transformation of economic and societal functions. According to a report produced within the framework of the European Regional Development Fund, sparsely populated rural areas in Europe rely on smaller enterprises, which require high-speed internet infrastructure to connect people and businesses to new markets (Interreg Europe, 2019). High-speed internet provides improved educational opportunities, increases productivity, enables new services and improves the delivery of public services. Greater access to information, transparency and openness lead to a higher level of public participation. The adoption of ICT to support, for instance, public transport, energy systems, waste and water management offers efficiency and effectiveness. According to a study conducted in rural Nordic regions, including Finnmark, digitalisation supports women's empowerment by reducing the digital gender gap and enhancing the inclusiveness of the ICT industry, as a job at the intersection of people and technology is more appealing to women (Nordregio, 2019). ICT and related innovations are, therefore, considered to be "supporting quality of life, higher standards of living, public services for citizens, better use of resources, less impact on the environment, and new opportunities for rural value chains in terms of products and improved processes", thereby complementing the realisation of the right to development (Interreg Europe, 2019). Digitalisation and sustainable development are mutually inclusive, as the former plays a role in promoting a healthy environment through, for example, reducing the need to travel and, thus, reducing energy consumption, carbon emissions and pollution.

Yet, as its development increases ever further, digitalisation also requires ever higher levels of energy. An increasing number of ICT means an increasing demand for electricity to keep the devices and underlying infrastructures running (e.g. Hodgson, 2015; Laakso & Terävä, 2019). Moreover, the production of ICT requires energy and a number of minerals, which have a global impact on both the environment and human rights. Many mineral mines are located in areas that are not known for cherishing human rights (e.g. Scott Jakobsson, 2019). At the other end of the ICT life cycle, the problem of electronic waste and its human rights implications are gaining increasing attention (e.g. Gardiner, 2014; Takkunen & Ortamo, 2018). Digitalisation's impact on the growing problem of space debris has also been acknowledged (e.g. Wei-Haas, 2019). Therefore, the relationships between digitalisation, environmental security and human rights are not as straightforward as many policy reports and much of the literature suggests. Another open question concerns the level of digital awareness and skills that the safe and secure use of ICT demands. The aforementioned Finnish, Swedish and Norwegian strategies highlight the need for improved digital skills among the citizenry. To bridge this recognised gap, for instance, more teaching hours have been allocated to digital skills and media literacy in the national curricula. In Finland, different kinds of awareness campaigns have been run, voluntary support groups established, courses on digital skills and training added to adult education centres' curricula, and obligations to provide help and support upon request placed on libraries. Yet, much remains to be done. The digitalisation of (public) services turns societies into "self-service societies" and increases the demand for skills, although no clear threshold for sufficient digital skills has thus far been established (Päläs & Salminen, 2019). This question is vitally important because some nations, such as Denmark in 2015, have imposed an obligation on all citizens to use digital means for all communication with public authorities, thereby banning the use of paper forms for official matters (Smith, 2012, p. 469). The Finnish and Norwegian strategies also point to digital services as the primary form of public service provision.

The empowering aspects of human rights entail the promotion of people's say in all matters that affect them. Therefore, certain aspects of human rights, such as taking part in decision-making processes, have become norms in recent years and been integrated into procedural practices. Environmental and social impact assessments represent measures taken by both state and non-state actors which is today operationalised through digital platform for instance, the conducting of surveys on the impact of extractive development practices. ICT may allow for more effective community involvement through improved inclusion (Hossain, 2019, p. 294). For the Sámi, this practice translates into an emerging international legal norm of "free, prior and informed consent", as embodied in the UN Declaration on the Rights of Indigenous Peoples. Acknowledging the positive, empowering impact of ICT on people's daily life does not negate the need to address the recognised non-equalising and harmful tendencies embedded within contemporary digital development in the European High North.

2.5 Conclusion

Placing the wellbeing of individuals and communities in their digitalising everyday life environment at the heart of the cybersecurity concept allows for a more comprehensive and detailed understanding of the effects of digital development in the European High North than the current mainstream approach. It provides a holistic view of both the positive, enabling developments and the negative, constraining trajectories. Moreover, it provides a chance for people to participate in the definition of security in the context of non-reversible digital development. By incorporating individuals' and communities' interests, wants and fears in the discussion, it supplements the technology- and/or national security-focused approaches. In addition, through anchoring itself to the human rights regime, it strengthens the application of this framework to the evolving cyber-physical life environment. Part of the problem with the mainstream approach concerns the divided discourse, that is, digitalisation promises to solve problems, including the realisation of human rights, through the further development and application of technology, whereas cybersecurity sees only threats and vulnerabilities and, further, can be used to justify the curtailment of human rights. The human security discourse brings these discussions together so that human rights and (national) security can support one another.

While this chapter, and indeed the entire volume, discusses digitalisation and cybersecurity in the European High North, many of the theoretical and methodological approaches can also be utilised in other contexts. However, the questions specific to the area, for instance, the digitalisation of reindeer herding or the rights of the Sámi, cannot be generalised. The intensification of issues such as network sustainability and the availability of digitalised (public) services due to long distances, the harsh climate and cost-efficiency pressures is not unique to the European High North but actually concerns rural areas in general. The challenges related to, for example, digital awareness and skills are global. A well-contextualised examination of digitalisation and cybersecurity that treats human beings as the referent objects of (and subjects contributing to) security can nonetheless provide insights and surprising perspectives and, thus, improve our understanding of the effects of digital development around the world.

References

Access Now. (n.d.). *Our work*. Retrieved from https://www.accessnow.org/.

Adrian, M., & Brown, D. (2012). Ambiguity and uncertainty in the "last mile": Using sense-making to explore how rural broadband networks are created. *The Journal of Community Informatics, 8*(3). Retrieved from http://ci-journal.net/index.php/ciej/article/view/587/940.

Alkire, S. (2002). *Conceptual framework for human security* (CRISE Working Paper 1(2)). Centre for Research on Inequality, Human Security and Ethnicity, University of Oxford.

BBC. (2017, May 13). NHS cyber-attack: GPs and hospitals hit by ransomware. *BBC News*. Retrieved from https://www.bbc.com/news/health-39899646.

Brown, D., Esterhuysen, A., & Knodel, M. (n.d.). *Briefing document: Cybersecurity policy and human rights*. Retrieved from https://www.apc.org/sites/default/files/brief8.pdf.

Buzan, B., Wæver, O., & de Wilde, J. (1998). *Security: A new framework for analysis*. Boulder, CO: Lynne Rienner.

Carr, M. (2012). The political history of the internet: A theoretical approach to the implications for U.S. Power. In S. S. Costigan & J. Perry (Eds.), *Cyberspaces and global affairs* (pp. 173–188). Farnham: Ashgate.

Cavelty, M. D. (2014). Breaking the cyber-security dilemma: Aligning security needs and removing vulnerabilities. *Science and Engineering Ethics, 20*(3), 701–715.

CBS News. (2016, August 16). Edward Snowden: Alleged latest NSA hack could be warning from Moscow. *CBS News*. Retrieved from https://www.cbsnews.com/news/edward-snowden-alleged-latest-nsa-hack-could-be-warning-from-moscow.

Commission on Human Security. (2003). *Human security now*. New York, NY: United Nations.

Council of Europe. (2001, November 23). *Convention on cybercrime*. Budapest, ETS No. 185. Retrieved from https://www.coe.int/en/web/conventions/full-list/-/conventions/rms/0900001680081561.

de Oliveira Albuquerque, R., Garcia Villalba, L. J., Sandoval Orozco, A. L., de Sousa, R. T., Jr., & Kim, T. H. (2016). Leveraging information security and computational trust for cybersecurity. *The Journal of Supercomputing, 72*(10), 3729–3763.

Dorn, W. (2001). *Human security: An overview*. Ottawa: Pearson Peacekeeping Centre. Retrieved from http://www.walterdorn.org/pub/23.

Dymet, M. (2019). Digital language divide in the European High North: The level of online presence of minority languages from Northern Finland, Norway and Sweden. *The Yearbook of Polar Law, 10*, 247–274.

Electronic Frontier Foundation (EFF). (n.d.). *Issues*. Retrieved from https://www.eff.org/work.

European Commission. (2018). *Digital Transformation Scoreboard 2018*. Retrieved from https://ec.europa.eu/growth/tools-databases/dem/monitor/scoreboard.

European Commission. (2019). *Digital Economy and Society Index (DESI) 2019*. Retrieved from https://ec.europa.eu/digital-single-market/en/desi.

European Digital Rights (EDRi). (n.d.). *Protecting digital freedom*. Retrieved from https://edri.org/.

European Parliament and the Council of the European Union. (2016, April 27). *The European Union General Data Protection Regulation (GDPR)*. EU 2016/679. Retrieved from https://eur-lex.europa.eu/legal-content/EN/TXT/PDF/?uri=CELEX:32016R0679.

Floyd, R. (2007). Human security and the Copenhagen school's securitization approach. *Human Security Journal, 5*, 38–49.

Franklin, M. I. (2019). Human rights futures for the internet. In B. Wagner, M. Kettemann, & K. Vieth (Eds.), *Research handbook on human rights and digital technology. Global politics, law and international relations* (pp. 5–23). Cheltenham: Edward Elgar.

Gardiner, B. (2014, December 8). The hidden environmental cost of the Internet of Things. *Computerworld*. Retrieved from https://www.computerworld.com.au/article/561064/hidden-environmental-cost-internet-things/.

Gasper, D. (2014). Human security: From definitions to investigating a discourse. In M. Martin & T. Owen (Eds.), *Routledge handbook of human security* (pp. 28–42). London and New York, NY: Routledge/Taylor Francis Group.

Gebeloff, R., & Russell, K. (2017, June 6). How the growth of e-commerce is shifting retail jobs. *The New York Times*. Retrieved from https://www.nytimes.com/interactive/2017/07/06/business/ecommerce-retail-jobs.html.

Gómez, O. A., & Gasper, D. (n.d.). *Human security guidance note. A thematic guidance note for regional and national human development report teams*. Retrieved from http://hdr.undp.org/sites/default/files/human_security_guidance_note_r-nhdrs.pdf.

Government Offices of Sweden, Ministry of Justice. (2016). *A National Cyber Security Strategy*. Skr. 2016/17:213. Retrieved from https://www.government.se/4970ac/contentassets/d87287e088834d9e8c08f28d0b9dda5b/a-national-cyber-security-strategy-skr.-201617213.

Hammine, M. (2016). *Sami language education in Sweden and Finland* (European Centre for Minority Issues [ECMI] Working Paper No. 92). Retrieved from https://www.ecmi.de/fileadmin/redakteure/publications/pdf/Working_Paper___92.pdf.

Handelsman, J., & Kalantar, E. (2014). *Life, liberty, and internet access: Is logging on a basic human right?* Retrieved from https://ublawjil.files.wordpress.com/2014/07/7-life-liberty-and-internet-access-kalantar.pdf.

Heinämäki, L. (2010). *The right to be a part of nature—Indigenous peoples and the environment*. Acta Universitatis Lapponiensis 180. Rovaniemi: Lapland University Press.

Heininen, L. (2013a). "Politicization" of the environment, and environmental politics and security in the Circumpolar North. In B. S. Zellen (Ed.), *The fast-changing Arctic: Rethinking Arctic security for a warmer world* (pp. 35–55). Calgary: University of Calgary Press.

Heininen, L. (2013b). Security in the Arctic. In N. Loukacheva (Ed.), *Polar law textbook II* (pp. 37–52). Copenhagen: Norden.

Hildebrandt, M. (2013). Balance or trade-off? Online security technologies and fundamental rights. *Philosophy & Technology, 26*(4), 357–379.

Hodgson, C. (2015, November 13). Can the digital revolution be environmentally sustainable? *The Guardian*. Retrieved from https://www.theguardian.com/global/blog/2015/nov/13/digital-revolution-environmental-sustainable.

Hoogensen Gjørv, G. (2012). Security by any other name: Negative security, positive security, and a multi-actor security approach. *Review of International Studies, 38*(4), 835–859. https://doi.org/10.1017/S0260210511000751.

Hoogensen Gjørv, G., Bazely, D. R., Goloviznina, M., & Tanentzap, A. J. (Eds.). (2014). *Environmental and human security in the Arctic*. London and New York, NY: Earthscan from Routledge.

Hossain, K. (2013). Securing the rights: A human security perspective in the context of Arctic indigenous peoples. *The Yearbook of Polar Law, 5*, 493–522. https://doi.org/10.1163/22116427-91000134.

Hossain, K. (2019). The evolving information-based society and its influence on traditional culture: Framing community culture and human security of the Sámi in the European High North. *The Yearbook of Polar Law, 10*, 275–296.

Hossain, K., Zojer, G., Greaves, W., Roncero, J. M., & Sheehan, M. (2017). Constructing Arctic security: An inter-disciplinary approach to understanding security in the Barents region. *Polar Record, 53*(1), 52–66. https://doi.org/10.1017/S0032247416000693.

Howard-Hassmann, R. E. (2012). Human security: Undermining human rights? *Human Rights Quarterly, 34*, 88–112.

Hutt, R. (2015, November 13). What are your digital rights? *World Economic Forum*. Retrieved from https://www.weforum.org/agenda/2015/11/what-are-your-digital-rights-explainer/.

Interreg Europe. (2019). *The challenges and necessity of rural innovation*. A Policy Brief from the Policy Learning Platform on Research and Innovation. Retrieved from https://www.interregeurope.eu/fileadmin/user_upload/plp_uploads/policy_briefs/2019-01-21_TO1_policy_brief_Rural_innovation_final.pdf.

Keskitalo, E. C. H., & Southcott, C. (2014). Globalization. In J. Nymand Larsen & G. Fondahl (Eds.), *Arctic human development report: Regional processes and global linkages* (pp. 397–422). Copenhagen: Nordic Council of Ministers.

Kilpeläinen, A. (2016). *Teknologiavälitteisyys kyläläisten arjessa. Tutkimus ikääntyvien sivukylien teknologiavälitteisyydestä ja sen rajapinnoista maaseutusosiaalityöhön* [Technology mediatedness in the everyday life of people living in rural villages. A study of technology mediatedness in ageing villages and interfaces with social work in the countryside]. Acta Universitatis Lapponiensis 316. Rovaniemi: University of Lapland Press.

Kotituomi, S., Huju, M., Kulmala, R., & Lahtela, A. (2019). *Selvitys Porokello-varoitusjärjestelmän vaikutuksista* [Impact evaluation of the Porokello alert service]. Traficomin tutkimuksia ja selvityksiä 9/2019 [Traficom Research Reports 9/2019].

Krause, K. (2004). The key to a powerful agenda, if properly delimited. *Security Dialogue, 35*(3), 367–368. https://doi.org/10.1177/096701060403500324.

Kulesza, J. (2014). *Protecting human rights online—An obligation of due diligence* (Jean Monnet Working Paper 24/14). New York, NY: NYU School of Law. Retrieved from http://jeanmonnetprogram.org/wp-content/uploads/2015/04/JMWP24Kulesza.pdf.

Kuner, C., Cate, F. H., Millard, C., & Svantesson, D. J. B. (2013). PRISM and privacy: Will this change everything? Editorial. *International Data Privacy Law, 3*(4), 217–219. https://doi.org/10.1093/idpl/ipt020.

Kurtti, T., & Berg, K.-M. (2018). Digitalisation among NGOs—The case of the Toimeksi.fi online service. In K. Hossain & A. Petrétei (Eds.), *Unconventional aspects of human security in the Arctic—Cases from the Barents Region* (pp. 100–128). Juridica Lapponica 45. Rovaniemi: University of Lapland. http://urn.fi/URN:ISBN:978-952-337-122-4.

Laakso, V., & Terävä, H. (2019, June 26). Netti syö kasvavalla tahdilla sähköä ja suurin syyllinen ovat nettivideot – "Epämiellyttävä totuus, josta ei haluta puhua" [Internet consumes electricity at an increasing rate and videos are the guilty party—"An unpleasant truth nobody wants to talk about"]. *YLE*. Retrieved from https://yle.fi/uutiset/3-10832413.

Lindell, J. (2018). Distinction recapped: Digital news repertoires in the class structure. *New Media & Society, 20*(8), 3029–3049. https://doi.org/10.1177/1461444817739622.

Maréchal, N. (2015). Ranking digital rights: Human rights, the internet and the fifth estate. *International Journal of Communication, 9,* 3440–3449.

Martin, M., & Owen, T. (2014). Introduction. In M. Martin & T. Owen (Eds.), *Routledge handbook of human security* (pp. 1–14). London and New York, NY: Routledge/Taylor Francis Group.

Mathisen, S. R. (2010). Indigenous spirituality in the touristic borderzone: Virtual performances of Sámi shamanism in Sápmi Park. *The Finnish Society for the Study of Religion, 46*(1), 53–72.

McGwin, K. (2019, July 11). In Swedish Lapland, drone deliveries may soon expand rural health services. *Arctic Today.* Retrieved from https://www.arctictoday.com/in-swedish-lapland-expanding-health-services-from-above/.

McSweeney, B. (1999). *Security, identity, and interests: A sociology of international relations.* Cambridge and New York, NY: Cambridge University Press.

Mearsheimer, J. J. (2014). *The tragedy of great power politics* (Updated ed.). New York, NY: W. W. Norton.

Mendez Garcia, O. P. (2011). Gender digital divide. The role of mobile phones among Latina farm workers in Southeast Ohio. *Gender, Technology and Development, 15*(1), 53–74. https://doi.org/10.1177/097185241101500103.

Mesch, G. S. (2012). Minority status and the use of computer-mediated communication: A test of the social diversification hypothesis. *Communication Research, 39*(3), 317–337.

Ministry of Justice, Finland. (2015). *The Criminal Code of Finland.* 39/1889 (amendments up to 766/2015 included). Retrieved from https://www.finlex.fi/en/laki/kaannokset/1889/en18890039.pdf.

Ministry of the Interior. (2019). *Kansallinen riskiarvio 2018* [National risk assessment 2018]. Publications of the Ministry of the Interior 2019:5. Retrieved from http://urn.fi/URN:ISBN:978-952-324-245-6.

Ministry of Transport and Communications. (2016, April 19). *Information security strategy for Finland. The world's most trusted digital business environment.* Publications 9/2016. Retrieved from https://julkaisut.valtioneuvosto.fi/bitstream/handle/10024/78416/Publications_9-2016_Information_Security_Strategy_for_Finland.pdf?sequence=1.

Naarttijärvi, M. (2018). Balancing data protection and privacy—The case of information security sensor systems. *Computer Law & Security Review: The International Journal of Technology Law and Practice, 34*(5), 1019–1038. https://doi.org/10.1016/j.clsr.2018.04.006.

Nam, S.-J., & Park, E.-Y. (2016). The effects of the smart environment on the information divide experienced by people with disabilities. *Disability and Health Journal, 10*(2), 257–263. https://doi.org/10.1016/j.dhjo.2016.11.001.

Nicol, H. N., & Heininen, L. (2014). Human security, the Arctic Council and climate change: Competition or co-existence? *Polar Record, 50*(1), 80–85. https://doi.org/10.1017/S0032247412000666.

Nordregio. (2019). *Governing the digital transition in the Nordic regions* (Policy Brief No. 8). Retrieved from http://norden.diva-portal.org/smash/get/diva2:1332131/FULLTEXT01.pdf.

Norwegian Ministries. (2012). *Cyber security strategy for Norway*. Retrieved from https://www.regjeringen.no/globalassets/upload/fad/vedlegg/ikt-politikk/cyber_security_strategy_norway.pdf.

Norwegian Ministry of Foreign Affairs. (2017, August 31). *International cyber strategy for Norway*. Retrieved from https://www.regjeringen.no/globalassets/departementene/ud/dokumenter/sikpol/cyberstrategy_2017.pdf.

Organisation for Economic Co-operation and Development (OECD). (2008). *OECD study on digital leaning resources as systemic innovation—Country case study report on Finland*. Retrieved from http://www.oecd.org/education/ceri/41951860.pdf.

Päläs, J., & Salminen, M. (2019). Alustan asiakkaan vastuusta ja vastuuttamisesta yksilöturvallisuuden tuottamisessa – sopimusoikeudellinen näkökulma kyberturvallisuuteen jakamistaloudessa [Responsibility and responsibilisation of the platform user in the production of individual security—A contract law perspective to cybersecurity in sharing economy]. In J. Päläs & K. Määttä (Eds.), *Jakamistalousjuridiikan käsikirja* [Handbook of sharing economy and law] (pp. 319–380). Helsinki: Alma Talent.

Paris, R. (2001). Human security: Paradigm shift or hot air? *International Security, 26*(2), 87–102. https://doi.org/10.1162/016228801753191141.

Perlroth, N., & Shane, S. (2019, May 26). In Baltimore and beyond, a stolen N.S.A. tool wreaks havoc. *The New York Times*. Retrieved from https://www.nytimes.com/2019/05/25/us/nsa-hacking-tool-baltimore.html.

PostNord. (2017). *E-commerce in the Nordics 2017*. Retrieved from https://www.postnord.com/globalassets/global/english/document/publications/2017/e-commerce-in-the-nordics-2017.pdf.

Rätti, O., & Wallén, H. (2017). *Porotalouden digitalisoituminen. Loppuraportti* [Digitalisation of reindeer husbandry. Final report]. Lapin yliopiston Arktinen keskus & Saamelaisalueen koulutuskeskus.

Romany, C., & Chu, J. (2004). Affirmative action in international human rights law: A critical perspective of its normative assumptions. *Connecticut Law Review, 36*, 831–870.

Rossini, C., & Green, N. (2015). *Cyber security and human rights*. Webinar Series Training Summaries. Retrieved from https://www.gp-digital.org/wp-content/uploads/2015/06/GCCS2015-Webinar-Series-Introductory-Text.pdf.

Salminen, M. (2018). Digital security in the Barents Region. In K. Hossain & D. Cambou (Eds.), *Society, environment and human security in the Arctic Barents Region* (pp. 187–204). Abingdon: Routledge.

Salminen, M. (2019). Refocusing and redefining cybersecurity: Individual security in the digitalising European High North. *The Yearbook of Polar Law, 10*, 321–356.

Salminen, M., & Hossain, K. (2018). Digitalisation and human security dimensions in cybersecurity: An appraisal for the European High North. *Polar Record, 54*(2), 108–118. https://doi.org/10.1017/S0032247418000268.

Sami Language Education. (2001). *The Sami language in education in Sweden*. Retrieved from https://www.mercator-research.eu/fileadmin/mercator/documents/regional_dossiers/saami_in_sweden.pdf.

Schou, J., & Hjelholt, M. (2018). *Digitalization and public sector transformations*. Cham: Palgrave Macmillan.

Scott Jakobsson, L. (2019, May 6). *Copper with a cost. Human rights and environmental risks in the mineral supply chains of ICT: A case study from Zambia* (SwedWatch, Report No. 94). Alice Blondel. Retrieved from https://swedwatch.org/wp-content/uploads/2019/05/94_Zambia_uppslag.pdf.

Secretariat of the Security Committee. (2013). *Finland's cybersecurity strategy*. Government Resolution 24.1.2013. Retrieved from https://www.defmin.fi/files/2378/Finland_s_Cyber_Security_Strategy.pdf.

Sen, A. (2014). Birth of a discourse. In M. Martin & T. Owen (Eds.), *Routledge handbook of human security* (pp. 17–27). London and New York, NY: Routledge/Taylor Francis Group.

Shane, S., Perlroth, N., & Sanger, D. (2017, November 12). Security breach and spilled secrets have shaken the N.S.A. to its core. *New York Times*. Retrieved from https://www.nytimes.com/2017/11/12/us/nsa-shadow-brokers.html?module=inline.

Smith, P. S. (2012). Imprisonment and internet-access human rights, the principle of normalization and the question of prisoners' access to digital communications technology. *Nordic Journal of Human Rights, 30*(4), 454–482.

Stevens, T. (2018). Global cybersecurity: New directions in theory and methods. Editorial. *Politics and Governance, 6*(2), 1–4. https://doi.org/10.17645/pag.v6i2.1569.

Subramony, D. P. (2007). Understanding the complex dimensions of the digital divide: Lessons learned in the Alaskan Arctic. *The Journal of Negro Education, 76*(1), 57–67. https://www.jstor.org/stable/40026330.

Taddeo, M. (2013). Cyber security and individual rights, striking the right balance. *Philosophy and Technology, 26*(4), 353–357.

Tadjbakhsh, S. (2014). In defense of the broad view of human security. In M. Martin & T. Owen (Eds.), *Routledge handbook of human security* (pp. 43–57). London and New York, NY: Routledge/Taylor Francis Group.

Takkunen, J., & Ortamo, S. (2018, December 22). Euroopan elektroniikkajäte kertyy Ghanan romupihoille – Baba Adi lajittelee tietokoneita ja vaarantaa terveytensä, koska romu tuo rahaa [Electronic waste from Europe piles up on the junkyards of Ghana—Baba Adi sorts computers and risks his life because scrap turns into money]. *YLE*. Retrieved from https://yle.fi/uutiset/3-10472211.

Tamnes, R., & Offerdal, K. (Eds.). (2014). *Geopolitics and security in the Arctic: Regional dynamics in a global world*. Oxford and New York, NY: Routledge.

Turvallisuuskomitean sihteeristö [Secretariat of the Security Committee]. (2019). *Suomen kyberturvallisuusstrategia 2019. Valtioneuvoston periaatepäätös 3.10.2019* [Finland's Cybersecurity Strategy 2019. Government Resolution 3.10.2019]. Retrieved from https://turvallisuuskomitea.fi/wp-content/uploads/2019/10/Kyberturvallisuusstrategia_A4_SUOMI_WEB_300919.pdf.

United Nations Development Programme (UNDP). (1994). *Human development report 1994*. New York, NY: Oxford University Press.

United Nations General Assembly (UNGA). (1986). *Declaration on the Right to Development (DRD)*. Retrieved from https://legal.un.org/avl/ha/drd/drd.html.

United Nations General Assembly (UNGA). (2016). *The promotion, protection and enjoyment of human rights on the Internet. Human Rights Council*. Thirty-second session. A/HRC/32/L.20. Retrieved from https://www.article19.org/data/files/Internet_Statement_Adopted.pdf.

United Nations Human Rights Committee. (1988, April 8). *CCPR General Comment No. 16: Article 17 (Right to Privacy), The right to respect of privacy, family, home and correspondence, and protection of honour and reputation*. Retrieved from https://www.refworld.org/docid/453883f922.html.

United Nations Human Rights Council (2018). *The promotion, protection and enjoyment of human rights on the Internet*. General Assembly. No. 38. A/HRC/38/L.10/Rev.1. July 4, 2018.

United Nations Population Fund (UNFPA). (2005). *Human rights principles*. Retrieved from https://www.unfpa.org/resources/human-rights-principles.

United Nations Secretary General. (2005). *In larger freedom: Towards development, security and human rights for all*. Report of the Secretary-General, A/59/2005.

United Nations Trust Fund for Human Security. (2009). *Human Security in theory and practice. An overview of the Human Security concept and the United Nations Trust Fund for Human Security*. Retrieved from http://www.un.org/humansecurity/sites/www.un.org.humansecurity/files/human_security_in_theory_and_practice_english.pdf.

United Nations Trust Fund for Human Security. (2016). *Human Security Handbook. An integrated approach for the realization of the Sustainable Development Goals and the priority areas of the international community and the United Nations system*. Retrieved from https://www.un.org/humansecurity/wp-content/uploads/2017/10/h2.pdf.

von Solms, B., & von Solms, R. (2018). Cybersecurity and information security—What goes where? *Information and Computer Security, 26*(1), 2–9. https://doi.org/10.1108/ics-04-2017-0025.

von Solms, R., & van Niekerk, J. (2013, October). From information security to cyber security. *Computers and Security, 38,* 97–102. https://doi.org/10.1016/j.cose.2013.04.004.

Waltz, K. N. (2010). *Theory of international politics.* Long Grove, IL: Waveland Press.

Webster, F. (2006). *Theories of the information society* (3rd ed.). Abingdon: Routledge.

Wei-Haas, M. (2019, April 25). Space junk is a huge problem—And it's only getting bigger. *National Geographic.* Retrieved from https://www.nationalgeographic.com/science/space/reference/space-junk/.

Zambianchi, M., Rönnlund, M., & Grazia Carelli, M. (2019). Attitudes towards and use of information and communication technologies (ICTs) among older adults in Italy and Sweden: The influence of cultural context, socio-demographic factors, and time perspective. *Journal of Cross-Cultural Gerontology*, 1–16. https://doi.org/10.1007/s10823-019-09370-y.

Zojer, G. (2019a). The interconnectedness of digitalisation and human security in the European High North: Cybersecurity conceptualised through the human security lens. *The Yearbook of Polar Law, 10,* 297–320. https://doi.org/10.1163/22116427_010010014.

Zojer, G. (2019b). Free and open source software as a contribution to digital security in the Arctic. *Arctic Yearbook*, 173–188.

CHAPTER 3

The New Frontier for Human Cybersecurity: Russia's Cybersecurity Policies in the Arctic

Nadezhda Filimonova and Mario Portugal-Ramirez

3.1 Introduction

The Arctic Zone of the Russian Federation[1] (AZRF) is strategically important in relation to the state's economic, military, and geopolitical interests (Konyshev, Sergunin, & Subbotin, 2017; Zhuravel, 2016). The Russian government links the development of energy and shipping projects in the AZRF to the country's socio-economic development, energy security, and global status as one of the leading exporters of mineral resources to the international energy markets.

[1]The AZRF is defined in the 2014 Presidential Decree № 296, which spelled out the land area of the Arctic.

N. Filimonova (✉) · M. Portugal-Ramirez
University of Massachusetts Boston, Boston, MA, USA
e-mail: N.Filimonova001@umb.edu

M. Portugal-Ramirez
e-mail: M.PortugalRamirez001@umb.edu

M. Salminen et al. (eds.),
Digitalisation and Human Security, New Security Challenges,
https://doi.org/10.1007/978-3-030-48070-7_3

In recent years, federal and regional governments in cooperation with business have elaborated plans to invest hundreds of billions of dollars in sectors of the regional economy such as mining, transportation, energy, and communications (Konyshev et al., 2017). In addition, the governments at various levels have also developed socio-economic programmes intended to maintain and bolster the shrinking population of the AZRF (Sokolova, 2016). The goal has been to guarantee the presence and implementation of economic projects in the AZRF.

Importantly, the growing significance of the AZRF requires the state to ensure that its national security is guaranteed. In addition to a military presence, telecommunications are key to Russia's security policy goals (Datyev, 2014). The policy targets prescribe the creation of modern information infrastructure that will allow the monitoring of the ice cover and the movement of ships, as well as providing hydrological, meteorological, and navigational services. In addition, the Russian government intends to bestow the AZRF with a better connection to the internet and more affordable mobile technology prices (Arctic Council Task Force on Telecommunications Infrastructure in the Arctic [TFTIA], 2017).

Alongside efforts to connect the AZRF to information networks, concerns about cybersecurity have arisen. In particular, the disruptions to critical infrastructure—the physical and cyber systems and assets—caused by hackers and cyberterrorism activities might have a tangible impact on the functioning of both governments and businesses (de Oliveira Albuquerque, García Villalba, Sandoval Orozco, de Sousa Júnior, & Kim, 2016). In 2012, in Russia, the total damage caused by cyberattacks amounted to some $2 billion (Tamitsky, 2013). In the AZRF, one of the primary targets of cyberattacks might be the telecommunication systems of oil rigs and ships that operate in the region.

The population of the AZRF is also vulnerable to cyberattacks. In 2017, around 86.9% of the total population of the AZRF reported using the internet (Federal Statistics, 2017). In a similar vein, an operation conducted in cyberspace—virtual computer networks—exposes individuals to cyber threats, including financial frauds and the loss of private data.

In general, illegal actions such as these are facilitated by a lack of efficient transnational and national security regulations and security structures (Shannon & Thomas, 2005). For this reason, since the early 2000s, Russia has been developing various strategies and introducing changes to the related legislation. Following the adoption of the 2008 Arctic Strategy, Russia has started to underscore the importance of measures intended to enhance the application of Information and Communication

Technologies (ICT), as well as to create a common information space for the Arctic territories (The Government of the Russian Federation, 2008).

This chapter examines how the cybersecurity narrative was framed by the Russian government in relation to the AZRF during the period from 2000 to 2017. The choice of this historical period is based on the fact that the Russian government adopted the key doctrines and legislation concerning information security and national Arctic strategies.

The prior academic literature regarding Russian Arctic security policy mainly focuses on environmental and military security in the Arctic (Gladun, 2015; Konyshev & Sergunin, 2014; Yagya & Sboychakova, 2016; Zagorski, 2018). This chapter fills the resultant research gap by contributing knowledge about the official Russian framing of cybersecurity. The chapter argues that a human security approach can contribute to making cybersecurity policies more inclusive to incorporate the needs and interests of people residing in the Arctic.

The present study has five sections. The first section presents existing academic literature on the notion of cybersecurity and identifies an analytical framework. The second and third sections present the case of the official Russian cybersecurity policy in the AZRF and outline the applied methodology, respectively. In the findings and discussion section, the chapter presents the major results within the frames of the three investigated themes (political security, community security, and economic security). Finally, in conclusion, the chapter explains how the framed Russian cybersecurity policy is related to the human security of individuals living in the Arctic.

3.2 Literature Review and Analytical Framework

The academic literature underscores the existence of multiple conceptions of cybersecurity (Schatz, Bashroush, & Wall, 2017). Craigen, Diakun-Thibault, and Purse (2014) identify five main themes that are applied by scholars to the concept of cybersecurity, namely: "(1) technological solutions; (2) events; (3) strategies, processes, and methods; (4) human engagement; and 5) referent objects (of security)" (p. 15). In general, the academic definitions include elements such as threats, referent objects, and means of protection. The approaches to the operationalisation of cybersecurity encompass various types of threats, including cyberattacks (Priyadarshini, 2019; von Solms & van Niekerk, 2013), cyberterrorism (Nam, 2019), occurrences (intentional and accidental events, natural hazards) (Craigen et al., 2014), and information risks (Anderson, 2003). The definitions also include myriad types of referent objects, such as cyberspace and cyber-enabled systems (Craigen et al., 2014), cyber users (Schatz et al., 2017; von Solms & van Niekerk, 2013), interests of a person, society and nation (Reid & van Niekerk, 2014), the integrity, confidentiality and availability of data and assets (Schatz et al., 2017), and the information system's infrastructure (telecommunications networks, computer systems, embedded processors and controllers) (de Oliveira Albuquerque et al., 2016). Depending on the referent object, the means of protection include guidelines, policies, training, tools (Schatz et al., 2017), the organisation and collection of processes and structures (Craigen et al., 2014), and the assurance that risks and controls are in balance (Anderson, 2003).

Aside from academic studies, the content of cybersecurity has also been explored by certain international organisations and states. For example, the International Telecommunication Union (ITU, 2009), a specialised agency of the United Nations (UN), has defined cybersecurity as:

> [t]he collection of tools, policies, security concepts, security safeguard, guidelines, risk management approaches, actions, training, best practices, assurance and technologies that can be used to protect the cyber environment and organization and user's assets. (p. 2)

The ITU's definition reflects the human dimension of cybersecurity by portraying a person as a user of telecommunication technologies, as well as an owner of the information that is deployed in cyberspace.

Currently, more than 50 countries have officially adopted strategic documents related to cyberspace, cybercrime, and/or cybersecurity (von Solms & van Niekerk, 2013). In the case of Russia, cybersecurity is viewed as part of the concept of information security. More specifically, the access to, sharing and protection of public and private electronic information are regulated in accordance with Russia's official information security policy. In this regard, von Solms and van Niekerk (2013) identify similarities in the definitions of the two concepts, such as their common goal being to preserve the confidentiality, integrity, and accessibility of public and private information.

Overall, it is worth noting that the above-mentioned approaches to the concept of cybersecurity share a common limitation. That is, the deployment of cybersecurity is blind to individual and group characteristics. In fact, any cyber threat has the same effect on an abstract group of individuals; therefore, their specific needs and interests are not considered. To address the identified definitional gap, several scholars have argued for the redefinition of the concept of cybersecurity by incorporating human security dimensions, specifically the needs, fears, and interests of ordinary people (Dunn Cavelty, 2014; Salminen & Hossain, 2018). The notion of cybersecurity, therefore, needs to integrate analytical tools that can capture with greater accuracy the human dimension of cybersecurity, thereby encompassing the diversity of individuals and groups with security needs.

In this regard, the concept of human security represents an important analytical tool for redefining cybersecurity. It has emerged as a reaction to the inability of traditional security studies to look beyond a state's security and to incorporate individuals as a referent of security (Wibben, 2016). Human security was initially defined in the 1994 United Nations Development Programme (UNDP) report as a person-centric approach to security involving seven types of security, namely food, economic, health, environmental, community, political, and personal security. The concept thus moves away from military security by underscoring how other realms of reality can be threatened, such as economic, environmental, and health security. It also tackles the security of states, societies, groups, and individuals, which should all have same importance (Paris, 2001).

Human security evolved following the publication of the UNDP report, reaching new meanings and areas of interest. During an allocution in 2000, the former UN Secretary-General, Kofi Annan, described human security as the opportunities and choices that lead to an individual's full potential (Annan, 2000). In 2001, the International Commission on Intervention and State Sovereignty (ICISS) mentioned human security in a report entitled *The Responsibility to Protect*. Despite its military-oriented content, the report identified human security as "the security of people – their physical safety, their economic and social well-being, respect for their dignity and worth as human beings, and the protection of their human rights and fundamental freedoms" (ICISS, 2001, p. 15). Some countries have also attempted to operationalise the concept of human security. For instance, in 2002, Canada identified five foreign policy areas related to human security, namely the protection of civilians, peace support operations, conflict prevention, governance and accountability, and public safety (Department for Foreign Affairs and International Trade [DFAIT], 2002).

With regard to academia, scholars acknowledge the contribution of human security to the critique of classical security studies, which often focus on a state's security. However, critical scholars have noted its ambiguity and broadness (Paris, 2001; Wibben, 2016). They argue how it has quickly become a policy-oriented concept (Newman, 2016), and how it has been co-opted by governments, "allowing them to tick 'the good international citizen' box of foreign policy" (Booth, 2007, pp. 323–234). Other scholars have argued that its integration into international policymaking circles merely covers up the fact that they have little impact on policy outcomes, thereby turning human security into a sort of "dog that didn't bark" (Chandler, 2008, p. 427). Moreover, certain Foucauldian analyses of biopower have argued that human security helps capitalist socio-political regimes to extend their power and control over the corporeality of the individual (Newman, 2016).

However, the employment of the human security concept offers several advantages. As a security objective, it emphasises the primacy of the individual and her life; it broadens the scope of security analysis (Fukuda-Parr & Messineo, 2012) and accurately addresses transnational threats (Thomas & Tow, 2002). The definition of cybersecurity from a human security perspective, therefore, should overcome the state/non-state dichotomy and help to position cybersecurity as a concern that involves threats to the state and the individual. In addition, it highlights how even if threats occur in cyberspace, their effects might also be felt in the material world.

Thus, this chapter defines cybersecurity as a set of policies and actions implemented by cyber users (state, private, and public users) to protect cyberspace, ICT, confidentiality, and the integrity of data. It also shields cyber users' interests and rights with regard to the application, transmission, and storage of data. The threats that cyberspace deals with must be understood and tackled in light of their effects on three aspects of human security: (1) economic security (an assured basic income, usually from productive and remunerative work), (2) community security (the survival of traditional cultures and ethnic groups, as well as the physical security of such groups), and (3) political security (the enjoyment of civil and political rights, as well as freedom from political oppression).[2]

[2]We acknowledge that such threats can also affect the other dimensions of human security included in the UNDP's definition. However, the choice of these three dimensions is based on the following factors. They include the impossibility of accessing certain data (interviews) and time constraints placed on the research. In addition, these elements reflect Russian strategic interests in the Arctic.

The use of a general definition of cyber users and their respective interests and rights makes it possible to unpack their respective notions by examining the official Russian framing of cybersecurity and its elements. Further, the application of this definition is based on the intention to move away from the dominant state-centric approach to cybersecurity. The chapter assumes that the responsibility to provide cybersecurity protection is assigned to a wide range of cyber users, including states, businesses, and individuals.

With regard to the Arctic, the critical interests of the state are reflected in the protection from cyber threats of critical infrastructure such as military bases, monitoring systems, mining factories, and oil rigs. Meanwhile, the interests of the local population are represented by the availability of the internet and their capacity to use online services.

3.3 Methodology

This chapter employs a single case study methodology to investigate the official Russian cybersecurity policy with regard to the AZRF. A single case study provides the researcher with an opportunity to examine a wide range of intervening variables, as well as the impacts of contextual factors (George & Bennett, 2005). Additionally, this chapter also represents a theoretical/configurative idiographic case study that does not contribute directly to theory building but it does produce outcomes that might be used in future theoretical research (George & Bennett, 2005). The use of a theoretical/configurative case study reflects the research goal of examining the human dimension of cybersecurity at the official level in Russia with the intention of redefining it in future studies. In addition, the application of single case study allows to grasp the influence of contextual factors such as Russian national security, its economic interests, and socio-economic situation in the AZRF. Lastly, the analysis of a specific case study provides knowledge about whether and, if so, how the needs and interests of a certain group are incorporated into the official cybersecurity discourse.

The choice of the case study is based on several factors. First, the AZRF is critically important for Russia's economic security, energy security, and military security. The AZRF is responsible for over 15% of Russia's gross domestic product (GDP) and 20% of its exports (Hunter, 2017). Currently, the Russian government views its economic development to be tied to the implementation of ambitious energy and shipping projects in the AZRF. More specifically, in 2008, during the Security Council meeting, President Dmitry Medvedev (2008–2012) announced a plan to convert the AZRF into a twenty-first-century resource base so as to compensate for the dwindling resources in the West Siberia (Ilyn, 2008). The development of oil and gas resources in the AZRF is closely connected to Russia's second ambitious project of boosting shipping through the Northern Sea Route (NSR) for domestic and international commercial purposes. Primarily, the NSR is viewed by the Russian government as an important route for exporting oil and gas extracted in the AZRF to the European and Asian markets.

Second, by 2015 almost half of the Arctic population, around 2.4 million people lived in the AZRF (Efremova, Didenko, Rudenko, & Skripnuk, 2017). Specifically, more than 80% of the Russian Arctic population lives in towns and urban settlements (Fauzer, Lytkina, & Fauzer, 2016). In total, in the AZRF, there are around 58 large municipalities and 200 settlements, with some of them being geographically remote and distant from each other (Simakova & Kozyreva, 2018). This means that a significant portion of the population could be affected by cyber threats.

Third, the Russian government also views the importance of the Arctic region through the lens of its military security. Since 1933, the Soviet/Russian Northern Fleet base has been located on the Kola Peninsula. This geographical location provides the country with ice-free water access to the Barents Sea and the Atlantic Ocean. In anticipation of an increase in shipping and natural resource extraction in the region, Russia views military security to be connected to its economic security. It will require better protection of its oil rigs and the NSR, as well as the enhancement of both human safety and the state's capacity for emergency response. Overall, the Russian government perceives the realisation of economic and military activities and the maintenance of the population in the AZRF as a means of guaranteeing its presence in the region.

In fact, since the 1990s, the Russian part of the Arctic Zone has faced a critical technological and information gap. The networks of short-wave communication systems introduced into the Arctic during Soviet times have started to degrade and have now almost completely ceased to operate (Datyev, 2014). Moreover, "capabilities in radar coverage, aviation, and naval patrol were gutted" (Flake, 2014, p. 102). From the early 2000s, the Russian government began implementing certain policies intended to address the identified challenges. For instance, Russia planned to construct the Optical Trans-Arctic Submarine Cable System to connect the country with a 100 Gbit/s broadband to China, Japan, and Norway. The laying of this cable might have also provided the Arctic population with access to online education and telemedicine. This project had an estimated cost of $750 million (Arctic Economic Council [AEC], 2016). However, its implementation has been delayed for an unknown period due to economic reasons, including the high estimated cost of implementation.

In 2012, the Ministry of Communications and Mass Media of the Russian Federation announced the national broadband-related target of making "100 Mbit/s ultra-fast broadband (UFB) available to 80 percent of Russian residents by 2018, with the ultimate goal of providing affordable and accessible broadband to 95 percent of households by 2020" (n.d.). In the AZRF, the Russian government aimed to achieve 100% mobile broadband internet availability in settlements with a population over 10,000 people by 2018 (TFTIA, 2017). By 2017, around 86.9% of the total population of the AZRF were using the internet (Federal Statistics, 2017), whereas the average figure for Russia is 73.4%. The official figures highlight the high rate of internet users in the AZRF when compared to the state's average number of internet users. In sum, due to the presence of a high number of internet users and Russia's strategic interests in the Arctic, human security represents a central component in the framing of national cybersecurity policies.

3.4 Data Collection and Analysis

The major data source for this chapter is the official Russian documents.[3] These policy documents are cited individually where appropriate, although they are referred to collectively as "the documents" for

[3]The Fundamentals of the State Policy of the Russian Federation in the Arctic for the period up to 2020 and beyond (2008); the Development Strategy of the Arctic zone of the Russian Federation and national security up to 2020 (2013); the Federal Law "Information, Information Technologies and Information Protection" (2006); the state programme of the Russian Federation "Socio-economic development of the Arctic zone of the Russian Federation" (2017); the Information Security Doctrine of the Russian Federation (2000, 2016); the Strategy for the Development of the Information Society in the Russian Federation (2008); the Strategy for Development of Information

simplification purposes. These policy documents were published by the Russian government and, therefore, represent the primary source of information regarding the official Russian stance on the topic at hand. In addition, the documents are publicly accessible on official Russian websites.

It is important to note that this type of data has limitations that the present research takes into account. First, official policy strategies are usually declarative and general in nature, as they aim to achieve the identification of major policy directions and actions in a given policy area. Second, regardless of the fact that the most recent policy document concerning cybersecurity was published in 2017, there might have been recent changes to the related legislation that this present document does not reflect. To overcome these limitations, future research could also make use of data collected from interviews with Russian officials involved in the formation of cybersecurity policies in the AZRF.

For the purposes of data collection, the chapter employed a critical discourse analysis (CDA). The choice of a CDA is based on the fact that "[i]t offers interpretations of the meanings of texts rather than just quantifying textual features and deriving meaning from this; situate what is written or said in the context in which it occurs, rather than just summarizing patterns or regularities in texts and argue that textual meaning is constructed through an interaction between producer, text, and consumer" (Richardson, 2007, p. 15). Besides, a CDA allows one to identify explanatory reasons, that is to say, why the discourse is the way it is (Amer, 2017).

The analysis of the data derived from these documents involved a critical examination of the language embedded in the narrative of cybersecurity and its connection to human security. To examine this language, the chapter used narrative framing and positioning techniques (Stoskopf & Bermudez, 2017). The narrative framing technique identified how the cybersecurity was constructed by Russia and how it was used in connection to human security. Moreover, the positioning technique determined which individuals were included in the concept of cybersecurity.

Society for the period from 2017 to 2030 (2017); the Presidential Decree "Measures to Ensure the Information Security in the Russian Federation when Using Information and Telecommunication Networks of International Information Exchange" (2008); and the State Policy of the Russian Federation on International Information Security for the Period up to 2020 (2013).

Table 3.1 Examples of techniques and analytic questions

Technique	*Analytic question*	*Human security*	*Intersectionality*
Framing	How is the cybersecurity narrative defined and structured?	Are the notions of political security, economic security, and/or community security mentioned?	
	How is the notion of a cybersecurity agent framed?	Is the notion of an agent mentioned in relation to political security, economic security, and/or community security?	Are age, class, gender, and/or ethnicity mentioned in relation to the notion of an agent?
Positioning	Who is defined as a human in relation to cybersecurity?		How are gender, ethnicity, age, and/or class framed in relation to humans and to cybersecurity?

Source Author's own creation

These techniques were applied by posing several analytic questions to the official documents (see Table 3.1).

To identify the potential connections between cybersecurity and human security, the questions posed reflected the former's key elements, including political security, economic security, and community security. The next step was to examine the diversity within groups by applying an intersectional approach. For this purpose, the questionnaire encompassed questions about gender, age, ethnicity, and class. By doing this, the chapter aimed to catch "the heterogeneity of minority experiences" (Atewologun & Mahalingam, 2018, p. 150) of the people residing in the Arctic.

The first step involved preparing the data by identifying those sentences that contained answers to the posed questions. The second step involved breaking the identified sentences down into separate words or groups of words (two or more words united by a general meaning). The identified words and groups of words were then compared to each other. Afterwards, the words and groups of words with similar characteristics were merged and divided into properties (characteristics) and dimensions (the ways in which the characteristics vary). In addition, categories that

Table 3.2 Examples of categories, properties, and dimensions identified in the documents

Category	*Dimension*	*Property*	*Example*
Interests and needs	An individual and multi-ethnic population	Political interests and needs	Access, use, and protection of information
	An individual and multi-ethnic population	Economic interests and needs	A decent standard of living and employment
	An individual and multi-ethnic population	Community interests and needs	Preservation of moral and cultural values

Source Author's own creation

describe their contexts were constructed. The decisions to assign certain words and groups of words to categories, properties, and dimensions were made based on data saturation, "[a] point at which no new information or themes are observed in the data" (Guest, Bunce, & Johnson, 2006, p. 59). For instance, the dimension of an individual and multi-ethnic population was formed based on who is defined as a human in relation to cybersecurity. Correspondently, properties such as political, economic, and community interests and needs were constructed based on how the cybersecurity narrative was defined in relation to human security (see Table 3.2).

The third step involved the re-examination and comparison of the collected data in order to identify connections between the categories, properties, and dimensions. This examination was performed through the application of the concepts of human security and intersectionality and the analytic questions posed. In sum, the analysis of the data identified the presence of connections between cybersecurity and human security in the official Russian cybersecurity discourse.

3.5 Findings and Discussion

Based on the analysis, three major topics are identified with regard to the relationship between the framing of Russia's cybersecurity policy and human security. The identified themes are interests, needs, threats, and agency. Furthermore, the theme of needs and interests can be subdivided into political, economic, social needs, and interests.

Political security is primarily perceived in the documents as the enjoyment of constitutional rights and freedoms in terms of access to, use, protection, and receipt of objective and reliable information. Here, the official Russian framing correlates with the existing notions of cybersecurity as involving the protection of the integrity, confidentiality, and availability of data (Craigen et al., 2014; Schatz et al., 2017). In addition, it encompasses the human dimension of cybersecurity by including personal interests and rights (Reid & van Niekerk, 2014; von Solms & van Niekerk, 2013). However, there are several important elements of the Russian approach.

First, the documents underscore the importance of the providence and availability of reliable data for users. In this case, the state extends its functions by not only protecting the integrity and availability of information but also by trying to protect the content of that information. Russia's framing of cybersecurity thus correlates with existing approaches, whereby individuals are viewed by the state as victims and less often as beneficiaries of cybersecurity (Dunn Cavelty, 2014).

Cyberspace presents a risk to political security since the availability of large volumes of online data leads to its shallow perception by an individual, which facilitates its influence on an individual's attitudes and behaviour (The President of Russia, 2017). For instance, various criminal structures, terrorist groups, and foreign services apply information technologies to spread disinformation and to fuel social, radical, and religious strife (*Novaya Gazeta*, 2000; The President of Russia, 2016). Such an influence might result in the eradication of an individual's moral and cultural values, as well as her feeling of patriotism.

Although the documents try to address the needs of a diverse group of the population, they still follow a rather homogeneous approach. For instance, while talking about the referent object they refer broadly to a multi-ethnic population or a younger generation (*Novaya Gazeta*, 2000; The President of Russia, 2016). In general, the documents are not specifying their characteristics such as ethnicity, gender, and age, as well as their respective vulnerabilities and needs.

As part of the state policy to address the issue, during the period of 2021–2025, the Russian government aims to increase the share of the Arctic population who are aware of state activities by 45% (The Government of the Russian Federation, 2017). To achieve this goal, the documents propose the creation and financing of mobile cultural centres and libraries in small settlements. In this regard, the framing of cybersecurity also includes the creation of secure cyberspace for the fostering of community security, particularly, for the preservation of moral and cultural values and a feeling of patriotism.

Additionally, the documents specify the meanings of both access to and use of information. They are primarily discussed in relation to two subtopics, namely the provision of state services and the development of energy and shipping projects in the Arctic. The former, for example, encompasses "[t]he creation of a reliable system of communication, navigation, meteorological and information services, including coverage of ice conditions" (The Government of the Russian Federation, 2013, p. 7). Importantly, the documents are silent as to the users of such information. It can be assumed that the access to and usage of such information principally addresses issues of personal safety while shipping along the NSR, as well as during the extraction of mineral resources. The users of such information, therefore, might be workers on oil rigs and vessels that operate in the AZRF. However, the documents' silence about the human dimension in relation to the providence of access and usage of such information might signify the prevalence of the state-centric over the human-centric approach. Primarily, the documents highlight the prevalence of the state's national interests in the development of mineral and natural resources, as well as the Arctic marine routes as sources of revenue for the federal budget.

With regard to the second element, the Arctic Strategy (2013) refers to the provision of telecommunications infrastructure, online education, and telemedicine technologies. Simultaneously, the documents follow a rather homogeneous approach towards potential users. For example, the Strategy for the Development of Information Society (2017) declares as one of its goals the provision of internet connections to all settlements with populations ranging from 250 to 500 people. However, its homogeneous approach to the potential internet users' gender, class, and age might result in an unequal distribution of access to the internet and its usage. In general, in Russia, every third man over the age of 65 (33%) uses the web resources, only one-fifth of women (19%) in the same age group go online (TADVISER, 2018). Among users of other ages, the difference is not so significant, with approximately the same number of men and women using the internet (TADVISER, 2018).

Furthermore, in 2008, the share of high-income online users was 44%, although that group only accounted for 27% of the total population (Radkevich, 2009). Moreover, the price of telecommunications and internet services in the AZRF is much higher when compared to the price in other regions of Russia. In this case, the state's decision to focus on only the size of the population of Arctic settlements results in a limitation due to the failure to take into consideration the lower-income class and aging populations, who might face constraints with regard to the enjoyment of their constitutional right to access and use of electronic information.

In relation to education, the Strategy for the Development of Information Society (2017) underscores how a secure cyberspace should generate conditions for self-development and the acquisition of knowledge by individuals. For this purpose, the documents identify training, education, retraining of specialists, and raising awareness about potential cyber threats as the state's services. In general, these services target the following groups, namely indigenous children and students.

Regarding individuals, the documents take a homogeneous approach by specifying the targeted group as indigenous children. Specifically, they do not acknowledge their cultural and linguistic diversity. In fact, there are currently 40 different indigenous peoples officially recognised by the Russian government as indigenous small-numbered peoples of the North, Siberia, and Far East (Lukin, 2015). Such a homogeneous approach reflects the situation in Russian education whereby none of the languages of these groups is used in schools' instructions (Malysheva & Nabok, 2015). Further, the focus on indigenous children also represents a limitation, as it ignores the educational needs of non-indigenous children for online education, which is particularly relevant for children residing in remote Arctic settlements.

The reference to higher education is made in connection to such policy goals as the reduction of the technological gap and the preparation of specialists in ICT and cybersecurity. According to different estimations, Russia's dependence on foreign software and telecommunications technologies stood at 80–90% in 2015 (Makedonsky, 2016). The documents suggest that such a technological gap and that level of dependence on foreign technologies pose risks to Russia's national security. That would happen because the country becomes less resilient to address electronic attacks by foreign states against information, its processing, and telecommunication systems (*Novaya Gazeta*, 2000; The President of Russia, 2016).

In this case, Russia's approach to preparing ICT and cybersecurity specialists reflects discussions in the academic literature concerning cybersecurity as a set of skills that can be developed among young people in school (Jones, Namin, & Armstrong, 2018) and in college (Chou & Jones, 2018). However, Russia's framing of cybersecurity goes beyond the perception of an individual as a cyber defender of state security (Herr & Allen, 2015). It also underlines the importance of providing personal security (*Novaya Gazeta*, 2000; The President of Russia, 2016). The country connects the provision of personal security with the goal of raising awareness about existing cyber insecurities. Here, individuals play a distinct role as agents tasked with strengthening their personal cybersecurity. Such an approach reflects the emphasis placed by scholars and policymakers on the importance of raising awareness and educating the population about protecting themselves from cyber threats (Reid & van Niekerk, 2014).

With regard to economic security, the documents mention it in relation to the guaranteeing of employment and a decent standard of living. Russia's approach to cybersecurity reflects the notion of individual interests and rights seen in the academic definitions (Reid & van Niekerk, 2014). It encompasses the creation of secure cyberspace where an individual can realise and perform her professional competencies. In particular, the Strategy for the Development of Information Society (2017) stipulates the creation of conditions that promote remote employment by various organisations.

The documents take a general approach to remote employment, and there is no reference made to the AZRF. Currently, high levels of unemployment represent a significant problem in some regions of the AZRF. In 2016, the average unemployment rate in the Arctic was 8.6%, although it varied across the AZRF. Generally speaking, scientific studies identify a low satisfaction rate among working people with regard to labour conditions and remuneration in the AZRF (Zlenko, 2018).

At the same time, save for mentioning members of the military and their families, the documents follow a general approach to the potential targeted groups in relation to addressing socio-economic problems (*Novaya Gazeta*, 2000). The focus on this specific group relates to Russia's strategic military interests in the region. The latest version of Russia's military doctrine, which was adopted in 2014, assigns the protection of the state's national interests in the AZRF during peacetime to the Russian armed forces (Konyshev et al., 2017).

A number of scholars emphasise the high level of disunity of interests between the Russian government and the population residing in the AZRF (Kondral & Morozov, 2015). The core reasons for this are seen in the strengthening of the Russian state's "vertical power," that means the top-down approach that characterises the decision-making process, and the weakness of the civil society institutions in the country (Kondral & Morozov, 2015; Olompiev, 2015). More specifically, in 2008, Russian citizens rated the state of civil society by 4.48 points out of 10 (Ryabev, 2010). The identified disunity of interests might actually worsen with the aggravation of socio-economic conditions in the AZRF (Kondral & Morozov, 2015). Consequently, it might lead to a disbalance of interests between individuals and the state in the area of cybersecurity. Primarily, this disbalance will result from the state's intentions to enhance the state-centric and to constrain the human security dimension in its policies.

3.6 Conclusion and Further Research

This chapter explores Russia's cybersecurity policies in the AZRF based on publicly available official documents. It aims to determine how the cybersecurity narrative has been framed by the Russian state during the period from 2000 to 2017, with a focus on whose interests and needs have been incorporated into the official discourse on cybersecurity. Ultimately, this chapter sought to understand whether and, if so, how human security is connected to the official Russian framing of cybersecurity.

The data derived from the official policy documents show that the Russian government applies a rather broad approach to the concept of cybersecurity in relation to the AZRF. Although Russia does not mention the notion explicitly, the analysis reveals that the understanding of cybersecurity with regard to the AZRF encompasses human security, including political security, economic security, and community security. In particular, Russia goes beyond the traditional approach to cybersecurity by protecting the information in cyberspace (Schatz et al., 2017). In this regard, the cybersecurity narrative for the AZRF moves beyond traditional perceptions and the academic critique of the state-centric approach (Dunn Cavelty, 2014; Salminen & Hossain, 2018). Further, the concept also includes national security elements, such as the protection of the state's critical information infrastructure and integrated telecommunications networks. By having both national security and human security among its elements, Russia's approach to cybersecurity aims to balance the interests and needs of the state and the individuals (*Novaya Gazeta*, 2000; The President of Russia, 2016).

The process of unpacking the narrative reveals a strong connection between the state's national security interests and the human security categories. The major reason for this is that Russia views the AZRF as strategic for its military security and economic security. The connection between human security and the state's policy goals is reflected in the documents in several ways. For instance, political security and community security are linked to the state's national security. The former aims to preserve the unity of the population in order to maintain the territorial integrity of the country. In turn, the latter relates to the training and retraining of information and telecommunications specialists. In this case, the presence of these connections raises a question regarding the primary motivations behind the Russian state's cybersecurity policy, as well as whether it represents a state-or a human-driven approach.

Importantly, the incorporation of the human security dimension broadens the scope of potential threats. Aside from traditional cyber threats, such as hackers and cyberterrorists, Russia's framing encompasses the creation of secure cyberspace for self-development, as well as the acquisition of new knowledge and competencies by individuals. The Russian approach broadens the state's functions by including the protection of cyberspace and the availability of reliable information.

Regarding individuals, Russia follows a dual approach by perceiving them as both victims of potential cyber threats and agents that can identify such threats. Following this stance, the documents shift the concept of an agent from a state-centric approach to a human-centric one by picturing an individual as capable of raising awareness of cyber insecurities and the associated means of protection. In this sense, Russia's approach finds similarities with some academic works that propose individuals as being agents for cybersecurity (Chou & Jones, 2018; Jones, Namin, & Armstrong, 2018).

In general, and with certain exceptions, the documents follow a rather homogeneous approach to identifying vulnerable groups. However, the prioritisation of members of the military, young people, and indigenous population raises a question regarding the reasons for mentioning these groups and not others. One reason for this is their importance in relation to the Russian government's strategic interests and goals. More specifically, members of the military represent key actors in the Russian strategy for ensuring military security in the AZRF, while the younger population is important in terms of providing cybersecurity for the state in the future. In addition, both the younger generation and the indigenous population are important for the preservation of cultural and historic traditions, as well as for maintaining political stability and territorial integrity.

Overall, despite the incorporation of human security, it remains unclear whether there has been a shift from a state-centric to a more human-centric approach in Russia's official framing of the cybersecurity. There hence exists a need for further research, especially in the light of emerging academic discussions on the topic. Future studies could examine the multilevel framing of cybersecurity and human security in the AZRF by focusing on specific policies, such as online education and telemedicine. Such research could focus on elucidating differences in the discourses on cybersecurity and its connections to human security across different administrative levels in certain policy areas. Finally, future research could also focus on comparative studies of multilevel cybersecurity framing in several Arctic states. Such studies might provide an explanation for the varying levels of importance assigned to the human dimension in the official framing of cybersecurity.

References

Amer, M. (2017). Critical discourse analysis of war reporting in the international press: The case of the Gaza war of 2008-2009. *Palgrave Communications, 3*(13), 1–11. https://doi.org/10.1057/s41599-017-0015-2.

Anderson, J. M. (2003). Why we need a new definition of information security. *Computers & Security, 22*(4), 308–313. https://doi.org/10.1016/S0167-4048(03)00407-3.

Annan, K. (2000, May 8–10). Secretary-General salutes international workshop on human security in Mongolia. *Press Release SG/SM/7382*. Ulaanbaatar.

Arctic Council Task Force on Telecommunications Infrastructure in the Arctic (TFTIA). (2017). *Telecommunications infrastructure in the Arctic: A circumpolar assessment*. Arctic Council Secretariat. Retrieved from https://oaarchive.arctic-council.org/handle/11374/2001.

Arctic Economic Council (AEC). Telecommunication Infrastructure Working Group. (2016). *Arctic broadband: Recommendations for the interconnected*

Arctic. Retrieved from https://arcticeconomiccouncil.com/wp-content/uploads/2017/03/AEC-Report_Final-LR.pdf.

Atewologun, D., & Mahalingam, R. (2018). Intersectionality as a methodological tool in qualitative equality, diversity and inclusion research. In L. A. E. Booysen, R. Bendl, & J. K. Pringle (Eds.), *Handbook of research methods in diversity management, equality and inclusion at work* (pp. 149–170). Cheltenham: Edward Elgar.

Booth, K. (2007). *Theory of world security*. Cambridge: Cambridge University Press.

Chandler, D. (2008). Human security: The dog that didn't bark. *Security Dialogue, 39*(4), 427–438. https://doi.org/10.1177/0967010608094037.

Chou, T.-S., & Jones, J. (2018). Developing and evaluating an experimental learning environment for cyber security education. In *Proceedings of the 19th Annual SIG Conference on Information Technology Education* (pp. 92–97). Fort Lauderdale, FL: ACM Proceedings.

Craigen, D., Diakun-Thibault, N., & Purse, R. (2014). Defining cybersecurity. *Technology Innovation Management Review, 4*(10), 13–21. https://doi.org/10.22215/timreview/835.

Datyev, I. (2014). Развитие инфотелекомукационных систем арктических территорий [The development of information telecommunication systems in the Arctic]. *Труды Кольского Научного Центра* [Trudi Kol'skogo Naychnogo Tsentra], 41–63.

de Oliveira Albuquerque, R., García Villalba, L. J., Sandoval Orozco, A. L., de Sousa Júnior, R. T., Jr., & Kim, T.-H. (2016). Leveraging information security and computational trust for cybersecurity. *The Journal of Supercomputing, 72*(10), 3729–3763. https://doi.org/10.1007/s11227-015-1543-4.

Department for Foreign Affairs and International Trade (DFAIT). (2002). *Freedom from fear: Canada's foreign policy for human security*. Ottawa: Department for Foreign Affairs and International Trade.

Dunn Cavelty, M. (2014). Breaking the cyber-security dilemma: Aligning security needs and removing vulnerabilities. *Science and Engineering Ethics, 20*(3), 701–715. https://doi.org/10.1007/s11948-014-9551-y.

Efremova, I., Didenko, N., Rudenko, D., & Skripnuk, D. (2017). Disparities in rural development of the Russian Arctic zone regions. *Economics, 2*(2017), 189–194. https://doi.org/10.22616/rrd.23.2017.067.

Fauzer, V., Lytkina, T., & Fauzer, G. (2016). Особенности расселения населения в Арктической зоне России [Population trends in the Russian Arctic zone]. *Арктика: Экология и Экономика* [The Arctic: Environment and Economy], *2*(22), 40–50.

Federal Statistics. (2017). Доля населения, использующего информационно-телекоммуникационную сеть "Интернет", в общей численности населения Арктической зоны Российской Федерации [The share of the population using the information and telecommunication network "Internet" in the total population of the Arctic zone of the Russian Federation]. Retrieved from https://data.gov.ru/opendata/7708234640-fiveaoneasixafouranine.

Flake, L. E. (2014). Russia's security intentions in a melting Arctic. *Military and Strategic Affairs, 6*(1), 99–116.

Fukuda-Parr, S., & Messineo, C. (2012). *Human security: A critical review of the literature*. Leuven: Centre for Research on Peace and Development.

George, A., & Bennett, A. (2005). *Case studies and theory development in the social science*. Cambridge: MIT Press.

Gladun, E. (2015). Environmental protection of the Arctic region: Effective mechanisms of legal regulation. *Russian Law Journal, 3*(1), 92–109. https://doi.org/10.5539/jpl.v6n2p90.

Guest, G., Bunce, A., & Johnson, L. (2006). Field methods: How many interviews are enough? An experiment with data saturation and variability. *Field Methods, 18*(1), 59–82. https://doi.org/10.1177/1525822X05279903.

Herr, C., & Allen, D. M. (2015). Video games as a training tool to prepare the next generation of cyber warriors. In *Proceedings of the 2015 ACM SIGMIS Conference on Computers and People Research* (pp. 23–29). New York, NY: ACM. Retrieved from https://resources.sei.cmu.edu/asset_files/Presentation/2015_017_001_442344.pdf.

Hunter, T. (2017). Russian Arctic policy, petroleum resources development, and the EU: Cooperation or coming confrontation? In N. Liu, E. A. Kirk, & T. Henriksen (Eds.), *The European Union and the Arctic* (pp. 172–199). Leiden: Brill.

Ilyn, A. (2008). Арктике определят границы [One will define borders for the Arctic]. *Rossiiskaya Gazeta*. Retrieved from https://rg.ru/2008/09/18/arktika.html.

International Commission on Intervention and State Sovereignty (ICISS). (2001). *The responsibility to protect: The report of the International Commission on Intervention and State Sovereignty*. Ottawa: The International Development Research Centre.

International Telecommunication Union (ITU). (2009). *Overview of cybersecurity. Recommendation ITU-T X.1205*. Geneva: International Telecommunication Union. Retrieved from http://www.itu.int/rec/T-REC-X.1205-200804-I/en.

Jones, K. S., Namin, A. S., & Armstrong, M. E. (2018). The core cyber-defense: Knowledge, skills, and abilities that cybersecurity students should learn in school: Results from interviews with cybersecurity professionals. *ACM Transactions on Computing Education (TOCE), 18*(3), 11.

Kondral, D. P., & Morozov, N. A. (2015). Власть, бизнес и население в северных регионах России: проблемы и перспективы взаимодействия [Government, business and the population in the Northern regions of Russia:

Problems and prospects for cooperation]. *Арктика и Север* [Arctic and North] (20), 58–72.

Konyshev, V., & Sergunin, A. (2014). Is Russia a revisionist military power in the Arctic? *Defense and Security Analysis, 30*(4), 323–335. https://doi.org/10.1080/14751798.2014.948276.

Konyshev, V., Sergunin, A., & Subbotin, S. (2017). Russia's Arctic strategies in the context of the Ukrainian crisis. *The Polar Journal, 7*(1), 104–124. https://doi.org/10.1080/2154896X.2017.1335107.

Lukin, Y. (2015). Этнокультурный ландшафт Российской Арктики: от концептуализации знаний к управлению конфликтами [Ethno-cultural landscape of the Russian Arctic: From the conceptualization of knowledge to the management of conflicts]. *Арктика и Север* [Arctic and North] (21), 118–143.

Makedonsky, S. (2016). Анализ рисков зависимости от иностранных технологий [Risk analysis of dependence on foreign technologies]. *Символ Науки* [Symbol of Science], (4), 93–96.

Malysheva, E., & Nabok, I. (2015). Образование коренных малочисленных народов Арктики: проблемы и перспективы [Education of the indigenous small-numbered people of the Arctic: Problems and perspectives]. *Общество, Развитие, Среда* [Society, Development, Environment], (1), 139–144.

Nam, T. (2019). Understanding the gap between perceived threats to and preparedness for cybersecurity. *Technology in Society, 58*, 101122. https://doi.org/10.1016/j.techsoc.2019.03.005.

Newman, E. (2016). Human security: Reconciling critical aspirations with political "realities". *The British Journal of Criminology, 56*(6), 1165–1183. https://doi.org/10.1093/bjc/azw016.

Novaya Gazeta. (2000). Доктрина информационной безопасности Российской Федерации [Information Doctrine of the Russian Federation]. Retrieved from http://www.ng.ru/politics/2000-09-15/0_infodoctrine.html.

Olompiev, K. (2015). Модели государственной политики в сфере взаимодействия с институтами гражданского общества в современной России [State policy models in the sphere of interaction with civil society institutions in modern Russia]. *Федеральная Политика и Управление* [Federal Politics and Governance], 57–67.

Paris, R. (2001). Human security: Paradigm shift or hot air? *International Security, 26*(2), 87–102. https://doi.org/10.1162/016228801753191141.

Priyadarshini, I. (2019). Introduction on cybersecurity. In D.-N. Le, R. Kumar, B. K. Mishra, M. Khari, & J. M. Chetterjee (Eds.), *Cybersecurity in parallel and distributed computing: Concept, techniques, application and case studies* (pp. 3–39). Hoboken, NJ: Wiley. https://doi.org/10.1002/9781119488330.ch1.

Radkevich, A. (2009). Интернет аудитория в России: состояние, динамика, тенденции [Internet audience in Russia: Current situation, dynamics, tendencies]. *Знание, Понимание, Умение* [Knowledge, Understanding, Proficiency], *1*, 230–236.

Reid, R., & van Niekerk, J. (2014). From information security to cyber security cultures: Organizations to societies. In *Proceedings of the conference on Information Security for South Africa (ISSA)* (pp. 1–7). Johannesburg, South Africa. Retrieved from https://pdfs.semanticscholar.org/ae5d/9568064a2b66dd9fb7f8b69c6ff21dec3c2e.pdf.

Richardson, J. E. (2007). *Analysing newspapers: Critical discourse analysis*. New York, NY: Palgrave Macmillan.

Ryabev, V. (2010). Гражданское общество в современной России: проблемы и перспективы становления [Civil society in modern Russia: Problems of prospects for development]. *Вестник Мурманского Государственного Технического Университета* [Herald of Murmansk State Technical University], *13*(2), 439–445.

Salminen, M., & Hossain, K. (2018). Digitalization and human security dimensions in cybersecurity: An appraisal for the European High North. *Polar Record, 54*(2), 108–118. https://doi.org/10.1017/S0032247418000268.

Schatz, D., Bashroush, R., & Wall, J. (2017). Towards a more representative definition of cyber security. *The Journal of Digital Forensics, Security and Law, 12*(2), 53–74. https://doi.org/10.15394/jdfsl.2017.1476.

Shannon, J., & Thomas, N. (2005). Human security and cyber-security: Operationalising a policy framework. In R. Broadhurst & P. Graborsky (Eds.), *Cyber-crime: The challenge in Asia* (pp. 327–346). Hong Kong: Hong Kong University Press.

Simakova, A., & Kozyreva, G. (2018). Краудсорсинг как инструмент взаимодействия власти, бизнеса и общества в решении проблем местного самуправления в Арктической зоне России [Crowdsourcing as a tool for interaction between government, business, and society in solving problems of local self-government in the Russian Arctic Zone]. *Мир Экономики и Упраления* [World of Economics and Governance], *18*(1), 129–139. https://doi.org/10.25205/2542-0429-2018-18-1-129-139.

Sokolova, F. H. (2016). Миграционные процессы в Российской Арктике [Migration processes in the Russian Arctic]. *Арктика и Север* [Arctic and North], *25*, 158–172. https://doi.org/10.17238/issn2221-2698.2016.25.158.

Stoskopf, A., & Bermudez, A. (2017). The sounds of silence: American history textbook representations of non-violence and the Abolition Movement. *Journal of Peace Education, 14*(1), 92–113. https://doi.org/10.1080/17400201.2016.1230543.

TADVISER. (2018). *Интернет-доступ. Рынок России и СНГ* [Internet access. Russian and CIS market]. Retrieved from http://www.tadviser.ru/index.php/Статья:Интернет-доступ_(рынок_России).

Tamitsky, A. (2013). Информационная безопасность как фактор международных отношений в арктическом регионе [Information security as a factor of international relations in the Arctic region]. *Вестник Северного (Арктического) Федерального Университета* [Herald of the Northern (Arctic) Federal University], 63–70.

The Government of the Russian Federation. (2008). *Об Основах государственной политики России в Арктике на период до 2020 года и дальнейшую перспективу* [The fundamentals of the state policy of the Russian Federation in the Arctic for the period up to 2020 and beyond]. Moscow: The Government of the Russian Federation. Retrieved from http://government.ru/info/18359/.

The Government of the Russian Federation. (2013). *О Стратегии развития Арктической зоны Российской Федерации и обеспечения национальной безопасности на период до 2020 года* [Strategy for the development of the Arctic zone of the Russian Federation and national security for the period up to 2020]. Moscow: The Government of the Russian Federation. Retrieved from http://government.ru/info/18360/.

The Government of the Russian Federation. (2017). *О внесении изменений в постановление Правительства Российской Федерации от 21 апреля 2014 г. № 366 "Об утверждении государственной программы Российской Федерации "Социально-экономическое развитие Арктической зоны Российской Федерации на период до 2020 года"* [On Amendments to the Resolution of the Government of the Russian Federation from April 21, 2014 No. 366 "On approval of the state program of the Russian Federation Socio-economic development of the Arctic zone of the Russian Federation for the period up to 2020"]. Moscow: The Government of the Russian Federation. Retrieved from http://government.ru/docs/29164/.

The Ministry of Telecommunications of the Russian Federation. (2012). *Objectives of the Ministry of Telecommunications and Mass Communications of the Russian Federation for the years of 2012–2018.* Retrieved from http://2018.minsvyaz.ru/en/.

The President of Russia. (2016). *Доктрина информационной безопасности Российской Федерации* [Information security doctrine of the Russian Federation]. Moscow, Russia: The President of Russia. Retrieved from http://kremlin.ru/acts/bank/41460.

The President of Russia. (2017). *Стратегия развития информационного общества в Российской Федерации на 2017–2030 годы* [Strategy for Development of Information Society for the period from 2017–2030]. Moscow: The President of Russia. Retrieved from http://www.kremlin.ru/acts/bank/41919.

Thomas, N., & Tow, W. T. (2002). The utility of human security: Sovereignty and humanitarian intervention. *Security Dialogue, 33*(2), 177–192. https://doi.org/10.1177/0967010602033002006.

United Nations Development Program (UNDP). (1994). *Human development report*. New York, NY: Oxford University Press.

von Solms, R., & van Niekerk, J. (2013). From information security to cyber security. *Computers & Security, 38*, 97–102. https://doi.org/10.1016/j.cose.2013.04.004.

Wibben, A. T. (2016). The promise and dangers of human security. In J. Nyman & A. Burke (Eds.), *Ethical security studies* (pp. 114–127). New York, NY: Routledge.

Yagya, V., & Sboychakova, A. (2016). Политические аспекты экологической безопасности российской Арктики [Political issues of environmental security of the Russian Arctic]. *Наука Красноярья* [Krasnoyarsk Science], *4*(37), 104–121. https://doi.org/10.12731/2070-7568-2016-4-104-121.

Zagorski, A. V. (2018). Военное строительство в Арктике в условиях конфронтации России и Запада [The Arctic defense postures in the context of the Russia-West confrontation]. *Арктика и Север* [Arctic and North], *31*, 80–97. https://doi.org/10.17238/issn2221-2698.2018.31.80.

Zhuravel, V. P. (2016). Россия в Арктике. Будет ли госкомиссия ее рулевым? [Russia in the Arctic. Will the State Commission be its helmsman?]. *Арктика и Север* [Arctic and North], *23*(2), 5–16. https://doi.org/10.17238/issn2221-2698.2016.23.5.

Zlenko, E. G. (2018). Ограниченность прожиточного минимума трудоспособного населения как инструмента регулирования минимального уровня трудовых доходов в России и ее арктическая специфика [Limitations of the living working-age population as a tool for regulating the minimum level of labor incomes in Russia and its Arctic specificity]. *Уровень Жизни Населения Регионов России* [The Level of Life of the Russian Regions], *1*(207), 37–43. https://doi.org/10.24411/1999-9836-2018-10004.

CHAPTER 4

Critical Human Security and Cyberspace: Enablement Besides Constraint

Alan Collins

4.1 Introduction

Whether it is accusations that Russia has meddled in Western elections by shaping public opinion via social media, or fears that democracy is being undermined by algorithms that restrict news access through social media, or concern that social media is infected by trolls that engage in cyberbullying and spread misinformation, it is clear that social networking sites and, more broadly, the infrastructure that underpins information and communication technologies (ICT) have not proved to be the unmitigated success that early pioneers had hoped. Nevertheless, it is also clear that the digital revolution has ushered in an age of transparency that, at least initially, severed the elites' control over information and, therefore, knowledge and connected like-minded people on a scale not previously possible. In this chapter, I explain how conceiving cyberspace as an arena for human empowerment or human constraint represents a valuable contribution that a critical human security approach can make to

A. Collins (✉)
Swansea University, Swansea, Wales, UK
e-mail: a.collins@swansea.ac.uk

M. Salminen et al. (eds.),
Digitalisation and Human Security, New Security Challenges,
https://doi.org/10.1007/978-3-030-48070-7_4

the debate. The chapter will reveal the duality of trends in social media that espouse an emancipatory potential while embedding constraints on human agency.

Applying the concept of human security to cyberspace enables us to consider how different types of power prevail in cyberspace as well as what this means for human security. More specifically, a question arises regarding whether human interaction with ICT has enhanced or even enabled human security, or whether it has constrained the pursuit of human security by reinforcing pre-existing, oppressive power structures?

Cyberspace, according to Myriam Dunn Cavelty (2016), "connotes the fusion of all communication networks, databases, and sources of information into a vast, tangled, and diverse blanket of electronic interchange ... Cyberspace, however, is not only virtual, since it is also made up of servers, cables, computers, satellites, etc. In popular usage we tend to use the terms cyberspace and internet almost interchangeably, even though the internet, albeit the most important one, is just one part of cyberspace" (p. 401). In this chapter, cyberspace and the internet are used synonymously, while the degree of security or insecurity that humans experience in this environment is determined by the degree to which ICT enhance or constrain their agency. Thus, increasing human security equates to enabling agency, whereas human insecurity equates to ICT constraining human agency.

While there is much to examine with regard to how restricted access to the internet, whether in physical terms or due to a lack of the skills necessary to create and communicate online, can exclude people on a variety of social, economic and political grounds, this chapter focuses on how those who enjoy unfettered access to the internet and the world of social media can have their agency enabled and constrained. This means conceiving of human security as being less focused on people being the thing that needs to be secured (the referent of security) and more focused on people being the thing that determines the degree of security being felt (the agent). This entails conceiving of human security as the examination of human agency in terms of creating a secure, or insecure, world.

4.2 Human Security: From Referent to Agent

The term "human security" has its origins in the United Nations' 1994 *Human Development Report*. It emerged at this time as a result of middle-sized powers embracing the opportunity afforded by the end of the Cold War to shift the international focus from superpower confrontation to the multitude of problems confronting the post-Cold War world. These problems, which are now referred to as non-traditional security threats, encompass a wide range of issues, from environmental catastrophes to pandemics to economic insecurity. All such problems share a focus on people and how, by putting human life at the centre of the security debate, it is possible to focus state resources on the multitude of threats that negatively impact people's quality of life (see Peou, 2014; Tadjbakhsh & Chenoy, 2009). Given the vast array of problems that can affect "quality of life", the discourse on human security has since been split between those who see worth in a broad definition of human security and those who prefer a narrower focus. The use of a broad definition has the advantage of capturing the structural causes of insecurity that underpin many of the freedoms from want (poverty, food insecurity, underdevelopment), although it has been criticised for being too broad to offer any meaningful analytical insights (Paris, 2004). The narrow definition focuses on the threat of political violence against people on the part of the state (freedom from fear), but while this provides more conceptual clarity and rigour than the broad definition, it overlooks the insidious threats that stem from the exploitative nature of the political, social, and economic structure of everyday life. I do not intend to revisit this debate (for details of the two schools of thought, see Kerr, 2013). It is worth noting, however, that while both conceptions of human security position humans as the referent object of security—the thing to be secured—the responsibility for providing that security still rests with the state. The holders of political power are thus considered part of the solution to the problem of human insecurity, which is no surprise given that "human security started as a policy-making project" (Hynek & Chandler, 2011, p. 1). It is this conventional understanding of human security that has come to dominate the literature, and it is unsurprising that as human security has evolved, it has been co-opted by states so as to render themselves not only the securitising agents but also the referent object (see Breslin & Christou, 2015, p. 7).

The twin colloquialisms of freedom from fear and freedom from want, which capture threats to human security, also provide a means of reconsidering human security from a critical perspective. The work that aligns most closely with the notion of freedom is Ken Booth's (1991) work on emancipation. His work predates the arrival of the term human security and, indeed, his emphasis on emancipation does not underpin the notion of human security contained within the *Human Development Report*. Booth (1991) writes,

> 'Security' means the absence of threats. Emancipation is the freeing of people (as individuals and groups) from those physical and human constraints which stop them carrying out what they would freely choose to do. War and the threat of war is one of those constraints, together with poverty, poor education, political oppression and so on. Security and emancipation are two sides of the same coin. Emancipation, theoretically, is security. (p. 319)

Emancipation is thus the freeing of people,[1] both freeing them from political oppression (human security's freedom from fear) and freeing them from economic deprivation (human security's freedom from want). Freeing people means freeing them from those institutions that cause the identified threats, which includes the state but also refers more broadly to an international system (and, especially, an international economic system) that prevents those in deprived areas of the world from enjoying the prosperity on offer in the richest parts of the world. Underpinning Booth's (1991) notion of emancipation is a clearly post-Marxist interpretation of the world. The identified threats are not simply the outcome of repressive regimes, but rather they stem from the holding of political power by elites that pursue policies that benefit their own constituencies (the bourgeois) at the expense of the common people (the labourers, the peasants, the working class). This can be considered a "critical" approach to human security because it is concerned with exposing where political power lies—whether it be in the state or in the institutions that govern the international system (United Nations [UN], World Bank, World Health Organization [WHO])—and thus revealing that the social contract, whereby people give up their freedoms to a leviathan in return for security, is being abused. The holding of power by the elite has not only failed to bring security to the common people but has resulted in their exploitation. Whether that exploitation comes in the form of a denial of rights (social and economic or civic and political) or, as was the case with nuclear deterrence, deliberately leaving them exposed to nuclear annihilation, human security has been sacrificed to achieve national security.

[1] Booth (2007) has subsequently added to this original definition of emancipation by including the caveat that one's freedom cannot come at the expense of others and, further, that emancipation serves as a means of understanding the world we inhabit and as a framework for actualising change. I quote Booth's original definition here to show that a critical approach to human security, even though it was not self-identified as such, predates the UN *Human Development Report*.

The consequence of this interpretation of human security is that humans must be the referent of security because the alternatives have not only failed to provide security for the people but are actually part of the reason why people are insecure. This speaks directly to the accusation levelled against the elites within the European High North that they maintain a colonial mindset with regard to the Arctic's indigenous people (see Benjaminsen, Svarstad, & Eira, 2018). The inequality inherent within the relationship between modernity and indigenous people is also prevalent where economic interests in resource extraction trump local and indigenous concerns about the conservation of nature (Hossain, Zojer, Greaves, Miguel Roncero, & Sheehan, 2017, p. 60). The holders of political power are thus part of the problem, not part of the solution. Emancipation is, therefore, a clarion call for change. Since the holders of political power are part of the problem, the change envisaged does not just involve considering humans as the referent object of security but also empowering them so that humans become the agents that achieve security. Humans, through the exercise of their agency, become the securitising actor.

It was previously noted that the traditional understanding of human security came to dominate the human security approach, with the state being conceived of as part of the solution to human insecurity. This was, at least in part, a consequence of the difficulties associated with establishing emancipatory agency separate from the current holders of political power. The problem for emancipation was the question of how humans could become the securitising actor and realise this changed, alternative, world. The answer lay in the alternative world, as championed by critical security scholars, emerging from engagement with states and intergovernmental organisations that would serve as the conduits through which emancipation could be achieved. Nik Hynek and David Chandler (2013) have argued that this was the beginning of the end for "critical" security studies and, indeed, the co-opting of the emancipatory discourse by the traditional holders of political power, essentially those promoting and maintaining the neoliberal order, have hollowed out any criticality in relation to emancipation. Edward Newman (2010) concurred with this view and added that because critical approaches are not problem-solving theories but are instead concerned with exposing where power lies, a critical security studies approach should ask different questions. Therefore, being "critical" is less about how to solve problems in the existing world than it is about questioning the "reality" of the existing world and why such as reality exists at all (Newman, 2010). With a different goal, critical security studies hence left the human security agenda free for the traditional approach to dominate.

In this chapter, I utilise the original critical insights that emancipation offers—the current ordering of political, social, and economic power needs exposing and examining so as to understand both the opportunities for enabling human agency and the constraints it imposes on human agency—to explore the impact of the digital revolution on peoples' security. Since emancipation concerns the freeing of people from constraints, it is an examination of whether ICT is an enabler of emancipation or a constrainer. In essence, does the availability of ICT empower people and render them agents of change, or does it reinforce and establish power structures that exploit humans? Are such technologies a conduit for emancipation?

In order to answer this question, it is useful to utilise the division of power—power being the concept that explains why people are empowered or constrained from carrying out what they would otherwise freely choose to do—into its three dimensions (or "faces"). This division was originally made by Steven Lukes (1974). He identified the first dimension as Robert Dahl's (1957) conception of power as a relationship between people in which the compliance of one person is brought about by another through a process of coercion or reward. Regarded as the "open face" of power, this conception captures how behaviour is manipulated by the tangible exertion of power. It is often linked to the power of a government to make legislation and to the state apparatus that enforces the rule of law. The second dimension is Peter Bachrach and Morton Baratz's (1962) idea that power is exercised within a complex system and, further, that it comes in the form of agenda setting. Regarded as the "secretive face" of power, this conception captures how behaviour is manipulated by establishing agendas determining what is, and is not, permissible. It is often linked to the notion of decisions being made "behind closed doors" and in the "corridors of power". The third dimension is Lukes' (1974) "radical view", in which power lies in the normalcy of everyday behaviour. Regarded as the "deceptive face" of power, this conception captures how behaviour is manipulated by establishing what appropriate, or normal, behaviour is and, therefore, what deviant behaviour is. It is often linked to the Marxist view of ideological power, according to which power is the control of what people think is right and thus they act within a belief system without questioning its core assumptions.

Conceiving of power in its dimensions, or "faces", is helpful because in this framework power is posited to lie in relationships. It is through power that someone or something can achieve a goal through either coercion/reward or framing discussion. Emancipation is thus concerned with exposing the exercise of power in this relationship and with freeing humans by empowering them. It is for this reason that I have not included the fourth dimension of power, namely Michel Foucault's notion of disciplinary power (see Digesar, 1992). While the other three dimensions are concerned with how power is used by an elite to control or oppress, and thus are directly relevant to the notion of emancipation, an explicit actor is missing from the fourth face, where power is not exercised by someone for some purpose but rather exists in the everyday activities of conformity and resistance. The fourth face treats power as omnipresent. There is no emancipation because it is impossible to be free of this conception of power (see Hayward, 2000). It shares with the third face or "deceptive face" of power the sense of ordering society into what is normal and abnormal, but unlike the third dimension, it is not a construct for a particular elite. The "disciplinary" nature of this conception of power emerges from society's notion of what is normal behaviour and, therefore, what is deviant. The ubiquitous ownership and use of the mobile phone is good example of disciplinary power. When first introduced into the UK, the mobile phone was a status symbol for young, upwardly mobile, professional people (commonly known as yuppies). Mobile phones were shunned by members of wider society, who regarded them as an irritant. Over time, society's attitude changed until there came a point at which mobile phone ownership was commonplace. This transition was marked by a changing reaction to those who did not own a mobile phone. This can be seen in the reaction to a common question asked about another person's contact details. Initially, the question was whether or not the other person had a mobile phone. The acceptable answer was either "yes" or "no". This question then changed to what the person's mobile phone number was, with the underlying assumption being that they must own one. While "I don't have a mobile phone" was initially an acceptable answer, it was later received with surprise, until eventually such an answer garnered incredulity. It was simply no longer "normal" to not own a mobile phone. The pressure, or disciplinary power, to own a mobile phone did not come from mobile phone manufacturers or service providers, but rather from society's attitude that it was abnormal to not own a mobile phone. A similar trajectory can be discerned with regard to social networking. In modern society, human agency is enabled by being an active participant in the world of social networking. To be part of modern society—to be disciplined—is to not just experience an event but also to share that experience by sending a Tweet, uploading a photograph or video to Instagram, or posting a Facebook message. To not have a Twitter, Instagram and/or Facebook account prevents such sharing, and this inability to share, perhaps instantaneously during the event itself, is abnormal because it means not being part of a modern society in which online and offline behaviours coexist. The rejection of the fourth face in this chapter does not dismiss the context that disciplinary power provides (see van Dijck, 2013, pp. 172–175), but rather sees it as a culture (or ideology) underpinning social media's economic model. This places it more firmly within the third face, as this kind of power is a construct by and for the commercial elites.

4.3 From Referent to Agent: The (False?) Promise of the Digital Revolution

The widespread availability of ICT have led to a truly remarkable global revolution in terms of how people access information and communicate, from the prevalence of the personal home computer to access and communicate via the internet, to the variety of mobile platforms that enable a continuous online existence, to the interconnectedness of home gadgets and wearable devices. We live in a world of self-monitoring, analysis and reporting technology (SMART), a world in which physical devices, from vehicles to home appliances, with an internet protocol (IP) address can connect, exchange data and be controlled remotely. This has brought about the Internet of Things (IoT), whereby an online and offline existence today coexist and reinforce each other. This ICT revolution has also enabled users to become more than simply consumers or end-users of information, but also producers of content, which is known as produsage in Axel Bruns' terminology (2008), as the internet is conceived as a platform for user-led content (2008). Web 2.0 epitomises this enablement of human agency to produce and disseminate content via blogs and sites such as Wikipedia and YouTube.

It is this ability on the part of anyone to access information, create information and share information widely, thereby severing the elites' control over information and, therefore, knowledge, that lies behind the claim that ICT is an enabler of human emancipation, from being a source of revenue via crowdsourcing, without which little can be accomplished, to exposing corruption in political office via, among others, WikiLeaks, to lifting the lid on societies' inherent patriarchy as unmasked by the #MeToo movement. The power of the digital revolution lies in the fact that it gives people the agency to instigate transparency in everyday life, thus exposing the power asymmetries that create pernicious insecurities and then doing something about it. This speaks directly to Booth's (2007, p. 112) call for emancipation to entail political action. The promise of the digital revolution as an enabler has, however, been questioned. Far from being an enabler, the "culture of connectivity" (van Dijck, 2013) that pervades social networking sites actually constrains people. According to this line of argument, cyberspace is not a free-for-all of opinions intended to inform and enlighten, but rather a collection of silos or "echo chambers" that, due to "filter bubbles", reaffirm ignorance by reinforcing convictions via congruence or "likes" and rejecting rebuttals as "fake". Far from exposing inequalities, ICT has been co-opted by political and commercial elites to maintain their hold over information and knowledge. Technical manipulation via algorithms and discursive beguilement, in which sharing, "liking" and having "followers" have become sources of self-esteem, insidiously patterns human behaviour so that agency is curtailed. Is cyberspace an enabler or a constrainer of human security? The answer lies in whether it empowers or constrains human agency, and determining this means applying the three faces of power. In Table 4.1, examples of enablement and constraint are provided under each of the three faces.

The separation of power into its three dimensions is not intended to categorise specific examples of internet usage, but rather to show how power is exercised during these activities. Therefore, important features of internet use, such as personalisation, will be evident across all three faces. Likewise, examples such as data processing and protection can appear in both the first and second dimensions, depending upon whether the issue is enforcing protection (first face) or establishing what can and cannot be done with peoples' data (second face). The separation of enablement and constraint in Table 4.1 is not intended to show that in some instances ICT enables and in others constrain. ICT actually do both concomitantly; it is enablement *besides* constraint. The division of power is not only helpful in terms of revealing how human agency is enhanced and constrained by the modalities of the internet, as it also reveals which elites control the power. For the purposes of this chapter, it is the state

Table 4.1 Examples of enablement and constraint

Human security	*Faces of power*		
	Open face	*Secretive face*	*Deceptive face*
Enablement	• Cyber-optimism – Removal of gatekeepers – Organise protests	• Relevance of information – Filter search engine results – Make recommendations	• "Produsage" • Human connectedness • SMART
Constraint	• Cyber-pessimism – Restrict access using firewalls – Enhance surveillance techniques	• Default privacy settings • Terms and conditions	• Culture of connectivity/manipulation of "sharing" • Filter bubbles/echo chambers

Source Author's own creation

elite in the open face and the commercial elite in the other two faces. However, this need not be the case; indeed, revealing the elites is one advantage of this taxonomy of power.

4.3.1 Open Face

The dichotomy of the internet as an arena for human enablement or constrainment is captured within the literature on digital activism by the terms "cyber-optimists" and "cyber-pessimists". This distinction relates to the open face of power, whereby enabling human agency frees people from the control exerted by the holders of political power, while constraining human agency involves the use of cyberspace to reinforce that control. The elite exercising power in the open face comprise, therefore, the state elite and the power that they exercise pertains to the freedom from want and fear. The latter is readily associated with political oppression, while the former can include digital infrastructure projects that primarily serve national commercial interests at the expense of people who are unable to access services online. For example, despite the significant investment in digital infrastructure in the European High North, digital divides endure, which leave ordinary people in want (see Salminen & Hossain, 2018).

Cyber-optimists perceive the digital revolution as empowering citizens in social and political arenas and thus speak directly to the notion that social media enables human agency. A combination of providing alternative avenues to traditional media outlets for the accessing of information and thus providing alternative sources of knowledge, combined with the ability to connect with like-minded individuals and to hence create a collective that shares a common desire, forms the basis for the cyber-optimists' view that social media can promote democracy and undermine authoritarianism. This view of social media as an arena that can bypass state-controlled media, organise demonstrations, engage with a global audience and instigate change was palpably demonstrated during the Arab Spring. On a smaller scale, human empowerment via social media has been seen in the European High North. For example, by utilising social media, in 2017, a Sámi activist group—Ellos Deatnu—occupied an island and declared a moratorium on fishing regulations in the Tana River Valley, which runs along the border between Finland and Norway (Lakkala, 2017). In 2013, at Gállok in Sweden, the Sámi community was able to use social media to raise awareness and organise a protest against a British mining company whose operations could jeopardise reindeer herding and other traditional livelihoods (Lindgren & Cocq, 2017). While such activities add weight to the notion of the empowering potential of social media's connectivity, they can also provide evidence for cyber-pessimists.

For cyber-pessimists, the kind of digital activism described above is less reflective of a wider audience than the cyber-optimists present, and it is also vulnerable to countermeasures on the part of traditional holders of power (see Morozov, 2011). For all the coverage of the work of Egyptian social media activists during the 2011 Arab Spring, who were invariably young people with liberal outlooks, when the populace was able to vote for Mubarak's replacement, they elected the Muslim Brotherhood, while the hard-line Salafist Nour Party came second. In fact, the Islamic parties convincingly outshone the liberal and secular parties (see Hermida, 2014, pp. 97–115). Within a year, a military coup restored authoritarian rule and President Abdel Fattah al-Sisi's government cracked down on social media, citing security concerns (Amnesty International, 2014; Human Rights Watch, 2017). Since the Arab Spring can serve as an example for both cyber-optimists and cyber-pessimists, the dichotomy is not enabler or constrainer but rather both. This gave rise to the present chapter's title: enablement besides constraint. In this case, it is interesting to determine which of these two approaches is in the ascendency.

In authoritarian states, control over the flow of information has always been a mechanism for maintaining social control and curtailing opposition. The drift towards authoritarian rule in Cambodia, for example, has been marked by the tightening of controls over the internet, with the technology to do this reportedly having been acquired from China (Murdoch, 2018). China is notorious for restricting its peoples' access to the internet by means of the "Great Firewall", blocking access to websites through the denial of service attacks (colloquially known as the Great Cannon) and conducting surveillance of its peoples' use of ICT (Marczak et al., 2015). China's strategy involves using human surveillance, coupled with peer pressure and self-censorship, to achieve social control (see Shahbaz, 2018). Yet, restricting access to political (mis) information is not just the preserve of authoritarian states. Western democracies have also sought to remove extremist material from the internet on the grounds that it can incite terrorism. The UK's Home Affairs Committee (2016) has accused social media giants such as Facebook, Twitter and YouTube of "consciously failing" to combat the use of their sites to promote terrorism. Further, it has recommended that the "Metropolitan Police's Counter Terrorism Internet Referral Unit (CTIRU) should be upgraded into a high-tech, state-of-the-art, round-the-clock, central Operational Hub which locates the perils early, moves quickly to block them and is able to instantly share the sensitive information with other security agencies" (Commons Select Committee, 2016). Whatever the merits (or otherwise) of this type of surveillance, it constitutes evidence that the use of ICT globally is increasingly monitored and controlled by the traditional holders of political power. The trend is thus one of the state authorities constraining human agency in cyberspace wherever that agency challenges state control. The effective internment of Julian Assange, the editor of WikiLeaks, in the Ecuadorian Embassy in London between June 2012 and April 2019, is a salient reminder of the tangible manifestation of power's open face in relation to those who challenge the state (see Taylor, 2018).

4.3.2 Secretive Face

In terms of the open face, power is exercised either by the state elite restricting access and co-opting the internet into a surveillance device or by human agency using the enabling characteristics of the internet to challenge the state elite. With regard to the secretive face, and this is also true of the deceptive face, the elite are not entities such as the Chinese Communist Party or the CTIRU, but rather those that influence how social media has evolved, namely commercial elites such as Facebook and Google.

The availability of copious amounts of information does not in itself empower and enlighten. Without the tools necessary to filter the available information, users would be unable to locate the particular information that they wanted. The availability of search engines that record an individual's browsing habits so that personalised searches are produced renders the vast swathes of data that exist both accessible and meaningful. It is also why search engines produce different results for the same search when it is conducted by different people. This is examined below as part of the personalisation of the internet, although it is captured by phrases such as "Customers who bought this item also bought", which make users aware of other products that may be of interest. This appears as the second dimension of power because the ways in which the algorithms operate so as to enable the personalisation of the internet are closely guarded commercial secrets. They can be thought of as agenda setting because the algorithms that underpin the tools humans use to access data on the internet determine what information is accessible and what is likely to be relevant. This is related to the previous dimension of power, whereby state authorities block access to certain sites, and, as will be discussed later, it also has implications for the third dimension, whereby the manipulation of social and political views is sought. However, in this case, it can be seen as a secretive enabler. Tracking users' internet and purchasing histories produces a rich data set concerning each individual's needs, and this data set can be used to create a digital experience that is relevant to the user and, therefore, empowering for them. In essence, without personalised filtering tools, the information available online would be overwhelming, which would hinder human agency.

It is worth noting that, in this dimension, human enablement is not an active human action, as it is with the open face of power. Agency in the case of the secretive face is enabled by the modalities of accessing and distributing internet content in a passive form. Thus, while the open face captures power's coercive capacity (power over someone), the second dimension captures power's constitutive capacity (power to frame). That is, the emancipatory potential of human agency is enabled not by exposing the algorithms that exert power over what is accessible, but rather by relying on them to exploit the internet's potential.[2] Enablement in this dimension is hence concerned with exploiting the internet's ability to enhance human security by utilising its inherent personalisation potential, which empowers human agency by rendering the wealth of available information accessible. The constraint in this second dimension is not how this personalisation can restrict access, although it certainly can (which explains the cyber-pessimists' fear noted above and the notion of filter bubbles discussed below), but rather how the agenda is established with little user awareness. When using social networking sites (SNSs) or downloading apps, users have to agree to certain terms and conditions, and they are encouraged to accept the tool's default privacy settings. This "encouragement" comes in the guise of an easier set-up and the assurance that privacy settings can be adjusted at a later point. However, because the economic model that underpins how SNSs and apps generate income requires users' data to be made available for commercial gain, there is every incentive to discourage the adjustment of the default privacy settings that make user information available. In their report, entitled *Why Johnny Can't Opt Out*, researchers at Carnegie Mellon University tested the skills of SNSs users with regard to controlling their settings, and they found that all the users encountered difficulties when trying to customise their privacy settings (see van Dijck, 2013, p. 172). While users can choose not to accept the stated terms and conditions, and have to give permission for apps to access their data, this "choice" is severely curtailed. Not accepting the terms and conditions means not being able to use the app or SNS in question, while if permissions are not granted, then either the app does not work or users are informed that the device might not work so effectively. This is often the case with apps and SNSs that require location-based services.

The constraint on human agency associated with the secretive face of power is that using the internet comes with conditions. These conditions ensure that the sharing of users' data is as prominent as the protection of users' data. It is not secretive in the sense that it is hidden from view. Indeed, users are explicitly required to accept the terms and conditions, and they are encouraged to adopt them by not adjusting

[2]Once algorithms are conceived of as restricting access and constraining human agency, challenging those elites who use algorithms for this purpose becomes an example of the open face of power.

the default privacy settings. These conditions are, however, obscured. They are concealed in difficult to understand terms of service (Calver & Miller, 2018), which can be changed by platform owners, and they are preserved by privacy settings that require significant technical expertise to customise so that tracking is negated. Users are thus generally unaware of what they have agreed to and how available their data is to third parties, although for the purposes of constraining human agency, the point of the second face of power is that these conditions are mandatory. Sharing users' data, in addition to protecting it, is the ***modus operandi*** of internet use, and agreeing to it is a prerequisite for accessing cyberspace.[3] A right to data privacy does not mean that data cannot be shared, but rather it is concerned with how and for what purpose it can be shared.

[3]On 25 May 2018, the General Data Protection Regulation (GDPR), a new EU law, came into effect, which changes how companies can collect personal data and how that data can be used. It applies to all companies that offer their services to EU citizens, even if they are based outside the EU. Failure to abide by the requirements of the GDPR can result in a fine of 20 million euros or 4% of a company's global turnover. The severity of this punishment explains the deluge of emails users received from companies that hold their data requesting that the users review their data privacy policies. In terms of the secretive power of framing how humans can use the internet, it makes the ways in which companies intend to use personal data more explicit (consent must be given by ticking a box, for example) and this consent must be clearly distinguishable from other matters in the terms of service. It is also not legal to require additional personal data in return for extra or premium services, as this does not constitute freely given consent. Users also have the right to access the data held about them and to require errors to be rectified and erased. The right to erase accurate data can also be requested.

Within hours of the GDPR coming into effect, a privacy protection group called none-of-your-business (noyb) filed four complaints against Facebook, Google, Instagram and WhatsApp. They accused these companies of forcing users to consent to targeted advertising to use their services; if that consent was not given, they prevented users from using their services. Max Schrems, chairman of noyb, claimed that this did not amount to people being given a free choice (noyb, 2018). Whether or not the complaint will be upheld is likely to rest upon whether social networking sites or instant messaging services can make the argument that processing users' data for targeted advertising is a necessary requirement for the service they provide. That is, that processing data for targeted advertising is a core feature of what it means to share in cyberspace. If this argument is not persuasive, then their processing of data for targeted advertising will require separate consent that the user can decline without being denied use of the service (Hern, 2018). This could have significant consequences for the economic model that underpins SNS, but in terms of the secretive power, because consent is more explicitly required, it is likely to make users more aware that their data are as much shared as protected.

4.3.3 *Deceptive Face*

The esoteric nature of terms and conditions as well as the abstruse character of default privacy settings actually do more than establish the mandatory conditions for internet usage. They also provide the background for understanding that the internet has its own culture and, further, that this culture has the power to manipulate users. The fact that such manipulation is done under the guise of empowering users makes it deceptive. It encourages an understanding of what is normal behaviour without an explanation that it is exploitative.

The deceptive face of power concerns the existence of a prevailing ideology that frames what is, as well as what is not, appropriate behaviour. As with the previous two faces of power, it can both enable and constrain human agency. In the case of internet usage, this ideology (or culture) is based on the human propensity to share. Sharing is both an enabler and a constrainer. It is, for example, the capacity both to collect data from citizens and devices as they interact with the environment and to use that data to manage assets and resources efficiently that lies at the heart of the concept of sustainable development via smart cities. It is this collecting and disseminating of data so as to enhance human well-being that informs Duncan McLaren and Julian Agyeman's (2015) assertion that,

> We believe that the world's cities, where the majority of people now live, could become more socially just, more environmentally sustainable and more innovative through the twenty-first-century reinvention and revival of one of our most basic traits: sharing. (p. 1)

The potential that sharing can have for human innovation, freedom, prosperity, and welfare points to its ability to enable human security, and it certainly captures the promise associated with social media and the arrival of Web 2.0 that the digital revolution would augur well for humanity. In particular, the ability of citizens to create web content and to establish online community communication was captured in the idea that these interactive platforms would be "user-centred", would enable more "collaboration" and would facilitate greater "participatory" activity. This notion reached its apex in 2006 when *Time* magazine chose "You" as its Person of the Year, with the clearly emancipatory assertion that, "It's about the many wresting power from the few and helping one another for nothing and how that will not only change the world, but also change the way the world changes" (Grossman, 2006, p. 38). By 2010, *Time* had replaced "You" with the Facebook founder, Mark Zuckerberg, and the conflation of transparency and empowerment was equated to "sharing". Zuckerberg said, "The thing that I really care about is making the world more open and connected ... Open means having access to more information, right? More transparency, being able to share things and have a voice in the world" (Grossman, 2010). Here, Zuckerberg was echoing the optimism of the early pioneers of the internet. For instance, the creator of the World Wide Web, Tim Berners-Lee, was inspired by the twin goals of opening up access to information and enabling people to share that information. However, the meaning of sharing has changed, and it is this evolution of meaning that has problematised the internet as an enabler and heightened its potential as a constrainer.

Here, the work of José van Dijck and her "culture of connectivity" is instructive. van Dijck (2013) juxtaposes the human connectedness espoused by advocates of social media with social media's prime requisite for connectivity. Thus, while social media offers humans the technology to facilitate connections (human connectedness), it does so through algorithms and discourse adaptations that engineer and manipulate those connections, thereby creating, surreptitiously, a culture of connectivity. Hence, while social media giants such as Facebook and Twitter encourage connections by utilising such emancipatory terms as "co-developers" and "empowering community leaders", or by offering the ability to create topics that garner global curiosity, or if you prefer, "trending", these platforms direct this human sociability through both algorithms that personalise an individual's online experience, which obscures as much as it reveals, and discourse adaptations that change the meaning of those phrases that underpin self-esteem, such as "friends" and "friending", "liking" and "following" creating "followers", which van Dijck (2013) refers to as the popularity principle.

The popularity principle arises because the culture of connectivity takes advantage of the human desire to share. Sharing, that is, the exchange of goods and information, fosters social capital, which enables communities to form and human security in the guise of material and emotional well-being to be enhanced. Online networks build on social capital by enabling users to garner what Pierre Bourdieu (see Swartz, 1997) refers to as cultural and symbolic capital. Cultural capital is intellectual knowledge, while symbolic capital is prestige or recognition. The sharing of knowledge via online discussions or posting links gives the provider recognition (symbolic capital) and the audience greater knowledge (cultural capital). So far, so empowering. The constraint on human agency arises because this connectedness is manipulated by the online process of connecting users.

This is, at least in part, because of algorithms that filter information based on individual users' preferences, and it is examined below as part of the personalisation of the online world. It also occurs, however, via the shifting meaning attached to phrases tied to self-esteem. van Dijck (2013) notes that in the offline world, people who are "well-connected" know a few, albeit very important, people. It is the status of the people they know that determines their sense of being well-connected. This quality signifier is replaced in the social media world with a quantity signifier. "Friends" and "friending" denote strong and weak ties, with no distinction being made between intimate contacts and strangers. The same is true of "followers". Before social media adopted the term, "followers" had religious overtones and the quality signifier was found in terms such as "devotees" and "believers" for those who followed the teachings of a preacher. Being well-connected on social media is quantifiable. It is synonymous with having many "friends" and "followers" without actually knowing them. The more you have, the more other people will think you are important, or well-connected, and thus the more they will want to connect with you. It is not dissimilar to "liking" and "retweeting", whereby the click of a button determines the popularity of an idea or thing. There is no quality assessment; instead, it is an accumulation of "likes" that determines approval and, by implication, disapproval. Once liking becomes quantifiable, it can be manipulated. Coding technologies ensure that what is shared between users is also shared with third parties that have an interest in promoting something or someone. Users are encouraged to "like", "comment" or "retweet", and participating in such activities can bestow symbolic capital. As van Dijck (2013) asserts,

> People who have many friends or followers are touted as influential, and their social authority or reputation increases as they receive more clicks. Ideas that are "liked" by many people have the potential of becoming trends. Friending, following, and trending are not the same functions, but they derive from same popularity principle underpinning the online economy of social media. (p. 13)

Popularity and, consequently, self-esteem are determined online by quantification: the more the better. While it is particularly noticeable in relation to many teenagers, where high quantities of "friends" and "followers" are considered a badge of honour, this sense of self-esteem is not unique to young people. Social media has thus co-opted phrases that have social meaning (liking, following) and replaced the quality previously associated with them with quantity. Further, self-esteem has been tied to this quantifiable metric. Sharing has become a good in itself rather than the value lying in what is shared, and the likelihood of sharing being increased is linked to the more time people spend on internet platforms. In their simplest manifestation, SNSs have become an essential feature of many people's everyday life, with the sharing of posts and links now being commonplace. It is estimated that as "of 2017, daily social media usage of global internet users amounted to 135 minutes per day" (Statista, 2018). It becomes, therefore, important to keep people online and to encourage them to share stories. This is captured by the phrase "attention economy", which refers to the fact that platform owners seek to maintain the user's focus on the information found on their site, using "clickbait" to encourage more online surfing. It is this manipulation of human behaviour that raises the spectre of constraining human activity. To appreciate this, we need to understand how the personalisation of the internet has restricted access to information and, further, that it has been done under the guise of human emancipation, which makes it particularly invidious.

The driver of this culture of connectivity is commercialisation and the increasing personalisation of the online world, coupled with social media's need for people to share data about themselves. Without people taking the time each day to share photographs of their nearest and dearest with their "friends" and "followers", or to "like" or "retweet" funny, disgusting or annoying stories, there would be no social media. This sharing has ballooned in recent years. This is less to do with the number of new people tweeting or creating a Facebook account than it is with the ease with which internet platforms enable users to share. This is most evident in the case of websites carrying the logos of the major social media companies. You can share with just the click of a button. Sharing is not, however, a two-way activity between the sharer and the receiver(s). The information that is shared is also made available to third parties that can use such data for commercial purposes. Put succinctly, your online search for information and those to whom you communicate that information are not a private affair between you and your browser. This is reflected in the phrase, "If you're not paying for something, you're not the customer; you're the product being sold" (Pariser, 2011, p. 21). Our preferences are known, and this directs advertisers to show us products and services that are meaningful and useful to us. Using the internet makes humans the commodity that commercial companies purchase so as to more efficiently target their advertising. These footprints that we leave in cyberspace are not simply trails followed by advertisers, as our pathways are increasingly being determined for us. This is reflected in the increasing personalisation of the online world.

Eli Pariser (2011) introduced the term "filter bubble" to explain the personalisation of the internet that began on 4 December 2009. He writes,

> Starting that morning, Google would use fifty-seven *signals* – everything from where you were logging in from to what browser you were using to what you had searched for before – to make guesses about who you were and what kinds of sites you'd like. Even if you were logged out, it would customize its results, showing you the pages it predicted you were most likely to click on (emphasis in the original). (Pariser, 2011, p. 2)

That knowledge about individuals' online behaviour has, as previously discussed, manifested in people receiving different search results for the same subjects, led social media sites to encourage users to connect with "People you may know", and established a purchasing history that leads to "Recommendations for you". As noted in relation to the secretive face of power, this does not in itself restrict human agency. Indeed, filtering the vast quantities of information available on the internet so as to present material of individual relevance can empower human agency. However, it can also have the effect of limiting exposure to contrasting opinions. The likeliness of this is exacerbated when SNSs, as networks of like-minded friends, peers, colleagues and family members, become the primary source of information. Personalisation, Pariser (2011) argues, creates filter bubbles that enclose the user in "an antiseptically friendly world" (p. 150) where the only online information received ascribes to their present view of the world. The concern that such personalised news, as epitomised by the phrase "Daily Me", can lead to society bifurcating or dividing into different world views underpins the argument that filter bubbles create echo chambers.

The term "echo chamber" signifies that the information shared by users conforms to a particular way of viewing the world. The idea draws upon social psychology studies that show how people tend to avoid dissonance and to prefer agreement, and it is captured by concepts such as groupthink and the favouritism shown to group members in social identity theory. As they are associated with filter bubbles, echo chambers are often thought of as forming and being maintained because people are only exposed to information they already agree with. This is only partially true, however, as an echo chamber can contain alternative views, although they may be presented in such a ways as to discourage agreement with them. The extent to which online activity has created echo chambers that effect political debate remains contentious. The surprise outcomes of the 2016 US presidential election and the 2016 Brexit vote appeared to offer evidence that swathes of American and British society were not engaging with one another or, if they did engage, they did so only to lambast the other as the propagator of fake news (see Grimes, 2017). However, Elizabeth Dubois and Grant Blank (2018) argue that while "networks on Twitter are polarized ... and networks on other social media may be equally polarized" (p. 741), because people source their political information from a wide variety of different media channels, including offline print media and television, they are regularly exposed to different views. They conclude that "when we look at the entire media environment, there is little apparent echo chamber" (Dubois & Blank, 2018, p. 740). Likewise, Jacob Groshek and Karolina Koc-Michalska's (2017) analysis of the 2016 US presidential election found that support for populist candidates, such as Bernie Sanders and Donald Trump, did not confirm the filter bubble thesis. They note that "it is tempting but may be somewhat premature to suggest that social media is damaging democracy through filter bubbles or even that such bubbles are responsible for the growing wave of populism now evident in the US and elsewhere" (Groshek & Karolina Koc-Michalska, 2017, p. 1402). However, the key implication of both filter bubbles and echo chambers is not that people are unaware of alternative views, but rather they discourage enlightened debate. Some people actively seek out alternative views, albeit often not to engage with them or to expose themselves to alternative ways of thinking, but rather to reinforce their pre-existing mindset by denigrating them. This has been referred to as "trench warfare" and, while presented as being different from echo chambers, it augments the prevailing view that online activity reinforces cognitive bias because it is not an arena that challenges pre-existing mindsets (Karlsen, Steen-Johnsen, Wollebæk, & Enjolras, 2017).

The clearest manifestation of the constraining of human agency can be seen in the February 2018 US Justice Department's indictment of Russian agents who were trying to subvert the 2016 US presidential election (Apuzzo & LaFraniere, 2018). This may appear to be an example of the open face of power, as it involves state authorities exercising the power to manipulate others so as to achieve their desired outcome. However, it is more pertinently an example of the deceptive face of power because the Russian agents were making use of social media's, and more specifically Facebook's and Instagram's, culture of connectivity to spread fake news. It was the Russian agents' use of "sharing" that enabled such manipulation. Hence, Jonathan Albright writes that "Facebook built incredibly effective tools which let Russia profile citizens here in the US and figure out how to manipulate us" (as cited in Frenkel & Benner, 2018).

The point is not that the Trump campaign was in cahoots with the Russians, but rather that by using social media's need for people to share—its culture—they were able to utilise social media's filter bubble to reinforce prevailing beliefs and to exacerbate the polarisation of political opinion. Human security is the freeing of people from both want and fear, and this can only be accomplished by tackling the ignorance that underscores blindness to inequalities (want) and breeds intolerance and distrust (fear). This does not necessitate seeking consensus and harmonious agreement, but rather seeing the virtue in debate and discussion as well as recognising that contestation empowers. For social media to be an arena for emancipation, it needs to be one in which differences can be confronted in a respectful space. As van Dijck (2013) laments, we do not just need "like" buttons, we also need "difficult but important" buttons (p. 66). The constraint, as perceived through the deceptive face of power, is thus that ignorance rather than enlightenment is the consequence of the internet's culture of connectivity.

4.4 Conclusion

Has the ICT revolution, and the introduction of Web 2.0 and social media, in particular, empowered users so that they can be free of the physical and human constraints that stop them from doing what they would freely choose to do? In many respects it has. It can enable people to access services and goods, who for a variety of reasons, from disabilities to geographical remoteness, are impeded from doing so. It can promote social capital by connecting like-minded individuals. It has enhanced symbolic and cultural capital through the sharing of information. It has helped to ferment opposition in authoritarian states. It has, without doubt, demonstrated the potential to empower users to create an environment that is more conducive to fulfilling their wants and needs. It can, as the early pioneers of the internet and Web 2.0 hoped, remove the gatekeepers that restrict people's access to information and knowledge. All this optimism and potential continue to resonate in contemporary campaigns. It is an enabler.

It is, however, also evident that human manipulation is prevalent and that the sharing of user data has created a personalised online experience that obfuscates as much as it reveals. The commercialisation of the internet, which funds social media giants, has led to the widespread, albeit rarely acknowledged by companies or understood by their users, sharing of people's data. The revenue obtained from advertisers has prompted social media companies to encourage their users to share and, therefore, allowed such companies to gain more knowledge about their users' preferences. Hence, the favourable connotations attached to "liking", "tweeting", "retweeting", "following", etc. The constraints that arise from this are captured by the notion of a filter bubble, which limits users' exposure to differences and which can be seen as stopping people from doing something they would freely chose to do (i.e. because the filtering is done surreptitiously). While the prevalence of offline information currently ensures that we do not have "real world" echo chambers, there is evidence to suggest online polarisation. Indeed, the use of people's data to manipulate their interpretation of the world is currently a major concern for Western democracies (Cellan-Jones, 2018).

The recognition of ICT and Web 2.0 innovation as both a constrainer and an enabler, as well as the 2018 Cambridge Analytica controversy regarding the sharing of users' data and the political purpose for which this can be done (Greenfield, 2018; *The Guardian*, 2018), have helped to shed light on the less than transparent workings of social media giants and their algorithms. Hynek and Chandler's (2011) critique of the hollowing out of criticality in critical security studies was based on the idea that emancipation had become a tool for, rather than a critique of, state action. However, when the threat to human security is also a threat to the state, seeing the state as part of the solution is not in itself problematic. What is problematic is how the state reacts in such a case. The "threat" posed to Western democracy by the sharing of user data has placed Facebook and other social media platforms in the cross-hairs of state authorities. This could be seen in Mark Zuckerberg's testimonies before the US Congress and the European Parliament in April 2018. What regulation emerges, such as the General Data Protection Regulation (GDPR), can serve as an enabler of human agency by ensuring that (a) users are better informed about the footprints they leave in cyberspace and (b) their data is better protected from (mis)use by third parties. The state can be a conduit for emancipation by enhancing human agency. However, it could also be a constrainer if the regulation curtails user activity from exposing corruption by those in authority. There are already disturbing signs that some state authorities are using the prevalence of "fake" news, or "deliberate online falsehoods" to use Singapore's phrase, to curtail the freedom of expression (Community Action Network, 2018). In this case, the state is part of the problem. To determine whether enablement or constraint is in the ascendency entails watching the trends rather than watching what is trending.

References

Amnesty International. (2014). *Egypt's plan for mass surveillance of social media is a "recipe for disaster"*. Retrieved from https://www.amnesty.org.uk/press-releases/egypts-plan-mass-surveillance-social-media-recipe-disaster.

Apuzzo, M., & LaFraniere, S. (2018, February 16). 3 Russians indicted as Mueller reveals effort to aid Trump campaign. *The New York Times*. Retrieved from https://www.nytimes.com/2018/02/16/us/politics/russians-indicted-mueller-election-interference.html.

Bachrach, P., & Baratz, M. S. (1962). Two faces of power. *American Political Science Review, 56*(4), 947–952.

Benjaminsen, T. A., Svarstad, H., & Eira, I. M. G. (2018, August 29). Nykolonialisme på norsk [Neocolonialism in Norwegian.] *Klassekampen*. Retrieved from https://dagens.klassekampen.no/2018-08-29/nykolonialisme-pa-norsk.

Booth, K. (1991). Security and emancipation. *Review of International Studies, 17*(4), 313–326.

Booth, K. (2007). *Theory of world security*. Cambridge: Cambridge University Press.

Breslin, S., & Christou, G. (2015). Has the human security agenda come of age? Definitions, discourses and debates. *Contemporary Politics, 21*(1), 1–10.

Bruns, A. (2008). *Blogs, wikipedia, second life, and beyond: From production to produsage*. New York, NY: Peter Lang.

Calver, T., & Miller, J. (2018, July 6). Social site terms tougher than Dickens. *BBC News*. Retrieved from https://www.bbc.co.uk/news/business-44599968.

Cellan-Jones, R. (2018, March 19). Facebook data—As scandalous as MPs' expenses? *BBC News*. Retrieved from http://www.bbc.co.uk/news/technology-43458110.

Commons Select Committee. (2016). *Internet giants "consciously failing" to tackle extremism on the web*. Retrieved from https://www.parliament.uk/business/committees/committees-a-z/commons-select/home-affairs-committee/news-parliament-2015/radicalisation-report-published-16-17/.

Community Action Network. (2018, March 26). CAN: Select Committee on online falsehoods should engage all NGOs in good faith. *The Online Citizen*. Retrieved from https://www.theonlinecitizen.com/2018/03/26/can-select-committee-on-online-falsehoods-should-engage-all-ngos-in-good-faith/.

Dahl, R. A. (1957). The concept of power. *Behavioral Science, 2*(3), 201–215.

Digeser, P. (1992). The fourth face of power. *The Journal of Politics, 54*(4), 977–1007.

Dubois, E., & Blank, G. (2018). The echo chamber is overstated: The moderating effect of political interest and diverse media. *Information, Communication & Society, 21*(5), 729–745.

Dunn Cavelty, M. (2016). Cyber-security. In A. Collins (Ed.), *Contemporary security studies* (4th ed., pp. 400–416). Oxford: Oxford University Press.

Frenkel, S., & Benner, K. (2018, February 17). To stir discord in 2016, Russians turned most often to Facebook. *The New York Times*. Retrieved from https://www.nytimes.com/2018/02/17/technology/indictment-russian-tech-facebook.html.

Greenfield, P. (2018, March 26). The Cambridge Analytica files: The story so far. *The Guardian*. Retrieved from https://www.theguardian.com/news/2018/mar/26/the-cambridge-analytica-files-the-story-so-far.

Grimes, D. R. (2017, December 4). Echo chambers are dangerous—We must try to break free of our online bubbles. *The Guardian*. Retrieved from https://www.theguardian.com/science/blog/2017/dec/04/echo-chambers-are-dangerous-we-must-try-to-break-free-of-our-online-bubbles.

Groshek, J., & Koc-Michalska, K. (2017). Helping populism win? Social media use, filter bubbles, and support for populist presidential candidates in the 2016 US election campaign. *Information, Communication & Society, 20*(9), 1389–1407.

Grossman, L. (2006, December 25). 2006 person of the year. *Time Magazine*, pp. 38–41.

Grossman, L. (2010, December 27). 2010 person of the year. *Time Magazine*, pp. 44–75.

Hayward, C. R. (2000). *De-facing power*. New York, NY: Cambridge University Press.

Hermida, A. (2014). *Tell everyone: Why we share & why it matters*. Toronto: Doubleday Canada.

Hern, A. (2018, May 25). Facebook and Google targeted as first GDPR complaints filed. *The Guardian*. Retrieved from https://www.theguardian.com/technology/2018/may/25/facebook-google-gdpr-complaints-eu-consumer-rights.

Hossain, K., Zojer, G., Greaves, W., Miguel Roncero, J., & Sheehan, M. (2017). Constructing Arctic security: An inter-disciplinary approach to understanding security in the Barents Region. *Polar Record, 53*(1), 52–66.

Human Rights Watch. (2017). *Egypt: Events of 2016*. Retrieved from https://www.hrw.org/world-report/2017/country-chapters/egypt.

Hynek, N., & Chandler, D. (2011). Introduction: Emancipation and power in human security. In D. Chandler & N. Hynek (Eds.), *Critical perspectives on human security: Rethinking emancipation and power in international relations* (pp. 1–10). London: Routledge.

Hynek, N., & Chandler, D. (2013). No emancipatory alternative, no critical security studies. *Critical Studies on Security, 1*(1), 46–63.

Karlsen, R., Steen-Johnsen, K., Wollebæk, D., & Enjolras, B. (2017). Echo chamber and trench warfare dynamics in online debates. *European Journal of Communication, 32*(3), 257–273.

Kerr, P. (2013). Human security. In A. Collins (Ed.), *Contemporary security studies* (3rd ed., pp. 104–116). Oxford: Oxford University Press.

Lakkala, A. (2017, August 28). "Like terrorists at the bottom of an outhouse": Being Sámi in the 100-year-old Finland. *The Barents Observer*. Retrieved from https://thebarentsobserver.com/en/life-and-public/2017/08/terrorists-bottom-outhouse-being-sami-100-year-old-finland.

Lindgren, S., & Cocq, C. (2017). Turning the inside out: Social media and the broadcasting of indigenous discourse. *European Journal of Communication, 32*(2), 131–150.

Lukes, S. (1974). *Power: A radical view*. Basingstoke: Macmillan.

Marczak, B., Weaver, N., Dalek, J., Ensafi, R., Fifield, D., McKune, S., … Paxson, V. (2015, April 10). China's Great Cannon. *The Citizen Lab*. Retrieved from https://citizenlab.ca/2015/04/chinas-great-cannon/.

McLaren, D., & Agyeman, J. (2015). *Sharing cities: A case for truly smart and sustainable cities*. Cambridge: MIT Press.

Morozov, E. (2011). *The net delusion: How not to liberate the world*. London: Penguin Books.

Murdoch, L. (2018, February 14). Hun Sen's Cambodia slides even further into dictatorship. *The Age*. Retrieved from https://www.theage.com.au/world/

asia/hun-sen-s-cambodia-slides-even-further-into-dictatorship-20180214-p4z0b8.html.

Newman, E. (2010). Critical human security studies. *Review of International Studies, 36*(1), 77–94.

NOYB—European Center for Digital Rights. (2018, May 25). GDPR: noyb.eu filed four complaints over "forced consent" against Google, Instagram, WhatsApp and Facebook. *noyb.eu*. Retrieved from https://noyb.eu/wp-content/uploads/2018/05/pa_forcedconsent_en.pdf.

Paris, R. (2004). Still an inscrutable concept. *Security Dialogue, 35*(3), 370–371.

Pariser, E. (2011). *The filter bubble: What the internet is hiding from you*. London: Penguin Books.

Peou, S. (2014). *Human security studies: Theories, methods and themes*. Singapore: World Scientific.

Salminen, M., & Hossain, K. (2018). Digitalisation and human security dimensions in cybersecurity: An appraisal for the European High North. *The Polar Record, 54*(2), 108–118.

Shahbaz, A. (2018). Freedom on the net 2018: The rise of digital authoritarianism. *Freedom House*. Retrieved from https://freedomhouse.org/report/freedom-net/freedom-net-2018/rise-digital-authoritarianism.

Statista. (2018). Daily time spent on social networking by internet users worldwide from 2012 to 2017 (in minutes). *The Statistics Portal*. Retrieved from https://www.statista.com/statistics/433871/daily-social-media-usage-worldwide/.

Swartz, D. (1997). *Culture and power: The sociology of Pierre Bourdieu*. Chicago: The University of Chicago Press.

Tadjbakhsh, S., & Chenoy, A. M. (2009). *Human security: Concepts and implications*. London: Routledge.

Taylor, M. (2018, May 18). Julian Assange is suffering needlessly. Why not report that? *The Guardian*. Retrieved from https://www.theguardian.com/commentisfree/2018/may/18/julian-assange-media-detention.

The Guardian. (2018). The Cambridge Analytica files. *The Guardian*. Retrieved from https://www.theguardian.com/news/series/cambridge-analytica-files.

van Dijck, J. (2013). *The culture of connectivity: A critical history of social media*. New York, NY: Oxford University Press.

PART III

Human Rights and Digital Infrastructure

CHAPTER 5

Social Exclusion as Human Insecurity: A Human Cybersecurity Framework Applied to the European High North

Kristin Smette Gulbrandsen and Michael Sheehan

5.1 Introduction

Digitalisation is rapidly changing the way in which governments provide public services. The Nordic states, which are often considered to be forerunners in terms of digitalisation, represent a clear case of this. Additionally, digital services are thought to have special utility for geographically peripheral and sparsely populated regions such as the European High North, which comprises northern Norway, Sweden and Finland. The European High North is currently facing several challenges, including the need to maintain services for the ageing population despite emerging labour shortages. There exists hope that digital

K. S. Gulbrandsen
Lund University, Lund, Sweden
e-mail: kristin_smette.gulbrandsen@keg.lu.se

M. Sheehan (✉)
Swansea University, Swansea, Wales, UK
e-mail: m.sheehan@swansea.ac.uk

M. Salminen et al. (eds.),
Digitalisation and Human Security, New Security Challenges,
https://doi.org/10.1007/978-3-030-48070-7_5

technologies can provide relief for healthcare systems facing tremendous pressure and, further, that they can reduce the risk of social exclusion from hard-to-access services at the same time. Success in terms of this shift from traditional to digital services is contingent on the existence of adequate digital competence among the population as well as on the provision of the necessary broadband infrastructure. Without these components, rural villages and elderly people in particular are at risk of suffering disadvantage or, potentially, social exclusion.

To address this issue, this chapter presents a cybersecurity framework that borrows from both human security and social exclusion theory. It argues that social exclusion can be considered a component of human insecurity and, further, it outlines what this means for cybersecurity. Drawing on a qualitative analysis of 23 key informant interviews conducted in the European High North during 2018 and 2019, it unpacks the linkages between digital and social exclusion. This chapter finds that while digitalisation can provide clear benefits to remote areas, significant attention must be paid to whose voices and interests shape the current digitalisation agenda. The fact that many elderly people are less competent users of information and communication technologies (ICTs) means that "unwanted" digitalisation and the closure of traditional services has the potential to create exclusion where it did not previously exist.

5.2 The Evolution of Security Theory

Traditionally, approaches to security within the field of international relations have dealt with military threats to sovereign states (Mearsheimer, 2001; Morgenthau, 1973; Waltz, 1979). During the 1980s and following the end of the Cold War, this narrow definition of security was "broadened" through the emergence of a security agenda that encompassed new security sectors, including the environmental, economic and societal sectors (Barnett, 2001; Buzan, 1983, 1991; Buzan, Wæver, & de Wilde, 1998; McSweeney, 1999). Additionally, the subsequent emergence of "critical" approaches to security, as well as the United Nations' adoption of the concept of "human security", reinforced and accelerated this shift in the understanding of security as being a state-centric and military-focused concept (Booth, 2007; Holland & Jarvis, 2015; Sheehan, 2005; Wyn Jones, 2007). Taken together, these developments enabled not only a shift towards the recognition of multiple sectors of security, but crucially also the adoption of individuals and sub-state communities as its referent objects (Booth, 2005, p. 22).

The new approaches changed the nature and scope of security threats too, as they were recognised as being socially constructed rather than objective material phenomena (Booth, 2005; Peoples & Vaughan-Williams, 2010). Constructivist-inspired critical security theories began to question the ontological and epistemological foundations of security studies as a field, arguing that norms and ideas shape the identities, interests and behaviours of political actors (McDonald, 2013). As a consequence, more complex and comprehensive concepts of security were developed, with human security being one of the most prominent to materialise within the global political discourse as well as in numerous state and multilateral policies. For the purposes of this chapter, there are certain characteristics of the concept that need to be emphasised. These characteristics have been captured by the Commission on Human Security (CHS), which in 2003 defined human security as "[t]he protection of the vital core of all human lives in ways that enhance human freedoms and human fulfilment [... including] processes that build on people's strengths and aspirations. It means creating political, social, environmental, economic, military and cultural systems that together give people the building blocks of survival, livelihood and dignity" (CHS, 2003, p. 4). As this chapter will later argue, when defined in this way, the concept of human security embraces many of the dimensions that are prominent within the social exclusion literature.

5.3 Cybersecurity

The concept of "cybersecurity" was also initially narrow in scope, focusing on threats to the military and economic assets of the state. This included concerns about the ability of the state or commercial organisations to maintain the continuity of operations in the face of efforts to subvert or disrupt computer-based operations. The original emphasis was on identifying and protecting critical (information) infrastructures. The language of societal resilience was used, not in reference to ontological human security, but rather to signify the ability of key organisations to withstand malicious interference with their computer systems. This resulted in threat imagery characterised by references to computer hackers, hacktivism, cybercrime and espionage, as well as to the potentiality of cyberterrorism (Limnéll, Majewski, & Salminen, 2015;

Singer & Friedman, 2014). As the literature concerning cybersecurity expanded within the field of international relations, the somewhat misleading term "cyberwarfare" also gained credence (Arquilla & Ronfeldt, 1993; Carr, 2010; Demchak, 2011; Rantapelkonen & Salminen, 2013; Rid, 2013), thereby encouraging the placement of cybersecurity issues within the more traditional national security framework. This development was accompanied by increasing attention being paid to issues related to cybercrime (Clough, 2015; McAfee & Center for Strategic and International Studies [CSIS], 2013, 2014; Yar, 2013), which also encouraged the placement of cybersecurity within the state security domain.

A notable feature of the prior cybersecurity literature is that, unlike human security approaches, it takes the state, or at least the information systems and networks of key organisations or infrastructures within the state, as its referent object. The literature identifies insecurities relating to the confidentiality, integrity and availability of information, either stored or in transit, and it elaborates techniques for countering such threats. This approach is essentially technical and focused on protecting against external intrusion and exploitation. It examines vulnerabilities and threats, such as cracking, malware, botnets, exploit kits, denial of service attacks and injection attacks, and develops countermeasures for dealing with them. A common feature of almost all the cybersecurity conceptualisations is, therefore, that they take the inanimate aspects of cybersecurity as their referent objects. The things that need to be safeguarded include the "cyber environment/domain", "interdependent networks and information infrastructures" and "organizations". Consequently, the human being is excluded from being a potential referent object of cybersecurity, instead being allocated the status of the "weakest link", "threat" or "victim" (Dunn Cavelty, 2014, pp. 703–704).

5.4 Human Cybersecurity

Unlike traditional conceptions of cybersecurity, when conceived within a human security framework, it is only through the *use* of such technology to ensure the well-being of people that infrastructures, equipment, software, information and networks become meaningful or, still less, *critical*. The fact that access to the internet has become so important was powerfully signalled in 2016 when the United Nations passed a non-binding resolution defining access to the internet as a human right (AccessNow,

2016). Finland had already accepted this logic, having made internet access a legal right for all Finnish citizens in 2010 (ArcticStartup, 2009). It is appropriate, therefore, to apply the same broader and more critical lenses to cybersecurity that have been utilised in relation to security more generally in order to reframe the idea of cybersecurity so that it incorporates human security, a development that requires a shift in the referent object away from technology and towards the human being. Doing so enables the analysis of cybersecurity in terms of its empowering and constraining dimensions, rather than simply in terms of the problematic aspects of human cyber use. Shifting the referent object of cybersecurity to the human being in this way requires treating individuals and communities as both objects and subjects of cybersecurity.

From a critical security perspective, this approach involves invoking alternatives to the prevailing technical and national security conceptualisations, acknowledging that both digitalisation and cybersecurity have varying implications in different areas and among different communities, and analysing digitalisation and cybersecurity from the perspectives of the citizens and communities themselves, with the aim of making their preferences and fears heard. To redefine cybersecurity from the human security perspective, as well as to incorporate cyber issues into the human security agenda, it is necessary to determine what the prevailing concerns of human security—*freedom from fear* and *freedom from want* (Kaldor, Martin, & Selchow, 2007)—entail in the digital realm.

Proposing a framework for addressing the core of human cybersecurity requires a brief dissection of the idea of "freedom" in terms of human security. Within the human security paradigm, freedoms correspond to notions of negative and positive liberty, which are also known as "freedom from" and "freedom to" (Alkire, 2003). Both aspects of freedom were emphasised in the 1994 Human Development Report, which also stated that threats to human security can exist at any stage of development and at any level of gross domestic product (GDP) (United Nations Development Programme [UNDP], 1994, pp. 23–24). Human security is, therefore, not just a concern of less developed states, but also of highly developed and technologically advanced societies. Following this conceptualisation, in relation to human security, the freedom from fear is concerned with the basis for survival, which is a component of the CHS's definition of human security outlined above, and it is protected by civil and political rights. From a human cyber perspective, threats to the "freedom from fear" include insecurities such as threats to privacy, personal data and the freedom of expression, as well as theft and fraud. The "freedom from want" component is concerned with the basis for livelihood and dignity, as realised through economic, social and cultural rights. In the cybersphere, this includes issues such as economic and educational opportunities and access to adequate services. When defined in this way, the latter dimension of human security overlaps with prominent aspects of the concept of social exclusion. Indeed, human security has been described as "a comprehensive approach that integrates the notion of social exclusion and links it to an extended framework that includes economic security, health, education, conflict, governance and migration perspectives" (Sokoloff & Lewis, 2005, p. 6).

5.5 Social Exclusion

A closer look at the commonly used definitions of social exclusion highlights further connections between these concepts. Burchardt, Le Grand, and Piachaud (2002) offer a fairly broad definition whereby an individual is considered to be socially excluded if they do not participate in "key activities" within the society in which they live. Sen (1999) notes that the lack of participation in key activities may stem from a lack of the required capabilities. This emphasises the relative nature of deprivation as well as why social exclusion can take place in any society. For this reason, Levitas et al. (2007, p. 9) define social exclusion as involving "the lack or denial of resources, rights, goods and services, and the inability to participate in the normal relationships and activities, available to the majority of people in society, whether in economic, social, cultural, or political arenas". The concept is also sometimes explicitly associated with citizenship rights (Berghman, 1998, as cited in Noll, 2002, p. 56) and with access to social goods, for example, public services (Byrne, 1999). Access to social goods is connected to a spatial aspect of social exclusion—area deprivation. In fact, when a geographical area faces relative (or absolute) deprivation, it may be referred to as *concentrated* exclusion (Miliband, 2006). Atkinson (1998, p. 14) makes the important point that exclusion may also be a "property of a group of individuals rather than of individuals" and, therefore, that it may manifest at a community level, both in terms of a geographic area and of a social community. These definitions highlight the overlap between the "freedom from want" dimension of human security and social exclusion when formulated as the deprivation of social goods.

In short, these definitions are all characterised by restricted access to opportunities and a limited capability to capitalise on such opportunities. As such, social exclusion indicates more than simply material deprivation. One way to approach this issue is through Bourdieu's (1986) concept of economic, social and cultural capital. These forms of capital can be thought of as essential to realising human freedoms. In this sense, social exclusion is a barrier to the freedom from both fear and want, thereby constraining an individual's or a community's ability to access and capitalise on opportunity and to achieve security.

The above definitions also emphasise the highly normative nature of social exclusion: it often refers to exclusion from the "normal" activities available to the "average" citizen or to the majority of citizens. It is, therefore, useful to view social exclusion through a constructivist lens. As Van Regenmortel et al. (2016, pp. 333–334) argue, the experience of social exclusion is context dependent and, as a consequence, universal, quantitative measures of social exclusion are not applicable or appropriate in all cases. To illustrate this point, they highlight how the labour market participation of people receiving old-age pensions may signify financial stress in some cases and a high degree of work satisfaction in others. The motivation behind the action, as well as the individual's perception of the situation, are thus critical to the experience of social exclusion. This has also been argued in relation to digital technologies. Helsper (2017) proposes a relative deprivation model of digital inequality, suggesting that an individual may be objectively relatively deprived and yet not experience subjective relative deprivation. In other words, if an individual does not see the value of digital technology, lacks the ability to use it and does not expect to adopt it in the future, they may not feel as if they are experiencing unjust disadvantage.

The emphasis on an individual's perception and agency further problematises the notion of social exclusion. A study of youth narratives in rural northern Finland illustrates this well. The young research participants contested the dominant discourse of social exclusion, wanting the researchers to recognise that "staying in a rural village can indicate a successful life" (Lanas, Rautio, & Syrjala, 2013, p. 393). Similarly, the growing body of literature concerning elderly people's relationships with digital technologies emphasises the importance of motivation in terms of explaining cases of non-use (Dahlberg, 2012; Lüders & Brandtzæg, 2016, 2017; Slettemeås, 2014). Some elderly people may choose to prioritise other activities, and they may not feel the need to use digital technologies. In these cases, non-use does not necessarily indicate social exclusion, but might instead reflect deliberate choices related to the achievement of individual well-being. This suggests that while "excluding" oneself from opportunities that are only accessible through adopting digital technologies or moving from a rural village to an urban centre may lead to forms of objective material deprivation, they do not necessarily result in a feeling of social exclusion. Indeed, Byrne (1999, p. 1) rejects the term "self-exclusion", arguing that "'exclusion' is something that is done by some people to other people", not by individuals to themselves. The same applies at the macro level: areas do not become marginalised on their own. It is something that often, however, "presents itself as inevitable, in the same way that market forces do: it just seems to happen, [...] cumulating in a phenomenon that can be called [...] 'modernization' or 'decay of the rural'" (Lanas et al., 2013, p. 385). This distinction between the "excluder" and the "excluded" has been identified with the "weak" and "strong" forms of the social exclusion discourse:

> In the "weak" versions of this discourse, the solutions lie in altering these excluded peoples' handicapping characteristics and enhancing their integration into dominant society. "Stronger" forms of the discourse also emphasise the role of those who are doing the excluding, and therefore aim for solutions which reduce the powers of exclusion. (Veit-Wilson, 1998, p. 45)

In other words, the former view focuses on modifying the behaviour of the excluded (such as moving to a bigger city or adopting ICT), while the latter focuses on the excluder and their actions (such as closing down public services in rural areas).

5.6 Digitalisation and Social Exclusion

During the 1990s, a new dimension of social exclusion emerged in both the academic literature and policy circles: the idea that social exclusion can result from limited ICT access, that is, the so-called *digital divide*. Initially, the digital divide was understood in terms of differentiated access to, and usage of, digital technologies, which created a gap between the "information rich" and the "information poor" (Wresch, 1996). As it "involves the unequal access and capacity to use ... [ICTs] that are seen as essential to fully participate in society" (Hope, Martin,

& Zubairi, 2016, p. 2), it clearly constitutes an element of social exclusion, as defined above. Recent research regarding this topic has focused more on ICT skills and usage rather than simply on inequality of access (Scheerder, van Deursen, & van Dijk, 2017).

As with social exclusion more generally, exclusion in the cybersphere can be thought of as being mediated by an individual's economic, cultural and social capital. Selwyn (2004) adopts the term "technological capital" to highlight the relevance of Bourdieu's (1986) traditional "capitals" to ICT usage. For example, in this model, economic capital refers to the "economic capacity to purchase ICT hardware and software", cultural capital to "participation in ICT education and training" and social capital to "networks of 'technological contacts' and support" (Selwyn, 2004, p. 355). One way of conceptualising the impact of an individual's capital on their ICT outcomes is through van Dijk's (2005) model of ICT access, which considers motivation, physical access, skills and usage as different stages of digital participation. For example, motivation concerns subjective lifestyle choices and individual perceptions of usefulness; access concerns things such as infrastructure, costs and accessibility; skills are to do with varying levels of competency and digital literacy and lastly, usage often distinguishes between use and "meaningful use" (Selwyn, 2004) or "capital-enhancing activities" (Hargittai & Dobransky, 2017). Each stage has implications for the utilisation of ICT and, importantly, they are all grounded in offline circumstances. By visualising this as a circular model, it is possible to emphasise its effects in terms of producing and reproducing the offline and online preconditions for access. Consequently, an individual's technological capital shapes their interaction with ICT, structuring their opportunities as well as their ability to capitalise on those opportunities, thereby producing unequal outcomes that have implications for social exclusion more broadly. Outcomes that enable participation in "key activities" and so enhance an individual's capital can facilitate social inclusion, while outcomes that simply reproduce offline inequalities can constrain human opportunity and, ultimately, security.

To fully unpack the idea of social exclusion in the cybersphere, it is also necessary to consider the nature of technology. This influences the perception of how an individual's technological capital enables or constrains their use of ICT. Three common views can be found in the literature. Henwood, Wyatt, Miller, and Senker (2000, p. 7) outline them as follows. The first view is one of technological determinism. It holds that technological developments inevitably lead to social progress. The second view states that while technology does not have straightforward and predictable effects, it is simply a "neutral" tool that can be appropriated for multiple purposes. The final view represents a constructivist understanding of technology, which recognises that technologies have cultural meanings and are intrinsically shaped by social values. According to this view, technologies are objects created by (certain) people and social groups. Lasch (2004) critiques the notion of technology as a value-free and neutral tool. He instead asks who shapes technologies and who those technologies serve, arguing that technologies, including the digital technologies of today, are shaped by struggles over production and power, ultimately serving certain interests over others.

There is an important connection between this view of technology as being socially produced and the previously mentioned debate regarding the weak and strong conceptions of social exclusion. The conventional literature in this regard rarely goes beyond viewing digital technology as neutral. This lays the groundwork for perspectives that consider only an individual's attributes when seeking to explain their ICT usage (or lack thereof), sometimes in quantifiable ways (Venkatesh, Thong, & Xu, 2012). According to this view, you either have or do not have the attributes required to adopt and exploit digital technologies for your own benefit. As noted above, similar views can be found concerning social exclusion: by possessing or acquiring certain attributes, you can modify your behaviour and so become socially included. However, by considering both sides of the coin, it is possible to move beyond the view that only an individual's capital matters in terms of their ability to reap the benefits of digitalisation. It is still useful to consider individual attributes in order to better understand the microprocesses that lead to adoption, although there is also a need to consider these factors given the realisation that the medium (i.e. digital technology) is socially shaped and serves certain interests. This allows for the consideration of what those interests are and who shapes them, as well as of which alternative voices, values and interests are excluded from the agenda.

5.7 The European High North

When discussing digitalisation and social exclusion in the European High North, it is necessary to first take a step back so as to make some conceptual connections, partly because the Nordic region is atypical among the case studies of social exclusion (Esping-Andersen, 1999). One study concerning social exclusion and welfare regimes indicates that the Nordic countries, given their high GDPs and strong social protection systems, have lower rates of social exclusion among the elderly than other types of welfare regimes in Europe (Ogg, 2005). Studies of social exclusion in the Nordic countries, such as those by Halleröd and Heikillä (1999) and Fløtten, Dahl, and Grønningsæter (2001), also suggest that problems are perceived and experienced differently when compared to the situation in central and southern European countries and, further, that multidimensional deprivation is less typical. Moreover, use of the term "social exclusion" still tends to build on assumptions related to issues that are typical of large urban areas, rather than of the problems more commonly associated with the sparsely populated areas of the European High North. However, as this chapter will show, there is still a case for focusing on peripheral areas. As Theobald (2005, p. 6) argues, social exclusion is focused on "the situation of members at the fringes of society", and the European High North represents a clear geographical "fringe" in relation to the population centres of the Nordic states. Its geographically peripheral location presents several challenges and potentials for cumulative disadvantage, which will be discussed below.

5.7.1 *Long Distances*

First, the remote location of the European High North, its vast area and the dispersed nature of the settlements result in an issue concerning transportation. Finnmark, Norrbotten and Lapland, the northernmost regions of Norway, Sweden and Finland, respectively, each have the largest area and the lowest population density of their countries (Teräs, Salenius, Fagerlund, & Stanionyte, 2018, pp. 5–6). Long distances and sparse settlements make it harder to provide adequate services, as explained by a project manager at the Centre of Excellence on Social Welfare in Northern Finland:

> [I]t's very hard to provide services there, and if you think about old people there, [... if they] need someone to come like many times a day [...] that's a problem, [...] the travelling takes so much time, and if there's only a few people and they have almost 50 kilometres to [the] next client, [...] then you have to travel a lot. So very often we have to move those people to [...] Ivalo, to provide them with the services they need. (Maarit, interview in Ivalo, Finland, 22 May 2018)

The costs associated with service provision in remote areas have resulted in more and more rural services, public or otherwise, being closed or scaled down (Autti & Hyry-Beihammer, 2014; Kilpeläinen & Seppänen, 2014). An example of this can be seen in the case of newspaper distribution in Norrbotten, Sweden. While the traditional medium will remain in production, the papers will not be distributed as regularly as before. The project leader at Norrbottens Media Distribution, owner of two local morning papers, explained that they want to digitalise their papers because:

> [...] we drive up this road and then back down the same road, empty; we travel 50 kilometres there and back for these papers. That's just over 60 papers. And it's enormous distances, it's enormous costs, both environmentally and money wise. (Elin, interview in Luleå, Sweden, 12 April 2019, author's translation)

In conditions such as these, restricted or missing mobility options can create a barrier to both opportunities and inclusion (Kenyon, Lyons, & Rafferty, 2002; Kilpeläinen & Seppänen, 2014). Long distances also make it harder to access services as they are moved to bigger towns. This has led policymakers and service providers to digitalise many key services in the hope of reducing costs and travel distances. A senior advisor at the Centre of Excellence on Social Welfare in Northern Finland, for example, noted that digital services have enabled more frequent meetings to be held between social workers and their clients, as well as between healthcare workers and their patients:

> [I]t saves time, and maybe you can see him or her more often because you don't have to drive to other places, but some of the meetings you can do [...] online. (Pia, interview in Rovaniemi, Finland, 21 May 2018)

Digitalisation also brings benefits when it comes to education, as the use of ICT can enable the provision of a specialist resource that would otherwise be uneconomical. An example of this is the provision of first language education to some primary school pupils in Luleå municipality through synchronous distance learning. This involves the use of digital tools to provide interactive and real-time education to pupils based at different schools simultaneously. According to a teacher at Flerspråkcentrum, the resource centre for language education in Luleå:

> [T]here are benefits [of distance education] when the municipality has schools spread far apart from each other. […] [I]f [a school] only has one pupil in one language, it feels like an unnecessary waste of resources to send a teacher there once a week. (Kristina, interview via e-mail, Luleå, Sweden, 20 May 2019, author's translation)

Synchronous distance education is also used in upper secondary education in Norrbotten county, Sweden and Finnmark county, Norway, as corroborated by additional interviews (Jonas, Lapland's Gymnasium, interview via Skype, Jokkmokk, Sweden, 9 May 2019 and Elisabeth, Nettskolen in Finnmark, interview via Skype, Vadsø, Norway, 4 September 2018). It enables schools to maintain a high quality of teaching despite small student groups, long travel distances and difficulties with teacher recruitment.

The regional digitalisation coordinator for Region Norrbotten also pointed out that digital services offer time and cost savings for individual citizens as well:

> [I]f we look, for example, at a patient who lives in Arjeplog, if it's 280 kilometres to Arjeplog, [they] can have a video consultation with their doctor at Sunderby Hospital, which takes five minutes, rather than having to travel 280 kilometres times two to see a doctor for five minutes. It's an incredible waste of time and resources, and not particularly environmentally friendly either, to transport yourself over such long distances for so little time. (Staffan, interview in Luleå, Sweden, 9 April 2019, author's translation)

Likewise, a senior lecturer at Luleå University of Technology explained that:

> […] digital technology should be able to solve a lot of problems, to avoid having to do these trips, to save both time and money, because it does have a cost for people. (Mari, interview in Luleå, Sweden, 11 April 2019, author's translation)

The digitalisation of key public services can, therefore, serve to reduce the costs associated with long distances—travel time, monetary costs and environmental impacts—and to render public services easier to access. Nonetheless, as traditional public services are withdrawn, people who cannot, or who will not, use digital technologies face the risk of social exclusion, as they are not in a position to access the replacement services (Hodge, Carson, Carson, Newman, & Garrett, 2017; Warren, 2007). This poses a particular challenge for areas with older-than-average populations and poorer-than-average connectivity, which will be expanded on below.

5.7.2 *Old Age*

Second, the European High North is facing a significant demographic challenge, as a small and decreasing labour pool will need to support an ageing population with complex needs in the coming years. This challenge is not unique to the region, although it is perhaps clearer there than elsewhere. Norrbotten, for example, is decades ahead of Sweden in terms of this demographic development, which has led to certain challenges, as outlined by the regional digitalisation coordinator for Region Norrbotten:

> [W]e have quite a large proportion of the population who are 65+and there we also have a demographic lead in Sweden, [... by] 2050 the rest of Sweden will catch up with us when it comes to demographic development, so we have a quite a big older population. That has consequences, [...] it's hard to find the labour needed to take care of [...] a population [...] that is older and more chronically ill. So, the increased number of elderly and the decreased number of people of working age is a great challenge for us. (Staffan, interview in Luleå, Sweden, 9 April 2019, author's translation)

Policymakers see digitalisation as a necessary intervention to address the challenges associated with service provision, labour shortages and an ageing population. There is hope that, through increased digitalisation, the public sector can become more efficient (Norwegian Ministry of Local Government and Modernisation, 2016). The growing use of digital public services and welfare technologies are examples of this. The process leader for regional development strategy for Region Norrbotten described digitalisation as a clear enabler for the region, especially if they are to successfully address its demographic challenges:

> I think, if we look at the challenges we have regarding demographics and so on, then we don't have a choice, we are going to be forced to work a lot with digitalisation and to look at how we can use it in different ways. [...] Digitalisation is the opportunity. We're sparsely populated, we have

> long distances, we have a declining population, we have more need for welfare services... So it's clear that somehow we have to provide services to people. (Kenneth, interview in Luleå, Sweden, 9 April 2019, author's translation)

A secondary effect of an ageing population can be seen in terms of digital competence. In highly digitalised societies such as the Nordic countries, digital competence has, in many ways, become necessary for participation. For example, using the internet has been identified as a "critical competency" in Norway (Staksrud, 2011, p. 69), while Sweden has likewise witnessed a rhetorical shift "from 'access' to citizens' skilful 'use' of digital resources" in government agendas (Reneland-Forsman, 2018, p. 336). Arguably, the relative skills gap becomes even more important in digitally advanced societies where digital competence has become vital to participation and inclusion. As Helsper (2017, p. 230) outlines:

> In a world where digital skills and engagement are fundamental for, for example, employment, it is not the absolute level but the relative level of skill or engagement that will make the difference for a person obtaining a job. If everyone else is highly digitally skilled, a person with medium skills will have a hard time finding a job but in a society where people have lower digital skills, the person with medium skills is king. The same level of access, skills, or engagement can thus indicate inclusion in one context and exclusion in another.

Despite the existence of high overall levels of digital competence, elderly people are considered to be the group most vulnerable to digital exclusion, as they constitute a majority of non-users and "weak" users (Fjørtoft, 2017; Slettemeås, 2014). They are thus at risk in terms of the "freedom from want" human security dimension noted earlier. This lack of digital competency and maturity is recognised as a challenge by several stakeholders:

> [T]hat's a problem because if we're going to be able to care for our elderly patients, maybe in their homes, with digital tools, then it's the digital immaturity or exclusion among the elderly population [that] is very much higher than among the younger population, and then it becomes [...] counterproductive. (Staffan, interview in Luleå, Sweden, 9 April 2019, author's translation)

> [W]hen [...] forms, for example, and services are explained in Finnish [on the] internet it's a challenge for older people [...] – first of all, they didn't learn to use [the] internet, and second of all they might use, for example, Sámi as their home language, [meaning] Finnish is not their first language, so when they have to use that kind of [...] official language [...], and then try to find [on the] internet all the services, it's getting really difficult for them. (Ristenrauna, SámiSoster NGO, interview with interpreter in Inari, Finland, 22 May 2018)
>
> [W]hen you talk with those who live in sparsely populated areas up here, we can't ignore that the average age is quite high, they aren't always digitally competent, they need help with an array of things, just understanding the technology [...]. (Mari, Luleå University of Technology, interview in Luleå, Sweden, 11 April 2019, author's translation)

It has also been found that, often, older people rely on networks of friends and family to help them use digital tools and to complete tasks online (Dahlberg, 2012; Koivunen, 2014; Rasi & Kilpeläinen, 2015; Selwyn, Johnson, Nemorin, & Knight, 2016; Slettemeås, 2014). This can prove challenging because small, remote villages tend to have an older-than-average demographic profile, as significant numbers of the younger working-age population have left for bigger towns and cities:

> Look, it's possible if your children are living here. But I think that mostly our children are going to [...] the south[ern] part of Finland, and they are studying there and they [are] working there, and then they have families there. (Marja, interview in Inari, Finland, 22 May 2018)

When on their own, many elderly people struggle with anxiety concerning the use of ICT, which limits the extent to which they are able to make use of a variety of digital services and platforms. According to the leader of the Pensioners' Association in Kirkenes, Norway:

> [W]e see that many of our members use Facebook if need be, which their grandchildren have taught them to do, but both banking services and other internet services they use very little of. [This is] because they have a fear and a fright about what it is, how dangerous it is. (Nils-Edvard, interview in Kirkenes, Norway, 5 September 2018, author's translation)

As a result, less confident users may avoid what some consider to be the most beneficial aspects of ICT, such as online banking or municipal services. In many cases these individuals instead have to pay user fees or travel long distances to continue accessing (non-digital) traditional alternatives.

The combination of long distances, poor mobility, the shift from locally available services to digital alternatives, the lower rates of internet use and the increasing competency requirements therefore raise questions regarding concentrated social exclusion, especially among the elderly, who are considered to be particularly vulnerable to social exclusion and multiple disadvantages (Van Regenmortel et al., 2016). Elderly populations in the European High North that lack ICT access or competence can be seen as experiencing the effects of concentrated exclusion in the geographical sense, as identified by Miliband (2006), which has a negative impact on their human security.

5.7.3 Connectivity

Finally, the geographic and demographic patterns of the European High North pose a problem in terms of internet connectivity. In the European High North, an adequate internet connection is seen as a fundamental precondition for delivering public services in sparsely populated areas, as explained by the process leader for regional development strategy for Region Norrbotten:

> It's really like a basic precondition for having digital services for the citizens. Otherwise, it becomes really difficult. Just imagine [...] if elderly people are going to live at home, and you might need services in the homes and you need to be able to provide services in a more efficient way, [...] then it's maybe digital services first and foremost... (Kenneth, interview in Luleå, Sweden, 9 April 2019, author's translation)

Indeed, among stakeholders in the European High North, it is commonly reported that an adequate internet connection is even more important in remote areas than in central areas, as the maintenance of adequate public services and the meeting of labour market challenges are seen as being dependent on efficient digitalisation. For example, a representative of Paliskuntain yhdistys, the Reindeer Herders' Association based in Rovaniemi, stated that:

> It's really helpful. I would say it's much more helpful for us in this area where we have, you know, [... such a] big area and so few people. It helps people on the, let me say, on the countryside. Much more than people in the cities. (Matti, interview in Rovaniemi, Finland, 21 May 2018)

The critical importance of digitalisation was further expressed by the digitalisation coordinator for Region Norrbotten, who emphasised that a digital shift simply *has* to happen:

> Norrbotten will be Sweden's most digitalised county. And why should we believe that? [...] [T]he driving force behind making it possible is that our county is such a big area and has such a small population. And in order for everyone to get great service, or access to services no matter where they live, [...] it requires innovative digitalisation. And the need for innovative digitalisation is greater in sparsely populated areas, or in counties with a lot of sparsely populated areas, than in more urban counties. That's why it's here that it's going to happen. It has to happen. (Staffan, interview in Luleå, Sweden, 9 April 2019, author's translation)

Yet, simultaneously, some parts of the European High North lag behind the national averages in their respective countries, which generally perform well in relation to digitalisation rankings such as the Digital Economy and Society Index (European Commission, 2018). Finnmark, for example, is the worst off Norwegian county in terms of both broadband speed and broadband adoption per household, despite there having been a gradual reduction in geographical differences (Statistics Norway, 2017a). While governments across the European High North have adopted strategies to improve broadband coverage (Eskelinen, Frank, & Hirvonen, 2008; Norwegian Ministry of Local Government and Modernisation, 2016; Randall, Berlina, Teräs, & Rinne, 2018), studies have identified a general trend whereby the gap in the digital infrastructure quality between commercially profitable and unprofitable areas persists due to the catch-up effect (Salemink, Strijker, & Bosworth, 2017). Simply put:

> [...] there aren't economic incentives to lay fibre everywhere. (Staffan, interview in Luleå, Sweden, 9 April 2019, author's translation)

According to the head of the ICT department of Sør-Varanger municipality in Norway, the high cost of providing broadband to remote areas is hard to shoulder without subsidies:

> [The challenge is] the distance and [that] people live very spread out. With a calculated fibre cost of 300 kroner per meter, when you talk about a stretch of 100 kilometres and around 40,000 additional cost per household, and [when] there are 700 households, then there goes the cost. (Anders, interview in Kirkenes, Norway, 5 September 2018, author's translation)

It is important to note, however, that the issue of poor internet connectivity is not limited to the European High North. Data from Statistics Norway (2017b) shows that the absolute number of people with a poor internet connection is much higher in more populous counties, including those with the highest average internet speeds. There is, therefore, no reason to overstate the importance of poor infrastructure as an issue in the more central areas of the European High North, as local stakeholders will attest:

> [...] I think that in Norway today there are very few who, well, there are of course some villages that have problems with it, but overall most people have more broadband than they can use. (Tove, Helse Nord IKT, interview via Skype, Tromsø, Norway, 5 September 2018, author's translation)

However, as suggested by the prior literature concerning rural social exclusion, long distances and the lack of proximity to public services leave rural populations at a particular disadvantage, meaning that a relative deficiency in broadband provision may have larger consequences. As Warren (2007) argues, this both heightens the benefits of being connected in rural areas and exacerbates the disadvantages of non-use. As public services become harder to access through traditional means, both internet connectivity and digital competency will become key to exploiting digital services and gaining the benefits of the ongoing digitalisation process. It is also crucial to note that due to the high cost associated with connecting the "final few", the most remote areas will remain the worst connected:

> [A] lot more [subsidies] are needed, because it's the last connections that are the most expensive, the ones that are the furthest away. (Kenneth, interview in Luleå, Sweden, 9 April 2019, author's translation)

5.8 Conclusion

By outlining the conceptual linkages between cybersecurity, human security and social exclusion, this chapter has demonstrated how social exclusion can be incorporated into a broader conception of human cybersecurity. This framework enables the analysis of the empowering and constraining dimensions of digitalisation for both individuals and communities through the language of security. A review of the prior literature showed that, in addition to representing more than poverty, social exclusion is a normative and relative concept because it refers to exclusion from "key activities" in a specific society. Moreover, there is a degree of subjectivity involved in whether or not social exclusion is experienced; it is not simply a material condition. The same is true for digital forms of exclusion. This chapter has shown, by reference to van Dijk's (2005) multilevel framework of access, that engagement with digital technologies and services is mediated by an individual's technological capital (Selwyn, 2004). Consequently, engagements that enable participation in "key activities" and enhance an individual's social, cultural and economic capital can facilitate social inclusion, while engagements that reproduce offline inequalities can constrain human opportunity and create insecurity. Moreover, recognition of the distinction between "weak" and "strong" conceptions of social exclusion (Veit-Wilson, 1998) has shifted the focus from purely changing the behaviour of the "excluded" to emphasising the actions of the "excluder". This has subsequently been connected to the idea that digital technology is produced in a social context (Henwood et al., 2000; Lasch, 2004), which gives rise to the question of whose voices, values and interests are represented in the digitalisation agenda.

By questioning the purpose of digitalisation and asking whose interests it is supposed to serve, light is shone on some of the structural problems involved in the pursuit of national and subnational digitalisation strategies, as well as on the potential mismatch of interests. It is no secret that efficiency, streamlining and cost reduction are powerful drivers of digitalisation. This contributes to the creation of a normative ideal regarding what it means to be a digitally competent citizen, an ideal that not everyone can fulfil. The result could, therefore, be digital services that mainly streamline internal processes rather than meeting the needs of the end user:

> I think what they sometimes miss [is that] they create e-services that simplify [things] for themselves organisationally and they forget to think about who the receiver is [...]. What does the customer need? What do they want? And I think that's a perspective that's missed a lot. One has slimmed down one's own processes within the organisation [and so] one has saved resources through providing some kind of e-service that is really turned outwards but that makes the organisation more efficient. And then we get these, we get these really diametrical stories, that's because one hasn't talked. (Mari, interview in Luleå, Sweden, 11 April 2019, author's translation)

When applied to the European High North, the human cybersecurity framework alerts us to several challenges and potentials for cumulative disadvantage concerning digitalisation, three of which are discussed in this chapter. First, the long distances and sparse settlements that characterise the European High North make it costly to provide services, which are often moved to bigger towns and cities or made available online. This creates the potential for exclusion among inhabitants with less digital competence and more restricted mobility. However, digital services also have an enabling potential, as the costs for individuals associated with accessing services are reduced. Second, the ageing population has led to challenges regarding service provision, not only in the European High North, but across Europe as well. In the European High North, this has had two effects. The first effect is the creation of labour shortages, as a small and decreasing labour pool attempts to support the ageing population. The second effect can be seen in the relationship between age and digital competence. Elderly people, an overrepresented demographic in the European High North, are generally less digitally competent than younger people. This relative skills gap becomes even more important in highly digitalised societies where digital competence has become critical to both participation and inclusion (Helsper, 2017). Lastly, the provision of adequate internet connectivity is problematic in the most remote parts of the European High North due to its geography and sparse settlements. The connection of the "final few" is thus expensive, but perhaps also critical in order to avoid concentrated social exclusion.

To conclude, it is clear that digitalisation can offer multiple benefits for peripheral areas such as the European High North, as well as for the security of communities and individuals, who gain easier access to services. Digital services are often described by local stakeholders as being more helpful to rural communities than to urban communities because they enable people to live, study and work in the periphery, which reflects the findings of prior studies concerning digitalisation in rural areas (Warren, 2007). However, this presupposes that all the inhabitants of the European High North want, can access and are able to use digital technologies to the extent that is demanded by the public sector and by society at large. In other words, successful digitalisation in the European High North and beyond is contingent on adequate digital competence and broadband infrastructure, but it is also dependent on the serious consideration of the needs and wants of affected populations. If these components are not put in place, rural villages and elderly people in particular are at risk of experiencing social exclusion. As such, while digitalisation provides solutions and opportunities that are crucial for the sustainability of areas such as the European High North, the benefits of such developments will be unevenly distributed if consideration is not given to the limitations of digitalisation and to whose voices, values and interests shape the current digitalisation agenda.

References

AccessNow. (2016). *U.N. passes landmark resolution condemning internet shutdowns*. Retrieved from https://www.accessnow.org/un-passes-resolution-condemning-internet-shutdowns/.

Alkire, S. (2003). *A conceptual framework for human security* (CRISE Working Paper 2). Centre for Research on Inequality, Human Security and Ethnicity, University of Oxford. Retrieved from https://assets.publishing.service.gov.uk/media/57a08cf740f0b652dd001694/wp2.pdf.

ArcticStartup. (2009). *Finland makes 1mbps internet connection a legal right—So what?* Retrieved from https://arcticstartup.com/finland-makes-1mbps-internet-connection-a-legal-right-so-what/.

Arquilla, J., & Ronfeldt, D. (1993). Cyberwar is coming! *Comparative Strategy, 12*(2), 141–165.

Atkinson, A. B. (1998). Social exclusion, poverty and unemployment. In A. B. Atkinson & J. Hills (Eds.), *Exclusion, employment and opportunity* (pp. 1–20). CASE Paper 4. Centre for Analysis of Social Exclusion, London School of Economics. Retrieved from http://sticerd.lse.ac.uk/dps/case/cp/Paper4.pdf.

Autti, O., & Hyry-Beihammer, E. K. (2014). School closures in rural Finnish communities. *Journal of Research in Rural Education, 29*(1), 1–17.

Barnett, J. (2001). *The meaning of environmental security: Ecological politics and policy in the new Security era*. London: Zed Books.
Booth, K. (2005). Introduction to part one. In K. Booth (Ed.), *Critical security studies and world politics* (pp. 21–25). London: Lynne Rienner.
Booth, K. (2007). *Theory of world security*. Cambridge: Cambridge University Press.
Bourdieu, P. (1986). The forms of capital. In J. Richardson (Ed.), *Handbook of theory and research for the sociology of education* (pp. 241–258). Westport, CT: Greenwood.
Burchardt, T., Le Grand, J., & Piachaud, D. (2002). Introduction. In J. Hills, J. Le Grand, & D. Piachaud (Eds.), *Understanding social exclusion* (pp. 1–12). Oxford: Oxford University Press.
Buzan, B. (1983). *People, states and fear: An agenda for security analysis in the post-Cold War era*. Brighton: Wheatsheaf.
Buzan, B. (1991). *People, states and fear: An agenda for security analysis in the post-Cold War era* (2nd ed.). Boulder, CO: Lynne Rienner.
Buzan, B., Wæver, O., & de Wilde, J. (1998). *Security: A new framework for analysis*. Boulder, CO: Lynne Rienner.
Byrne, D. (1999). *Social exclusion*. Buckingham: Open University Press.
Carr, J. (2010). *Inside cyber warfare*. Sebastopol, CA: O'Reilly Media.
Clough, J. (2015). *Principles of cybercrime* (2nd ed.). Cambridge: Cambridge University Press.
Commission on Human Security (CHS). (2003). *Human security now* (Report). Retrieved from https://reliefweb.int/sites/reliefweb.int/files/resources/91BAEEDBA50C6907C1256D19006A9353-chs-security-may03.pdf.
Dahlberg, Å. (2012). Användning av IT-baserade tjänster i de nordiska länderna [Use of IT based services in the Nordic countries]. In E. P. Göransson (Ed.), *Fokus på eldre i informasjonssamfunnet* [Focus on the elderly in information society] (pp. 12–24). Nordens Välfärdscenter. Retrieved from http://www.diva-portal.org/smash/get/diva2:706911/FULLTEXT01.pdf.
Demchak, C. C. (2011). *Wars of disruption and resilience: Cybered conflict, power and national security*. London: The University of Georgia Press.
Dunn Cavelty, M. (2014). Breaking the cyber-security dilemma: Aligning security needs and removing vulnerabilities. *Science and Engineering Ethics, 20*(3), 701–715.
Eskelinen, H., Frank, T., & Hirvonen, T. (2008). Does strategy matter? A comparison of broadband rollout policies in Finland and Sweden. *Telecommunications Policy, 32*(6), 412–421.
Esping-Andersen, G. (1999). *Social foundations of postindustrial economies*. London: Oxford University Press.

European Commission. (2018). *The Digital Economy and Society Index (DESI)*. Retrieved from https://ec.europa.eu/digital-single-market/en/desi.

Fjørtoft, T. O. (2017). *Digitale ferdigheter i ulike aldersgrupper: Unge og høyt utdannede er flinkest foran PC-en* [Digital skills in different age groups: The young and highly educated are best in front of the PC] (Report). Statistics Norway. Retrieved from http://www.ssb.no/teknologi-og-innovasjon/artikler-og-publikasjoner/unge-og-hoyt-utdannede-er-flinkest-foran-pc-en.

Fløtten, T., Dahl, E., & Grønningsæter, A. (2001). *Den norske fattigdommen: Hvordan arter den seg, hvor lenge varer den og hva kan vi gjøre med den?* [Norwegian poverty: How does it behave, how long does it last and what can we do about it?] (Fafo Working Paper 2001:16). Forskningsstiftelsen Fafo. Retrieved from https://www.fafo.no/media/com_netsukii/672.pdf.

Halleröd, B., & Heikkilä, M. (1999). Poverty and social exclusion in the Nordic countries. In M. Kautto, M. Heikkilä, B. Hvinden, S. Marklund, & N. Ploug (Eds.), *Nordic social policy: Changing welfare states* (pp. 185–214). London: Routledge.

Hargittai, E., & Dobransky, K. (2017). Old dogs, new clicks: Digital inequality in internet skills and uses among older adults. *Canadian Journal of Communication, 42*(2), 195–212.

Helsper, E. J. (2017). The social relativity of digital exclusion: Applying relative deprivation theory to digital inequalities. *Communication Theory, 27*, 223–242.

Henwood, F., Wyatt, S., Miller, N., & Senker, P. (2000). Critical perspectives on technologies, in/equalities and the information society. In F. Henwood, N. Miller, P. Senker, & S. Wyatt (Eds.), *Technology and in/equality* (pp. 1–18). London: Routledge.

Hodge, H., Carson, D., Carson, D., Newman, L., & Garrett, J. (2017). Using internet technologies in rural communities to access services: The views of older people and service providers. *Journal of Rural Studies, 54*, 469–478.

Holland, J., & Jarvis, L. (2015). *Security: A critical introduction*. London: Palgrave.

Hope, S., Martin, C., & Zubairi, S. S. (2016). *The role of digital exclusion in social exclusion* (Ipsos MORI Scotland & Carnegie UK Trust report). Retrieved from: https://d1ssu070pg2v9i.cloudfront.net/pex/carnegie_uk_trust/2016/09/LOW-2697-CUKT-Digital-Participation-Report-REVISE.pdf.

Kaldor, M., Martin, M., & Selchow, S. (2007). Human security: A new strategic narrative for Europe. *International Affairs, 83*(2), 273–288.

Kenyon, S., Lyons, G., & Rafferty, J. (2002). Transport and social exclusion: Investigating the possibility of promoting inclusion through virtual mobility. *Journal of Transport Geography, 10*(3), 207–219.

Kilpeläinen, A., & Seppänen, M. (2014). Information technology and everyday life in ageing rural villages. *Journal of Rural Studies, 33,* 1–8.
Koivunen, E. (2014). *Telecare and older people's social relations* (AKTIVE Working Paper 3). Leeds: Centre for International Research on Care, Labour and Equalities. Retrieved from http://www.aktive.org.uk/downloads/AKTIVE_Paper-3.pdf.
Lanas, M., Rautio, P., & Syrjala, L. (2013). Beyond educating the marginals: Recognizing life in northern rural Finland. *Scandinavian Journal of Educational Research, 57*(4), 385–399.
Lasch, C. (2004). The degradation of the practical arts. In F. Webster (Ed.), *The information society reader* (pp. 287–291). London: Routledge.
Levitas, R., Pantazis, C., Fahmy, E., Gordon, D., Lloyd, E., & Patsios, D. (2007). *The multi-dimensional analysis of social exclusion* (Report). Department of Sociology and School for Social Policy, Townsend Centre for the International Study of Poverty and Bristol Institute for Public Affairs. Retrieved from https://dera.ioe.ac.uk/6853/1/multidimensional.pdf.
Limnéll, J., Majewski, K., & Salminen, M. (2015). *Cyber security for decision makers.* Jyväskylä: Docendo.
Lüders, M., & Brandtzæg, P. B. (2016). Når alt sosialt blir flyktig: En kvalitativ studie av hvordan eldre opplever sosiale medier [When everything social becomes fleeting: A qualitative study of how older people experience social media]. *Norsk Medietidsskrift, 23*(2), 1–18.
Lüders, M., & Brandtzæg, P. B. (2017). 'My children tell me it's so simple': A mixed-methods approach to understand older non-users' perceptions of social networking sites. *New Media & Society, 19*(2), 181–198.
McAfee and Center for Strategic and International Studies (CSIS). (2013). *The economic impact of cybercrime* (Report). Retrieved from www.mcafee.com/us/resources/reports/rp-economic-impact-cybercrime2.pdf.
McAfee and Center for Strategic and International Studies (CSIS). (2014). *Net losses: Estimating the global cost of cybercrime* (Report). Retrieved from https://csis-prod.s3.amazonaws.com/s3fs-public/legacy_files/files/attachments/140609_rp_economic_impact_cybercrime_report.pdf.
McDonald, M. (2013). Discourses of climate security. *Political Geography, 33,* 42–51.
McSweeney, B. (1999). *Security, identity and interests: A sociology of international relations.* Cambridge: Cambridge University Press.
Mearsheimer, J. (2001). *The tragedy of great power politics.* New York, NY: W. W. Norton.
Miliband, D. (2006). *Social exclusion: The next steps forward.* London: ODPM.
Morgenthau, H. (1973). *Politics among nations: The struggle for power and peace* (5th ed.). New York, NY: Knopf.

Noll, H. H. (2002). Towards a European system of social indicators: Theoretical framework and system architecture. *Social Indicators Research, 58*, 47–87.

Norwegian Ministry of Local Government and Modernisation. (2016). *Digital agenda for Norway in brief: ICT for a simpler everyday life and increased productivity* (White Paper 27 [2015–2016]). Retrieved from https://www.regjeringen.no/contentassets/07b212c03fee4d0a94234b101c5b8ef0/en-gb/pdfs/digital_agenda_for_norway_in_brief.pdf.

Ogg, J. (2005). Social exclusion and insecurity among older Europeans: The influence of welfare regimes. *Ageing & Society, 25*, 69–90.

Peoples, C., & Vaughan-Williams, N. (2010). *Critical security studies: An introduction*. London: Routledge.

Randall, L., Berlina, A., Teräs, J., & Rinne, T. (2018). *Digitalisation as a tool for sustainable Nordic regional development: Preliminary literature and policy review* (Discussion Paper). Stockholm: Nordic Thematic Group for Innovative and Resilient Regions. Retrieved from http://www.nordregio.org/wp-content/uploads/2018/03/Digitalisation_Discussion-Paper_Jan-31.pdf.

Rantapelkonen, J., & Salminen, M. (2013). Introduction: Looking for an understanding of cyber. In J. Rantapelkonen & M. Salminen (Eds.), *The fog of cyber defence* (pp. 14–23). Publication Series 2, Article 10. Helsinki: National Defence University. Retrieved from https://www.cyberwar.nl/d/20130200_The%20Fog%20of%20Cyber%20Defence%20NDU%202013.pdf.

Rasi, P., & Kilpeläinen, A. (2015). The digital competences and agency of older people living in rural villages in Finnish Lapland. *International Journal of Media, Technology and Lifelong Learning, 11*(2), 149–160.

Reneland-Forsman, L. (2018). 'Borrowed access'—The struggle of older persons for digital participation. *International Journal of Lifelong Education, 37*(3), 333–344.

Rid, T. (2013). *Cyber war will not take place*. Oxford: Oxford University Press.

Salemink, K., Strijker, D., & Bosworth, G. (2017). Rural development in the digital age: A systematic literature review on unequal ICT availability, adoption, and use in rural areas. *Journal of Rural Studies, 54*, 360–371.

Scheerder, A., van Deursen, A., & van Dijk, J. (2017). Determinants of internet skills, uses and outcomes: A systematic review of the second- and third-level digital divide. *Telematics and Informatics, 34*, 1607–1624.

Selwyn, N. (2004). Reconsidering political and popular understandings of the digital divide. *New Media & Society, 6*(3), 341–362.

Selwyn, N., Johnson, N., Nemorin, S., & Knight, E. (2016). *Going online on behalf of others: An investigation of 'proxy' internet consumers*. Sydney, Australia: Australian Communications Consumer Action Network. Retrieved from http://eprints.lse.ac.uk/83436/1/Nemorin_Going%20online_2017.pdf.

Sen, A. (1999). *Development as freedom*. Oxford: Oxford University Press.

Sheehan, M. (2005). *International security: An analytical survey*. Boulder, CO: Lynne Rienner.

Singer, P. W., & Friedman, A. (2014). *Cybersecurity and cyberwar: What everyone needs to know*. Oxford: Oxford University Press.

Slettemeås, D. (2014). *Eldres bruk av digitale verktøy og internett: En landsdekkende undersøkelse av mestring, støttebehov, motivasjon og hindringer* [The elderly's use of digital tools and the internet: A country wide examination of mastery, needs for support, motivation and barriers] (Report 5). Oslo: Deltasenteret and Statens institutt for forbruksforskning—SIFO. Retrieved from http://www.hioa.no/extension/hioa/design/hioa/images/sifo/files/file79937_oppdragsrapport_5-2014_rev_1.pdf.

Sokoloff, C., & Lewis, R. (2005). *Denial of citizenship: A challenge to human security* (Issue Paper 28). Brussels: European Policy Centre. Retrieved from http://www.epc.eu/documents/uploads/724318296_EPC%20Issue%20Paper%2028%20Denial%20of%20Citizenship.pdf.

Staksrud, E. (2011). Norske barn på Internett: Høy risiko - liten skade? *Nordicom Information, 33*(4), 59–70.

Statistics Norway. (2017a). *Internett-målinga 20 November 2017* [The internet measurement 20 November 2017]. Retrieved from https://www.ssb.no/inet.

Statistics Norway. (2017b). *Ni av ti surfer på nettet hver dag* [Nine out of ten surf the web every day]. Retrieved from http://www.ssb.no/teknologi-og-innovasjon/artikler-og-publikasjoner/ni-av-ti-surfer-pa-nettet-hver-dag.

Teräs, J., Salenius, V., Fagerlund, L., & Stanionyte, L. (2018). *Smart specialisation in sparsely populated European Arctic regions* (Technical Report). Luxembourg: The Joint Research Centre (JRC), Publications Office of the European Union. Retrieved from https://s3platform.jrc.ec.europa.eu/documents/20182/201464/Smart+Specialisation+in+Sparsely+Populated+European+Arctic+Regions/6d39afc9-6b7d-442d-a4bd-d2c4f86d8092.

Theobald, H. (2005). *Social exclusion and care for the elderly: Theoretical concepts and changing realities in European welfare states* (WZB Discussion Paper SP I 2005-301). Berlin: Wissenschaftszentrum Berlin für Sozialforschung (WZB). Retrieved from https://www.econstor.eu/bitstream/10419/47367/1/482174676.pdf.

United Nations Development Programme (UNDP). (1994). *Human Development Report*. Oxford: Oxford University Press. Retrieved from http://hdr.undp.org/sites/default/files/reports/255/hdr_1994_en_complete_nostats.pdf.

Van Dijk, J. (2005). *The deepening divide: Inequality in the information society*. London: Sage.

Van Regenmortel, S., de Donder, L., Dury, A., Smetcoren, A., de Witte, N., & Verté, D. (2016). Social exclusion in later life: A systematic review of the literature. *Population Ageing, 9*, 315–344.

Veit-Wilson, J. (1998). *Setting adequacy standards: How governments define minimum incomes*. Bristol: The Policy Press.

Venkatesh, V., Thong, J. Y. L., & Xu, X. (2012). Consumer acceptance and use of information technology: Extending the unified theory of acceptance and use of technology. *MIS Quarterly, 36*(1), 157–178.

Waltz, K. N. (1979). *Theory of international politics*. New York, NY: Random House.

Warren, M. (2007). The digital vicious cycle: Links between social disadvantage and digital exclusion in rural areas. *Telecommunications Policy, 31*(6–7), 374–388.

Wresch, W. (1996). *Disconnected: Haves and have-nots in the information age*. New Brunswick, NJ: Rutgers University Press.

Wyn Jones, R. (2007). Message in a bottle? Theory and praxis in critical security studies. In B. Buzan & L. Hansen (Eds.), *International security* (pp. 299–319). London: Sage.

Yar, M. (2013). *Cybercrime and society* (2nd ed.). London: Sage.

CHAPTER 6

Mobile Internet Access as a Human Right: A View from the European High North

Stefan Kirchner

6.1 Introduction

Communication is essential—not only for the ability to work and to participate in social life, but also for human security. In sparsely populated areas, mobile communication can quickly become a lifeline in the truest sense of the word, allowing local residents to make a living and providing access to, for example, medical services, including telemedicine. Often, however, rural and remote areas, which could benefit hugely from access to sufficiently fast mobile internet, lack such access. This chapter will focus on mobile internet services, rather than on mobile phone services in general, due to the increasing practical relevance of mobile internet applications to the provision of services. Practical examples of the obstacles experienced by rural residents include the limited choice available to consumers (often there is only one mobile phone provider, leaving users with no backup in case of technical problems) and the frequent lack of or limited coverage in rural areas. Low speeds and

S. Kirchner (✉)
University of Lapland, Rovaniemi, Finland
e-mail: stefan.kirchner@ulapland.fi

M. Salminen et al. (eds.),
Digitalisation and Human Security, New Security Challenges,
https://doi.org/10.1007/978-3-030-48070-7_6

low bandwidths are also common, as is the absence of mobile internet services even though some mobile phone coverage might be available. This problem is not limited to less developed countries, as it is frequently encountered in rural areas of more developed countries too.

This chapter will show how mobile internet access is a concern from the perspective of international human rights law. The issue of access to communication infrastructure is distinct from the right to freedom of expression, although there is a significant practical overlap: freedom of speech does not require the state to provide a forum in which individuals can speak (different, for example, from judicial human rights, which require a functioning justice system to be provided by the state). While the right to internet access does not require the state to provide everybody with the devices necessary to make use of networks, it can be understood as the legal basis for a duty on the part of the state to ensure the existence of infrastructure which is sufficient for the needs of the local population: the right to internet access might be described as a *sui generis* right, which is more than an amalgamation of rights. Access to the internet is the practical prerequisite for the effective enjoyment of a number of other human rights, including the right to freedom of expression and the right to receive information. The same can be said with regards to the right to a private life and a whole range of other civil and political human rights. Mobile internet access is particularly relevant when it is necessary for the realisation of the right to health and, potentially, the right to life, for example, when sufficiently fast mobile internet connectivity is required for the provision of telemedical services. The degree to which the internet and related technologies have permeated society, especially in Northern Europe, has made the prior distinction between the online world and "real life" activities meaningless. This distinction is evaporating rapidly in a society in which mobile internet access is the norm rather than the exception, even in remote locations.

This chapter will begin by considering the availability of mobile internet services in remote regions. Afterwards, questions concerning the legal status of the access to such services in international human rights law will be approached in the context of contemporary communication needs. As an example of the Nordic approaches, Finland's contribution to this development will be focused on. Finally, the issue of the realisation of such a possible right to mobile internet access will be addressed. The legal focus will be on the two international treaties that are most relevant to the European High North from the perspective of general international human rights law, namely the International Covenant on Civil and Political Rights (ICCPR) and the European Convention on Human Rights (ECHR).

6.2 Mobile Internet Access in Rural Areas

6.2.1 Not an Empty Land

The European High North is often perceived as being nearly uninhabited (cf., albeit somewhat tongue-in-cheek, Clare, 2017, p. 1), although this is hardly the case and the region has been home to indigenous peoples for thousands of years. Yet, the population density in the area is significantly lower than that in most other parts of Europe (see Eurostat, 2018). This state of affairs impacts the availability of services in general as well as the availability of mobile internet services: "Remote areas with sparse population, disaster areas in the aftermath, and refugee camps all require communication that is not forthcoming from commercial vendors" (Brown & Mickelson, 2018, p. 1). This means that there exists a need for alternative pathways to universal connectivity. It is thought that even the least developed regions of the world will eventually be connected because the local population has a need to communicate that will eventually translate into commercial profits (Brown & Mickelson, 2018, p. 2; for an economic analysis see Brown & Mickelson, 2018, p. 4 et seq.). However, it remains debatable whether the same is true for regions such as those parts of the European High North where the population density in many mobile phone coverage cells is practically zero, save for the occasional reindeer herder. While there are different technical solutions available to provide internet access to rural villages, such as internet-by-satellite or small directional antennas operated by local companies that create neighbourhood-level wireless local area networks to cover gaps left by mobile phone service providers (see also Brown & Mickelson, 2018, p. 3), the situation outside settled areas can look very different. In Finland, however, the high volume of data usage among consumers and the high degree of competition between mobile phone (and mobile internet) service providers has resulted in significant investments in infrastructure (Robitzski, 2018). In addition, Finnish law requires those phone service providers that have been designated by the regulatory body as universal service providers to make their services, including mobile internet services, available to all customers nationwide, regardless of their place of residence (Electronic Communication Services Act, 2014, §§ 85 et seq.). As a result, the mobile internet coverage in Finland, even in remote areas far north of the Arctic Circle, exceeds that of, for example, rural areas near major population centres in central Europe. In addition to serving as an example of state practice concerning the determination of the existence of a norm of customary international law providing for a human right to mobile internet access, this approach (combining a free market with a minimum of regulatory effort on the part of the state) might serve as a model for other remote regions due to its positive impact on mobile internet coverage. Usually, there is a link between population density and mobile phone coverage (Brown & Mickelson, 2018, p. 3): In order for an investment in infrastructure, such as antenna towers, to be profitable, there have to be at least a certain number of users who will generate income for the investor (in this case, the mobile service provider). This link has been broken in the European High North, particularly in Finnish Lapland, because service providers made the necessary investments. In many other parts of the world, the continuing rural disconnect remains a serious problem. This is also a question of international human rights law. Efforts by the United Nations (UN) to promote human rights also extend to the internet (see, e.g. United Nations General Assembly, 2016), while the UN Human Rights Council has been "concern[ed] that many forms of digital divides remain between and within countries" (UN General Assembly, 2016, preamble).

6.2.2 The Human Right to Internet Access as an Idea

The question of whether or not internet access is a legally protected human right is not a new one. Given how important the internet has become in numerous spheres, from private life to business, culture to politics (Tully, 2014), the question needs to become whether internet access, as a human right, is still merely an emerging idea or whether it is already solidly rooted in existing norms of international human rights law (Tully, 2014). Initially, the debate regarding the right to internet access in general was mainly concerned with questions of infrastructure (Jørgensen & Marzouki, 2015, p. 7). While other issues, such as net neutrality and the prices of mobile data packages, have now also become relevant, the presence or absence of infrastructure and the so-called "digital divide" (Jørgensen & Marzouki, 2015, p. 7) remain relevant. The lack of infrastructure (cf. Jørgensen & Marzouki, 2015,

p. 7), especially in rural and thinly populated areas, remains a particular problem. Although these technical questions may have been answered in many urban regions in developed countries (and in the European High North in rural areas too), half of the planet's population remains offline (Jørgensen & Marzouki, 2015, p. 8). By contrast, the internet penetration rate in the Nordic countries is consistently above 90%. In fact, 94% of the population of Finland and 96.4% of the Swedish population are connected to the internet, while in Norway and Iceland the penetration rates are at 98.4% and 98.6%, respectively (Internet World Stats, 2019). At times, internet access is seen as such a fundamental part of daily life, at least in developed countries, that the discussion regarding deliberate state-created limitations on it has focused on special situations, for example, the limitations on internet use imposed on prisoners (cf. White, 2015). "Having access to the internet is increasingly considered to be an emerging human right. International organisations and national governments have begun to formally recognize its importance to freedom of speech, expression and information exchange" (Shackelford, 2017). This trend will continue as internet access becomes more and more important for the realisation of human rights: "In time, the tide will likely strengthen. Internet access will become more widely recognized as a human right" (Shackelford, 2017). However, some kind of internet access is only the first step (Shackelford, 2017), as cybersecurity, that is, the protection of users' data and property from online threats, is also necessary (Shackelford, 2017). Over time, other aspects, such as data protection, limits on roaming charges and protection against third-party, including government, access to personal data, are becoming increasingly important for many users. Likewise, fast mobile (rather than merely stationary) internet access is also increasingly seen not as an option but rather as a necessity. Therefore, it seems reasonable to assume that access to fast mobile internet is an emerging right as well. Indeed, the widespread adoption of mobile internet-capable devices and their permissive regulation by almost all states (for restrictions in this regard, see Feldstein, 2017) has contributed to the development of a norm of customary international law to the effect that mobile internet access is also a human right. It remains to be seen whether this development will result in the adoption of an international treaty norm to this effect. "So far, the human rights community has been unable to launch a consistent and successful campaign against repressive governments that restrict the Internet or to spur the creation of an international norm to protect Internet access" (Feldstein, 2017). In 2010, a poll conducted in the United Kingdom by the BBC (Global Scan, 2010) showed that around 80% of respondents considered internet access to be a human right (Jørgensen & Marzouki, 2015, p. 11). A 2014 poll showed that this number had grown to 83% (Canazza, 2018, p. 3). Mobile internet is quickly gaining in importance, not only for users but also from an economic perspective (Hosting Facts, 2018).

One problem with the notion of internet access as a human right concerns the continuing evolution of the technology and the associated uncertainties (Jørgensen & Marzouki, 2015, p. 8). The question under discussion here is no longer simply about some form of internet access but rather about access to mobile internet at the sufficiently high speeds required for navigation, videoconferencing, telemedicine, etc. It has been suggested that whether or not internet access is a human right should depend, *inter alia*, on whether "it is [...] concerned with very urgent interests" (Wang, 2016, p. 133). However, there is no basis for such an urgency requirement in the *lex lata*. Such a requirement would run counter to the very idea of human rights, which do not come into existence only when a need to make use of a right has actually arisen, let alone when an "interest" (Wang, 2016, p. 133) has become "urgent" (Wang, 2016, p. 133). Further, urgency can be very subjective and situation-dependent. While urgency can be relevant to proceedings concerning the application for a preliminary injunction from a court to prevent a violation of human rights, it should not be a consideration with regards to whether or not a right exists in the first place. The question of whether internet access is a mere interest or a legally protected right is not a question concerning the facts of a given situation but rather a question of law. It is, therefore, necessary to search for a legal basis for the protection of internet access as a human right according to international law.

6.2.3 Different Perspectives on Communication Rights

Particularly in those countries that are both economically developed and committed to the rule of law and human rights, communication rights are often considered from the perspective of privacy (cf. Bederman, 2008, p. 140 et seq.; Grabenwarter, 2014, p. 216 et seq.; Meyer-Ladewig & Nettesheim, 2017, p. 330 et seq.), although the fundamental question of access to communication systems, which is a much wider issue than the question of the availability of fast mobile internet service, must not be forgotten. What is often overlooked in debates held in economically developed societies is that the question of privacy (which predates the emergence of modern telecommunications technologies but which is gaining in importance for users as communication becomes more dependent on technology) only arises once communication has become technically possible. Yet this is only part of the problem. Inequality is often (cf. Rehbein & Schwengel, 2008, p. 210 et seq.) seen as a question of poverty, but this can obscure the different needs of residents in rural areas when compared to those who live in urban centres. Not only are their needs different, the low number of rural residents in the European High North also impacts their ability to make consumer decisions in such numbers that they have the potential to actually shape the market (cf. Walzer, 1995, p. 52). Residence choices are not perfectly free and, especially for the indigenous communities in the European High North, are often also culture-dependent. For a long time, the assumption has been that the provision of services is location-dependent and that it is necessary, or indeed expected, that one must move if one wants to access certain services. This notion, which has led to the continuing exodus from rural to urban areas and the myriad problems associated with rapid urban growth, has even been enshrined in international human rights law: Article 12 of the ICCPR allows one to choose where to live within a country (Yliruusi, 2014, p. 2). The proliferation of the internet and the widespread availability of mobile internet services might cause some to consider whether this paradigm will still apply in the future, as today it is possible to participate in a wide range of activities and to benefit from many services regardless of one's physical location. The mobile nature of internet access is an essential building block of this new world. When the idea of internet access as a human right was first discussed on a large scale a few years ago, internet access typically meant a desktop or tower computer connected to the internet via cables. Nowadays, for many users, mobile devices have become the primary, and in many cases only, way to access the internet. While some connectivity issues in settled areas can be overcome through the provision of wireless internet access (WiFi), through the provision of internet services by private actors ("internet cafes") or through the provision of such services by public libraries (Petri, 2018; on discrimination-free access to public libraries, see Jaeger, Wentz, & Bertot, 2015), rural residents, particularly those who live in remote areas, often require truly mobile internet.

The human right to internet access can be realised in many ways, although not all internet needs can be fulfilled by stationary and shared devices, which might only be accessible a few hours per week. This is especially true in terms of access to personal communication, navigational information, emergency services, telemedicine, the Internet of Things (IoT) and a combination of the latter two, which all require the use of personal devices that are provided with an internet connection with sufficient speed and bandwidth.

6.3 Characteristics of a Modern Mobile Internet Service

6.3.1 Speed and Safety

Communication is essential for all humans. Humans are social creatures and thus not well suited to isolation. It is, therefore, unsurprising that the legality of a state's attempts to limit communication is usually restricted to extreme cases, for example, in the case of the detention of organised crime or terrorism leaders. Being able to communicate is essential for the enjoyment of many human rights. In today's interconnected, global community, communication permeates society in many ways. Given the human need to communicate, it is not surprising that communication is widely accepted as a human right. Today, communication is no longer restricted to inter-human communication since long-distance communication is usually a kind of data transfer. This is not only true in the case of phone calls, which now frequently make use of Voice-over-Internet-Protocol (VoIP) technologies, and emails, but also in the case of the IoT. Communication, including mobile internet access, serves multiple functions, for instance ensuring personal safety, economic development and job creation in remote regions.

Today's (and tomorrow's) internet is not only concerned with those devices most commonly associated with communication but also with devices that play a big role in everyday life. As more and more devices are connected and thus form an IoT, the role of communication changes. To a large degree, communication now consists of the exchange of data, often without direct human involvement in the data release. The limited user role, however, does not reduce the need for human rights protection but instead increases the need for protection as individual users have less control over what happens to their data and devices. In fact, hacking attacks against the most personal aspects of the IoT, such as connected pacemakers, have already become a credible cybersecurity risk scenario (von Lutterotti, 2018).

As the technological developments continue, the mobile internet connection needs to evolve as well. What exactly constitutes a sufficiently fast mobile internet connection[1] is not only dependent on—evolving—technical standards but also on individual needs, such as increasing demands for bandwidth and new telecommunications standards, as well as on ethical and human rights considerations.

6.3.2 *Open Internet Architecture*

In addition to the lack of mobile phone and internet infrastructure and insufficient connection speeds, internet needs might also remain unmet due to restrictive internet architectures. So-called "walled garden" types of internet services, such as "Free Basics" by Facebook's Internet.org (Internet.org, 2015a), limit the ability of the user to freely access the internet since only selected websites are accessible using this free internet connection (Internet.org, 2015b). To qualify for inclusion in such services, websites must comply with a range of requirements, which exclude, for example, file transfer, video streaming and VoIP services (Internet.org, 2015c), all of which are essential for the fruitful use of the internet today. An open internet architecture is seen "as a driving force in accelerating progress towards development in its various forms" (UN General Assembly, 2016, para. 2). From the perspective of human rights, such a walled garden infrastructure can be considered a double-edged sword, as it provides some internet access but also maintains existing

[1]At the time of writing, in late 2018/early 2019, the technical standards include the GPRS (General Packet Radio Service), which is also referred to as 2G, with a download speed of up to 114 Kbps; the EDGE (Enhanced Data GSM Evolution), which is another form of 2G technology, albeit faster than the GPRS at speeds of up to 384 Kbps; 3G, with a connection speed of up to 3.1 Mbps, which is the minimum speed required for video streaming and video calls; HSDPA (High-Speed Down-Link Packet Access), which is an improved version of 3G with a speed of up to 14 Mbps; as well as the current standards HSPA+ (Evolved High-Speed Packet Access), commonly referred to as 4G, which provides users with a connection speed of up to 168 Mbps and 4G LTE (Long-Term Evolution), which is needed for HD video streaming and provides a speed of up to 299.6 Mbps.

social divides and makes it difficult for those who depend on walled garden types of access to overcome social inequalities using the very tool that promises global connectivity.

6.4 Mobile Internet Access as a Human Right

Given the importance of mobile internet access in relation to the enjoyment of fundamental human rights, including the right to life, it is important to consider whether there is a norm of international human rights law that guarantees rural residents access to sufficiently fast (i.e. usually 4G or 4G LTE) mobile internet.

6.4.1 *Is the Language of Human Rights Appropriate?*

It might be questioned, however, whether human rights is the right framework in which to search for a solution to the problem of insufficient connectivity in rural areas. Internet access can make the enjoyment of human rights possible (Jørgensen & Marzouki, 2015, p. 8; Sutherland, 2018, p. 3), although the use of the language of human rights to describe internet access has been criticised because internet access is no longer limited to humans since, given the IoT, it is also relevant to technical items and because machines cannot have a legal right to access the internet (Cerf, 2012; Sutherland, 2018, p. 4). The absence of a right for a machine (which does not enjoy a legal personality to begin with) does not necessarily mean that internet access cannot be a right enjoyed by people (Sutherland, 2018, p. 4). Additionally, the evolving nature of the internet does not per se prevent internet access from being a right (for the opposing view, see Sutherland, 2018, p. 10).

6.4.2 *Internet Access as a Way to Realise Other Rights*

The internet is a tool. It connects families around the world and it provides opportunities for education and income generation in unprecedented ways. It can be used to save lives. Within only a few years, life without mobile internet access has become almost unimaginable for a large part of the global population—while many people simultaneously suffers from poverty, war, famine and preventable diseases. Modern telecommunications systems, including mobile internet access, allow for the creation of a world in which the "other" is far less distant than it was in the past. Access to the internet allows for the realisation of a range of human rights (Katyal, 2016). As the internet is deemed to have a "valuable effect on the modern world" (Sergeev, 2017, abstract), it can be argued that it forms a central building block of the emerging global society. Further, there is hence a need to ensure its continued functionality and existence through legal tools. Due to the global nature of the internet, it follows that these tools have to come from international law (cf. Sergeev, 2017, abstract). At this point, it should be noted that walled gardens can be seen as attempts by nation-states and corporations to limit internet access for political or commercial purposes (Katyal, 2016; on the right to access the internet without such restrictions, see Jørgensen & Marzouki, 2015, p. 8). While it remains a serious issue in light of the human right to access information across borders and deserves in-depth treatment from the perspective of international law, covering this problem would exceed the limitations of the present chapter, as this text is only concerned with the question of the technical availability of mobile internet access (cf. also Katyal, 2016).

In the 2015 case of *Cengiz and Others v. Turkey*, the European Court of Human Rights recognised the importance of the internet for the enjoyment of the freedom of expression, which is protected by Article 10 (1) of the ECHR (European Court of Human Rights, *Cengiz and Others v. Turkey*, Judgment of 1 December 2015, para. 52). This conclusion was based on the Court's own earlier case law (European Court of Human Rights, *Times Newspapers Ltd v. the United Kingdom (nos. 1 and 2)*, Applications Nos. 3002/03 and 23676/03, Judgment of 10 March 2003, para. 27; European Court of Human Rights, *Delfi AS v. Estonia*, Application No. 64569/09, Judgment of 16 June 2015, para. 110).

The importance of mobile internet can be seen in examples such as the proliferation of mobile payment schemes in some African countries and the emphasis on internet-based forms of governance, for example, the e-Estonia concept (Orgad, 2018, p. 354). Although the classical conception of human rights is a vertical one, referring to the relationship between states and individuals and groups, corporations can also violate human rights (Teubner, 2008). In the case of company-dependent internet access, inequalities in terms of access are human rights concerns on multiple levels, including political participation rights. While it can be said that human rights apply offline as well as online (Livingstone, Carr, & Byrne, 2015, pp. 1, 11), the absence of equal access continues to provide a fundamental challenge to the participation of rural and poor individuals and communities in the political process.

Sparsely populated areas are known to be underserved when it comes to emergency medical services (Yliruusi, 2014, p. 2). This problem could be solved, at least in part, by the availability of telemedical services, including the permanent online monitoring of connected implanted devices. In the developed world, internet connectivity is often taken for granted, with the focus having shifted towards cybersecurity (cf. Goodman, 2016, p. 19 et seq.). As large parts of our lives have moved to the online sphere, cybersecurity is as important as real-world physical security. Indeed, the boundary between what was long referred to as IRL ("in real life") and cyberspace has been practically obliterated over the last decade. For this reason, connectivity cannot be taken for granted in rich countries either. Mobile phone coverage, including mobile internet coverage, is costly. The installation of the necessary infrastructure, particularly antenna towers, is cost-intensive. In a privatised system, the predictable decline in the number of residents in rural areas (see, e.g. for a prognosis for Germany until the year 2030, Hawes, 2017, p. 217) provides even less of an incentive for private companies to invest in infrastructure there. Even in a world dominated by technology, the spaces in which we live matter, particularly in terms of "the quality of everyday life" (Hilton, 2016, p. 338). If an otherwise positive space, providing sufficient housing, access to schools, clean air, etc., lacked an essential element, such as the ability to communicate on a global scale, the quality of life in that space would be severely restricted. This is the reality in rural communities in many parts of Europe, albeit much less so in the European High North, where mobile internet connectivity is the norm—literally.

This also follows from international human rights law. Under Article 2 of the ECHR, states have the obligation "to take appropriate steps to safeguard the lives of those within their jurisdiction" (Rainey, Wicks, & Ovey, 2014, p. 153), meaning that this legal obligation to protect human life is positive in nature (Grabenwarter, 2008, p. 137; Smith, 2014, p. 217 et seq.). Internet access allows for the realisation of human rights, including the right to health (on the right to health, see Office of the United Nations High Commissioner for Human Rights/World Health Organisation, 2008), which in turn includes the human right to emergency medical services (cf. Yliruusi, 2014, p. 48 et seq.). Indeed, access to emergency medical services is a human right (cf. Nissen, 2017), as part of the right to health, which is protected under the broadly phrased Articles 11 (Yliruusi, 2014, p. 21 et seq.) and 12 (Yliruusi,

2014, pp. 2, 21) of the International Covenant on Economic, Social and Cultural Rights (ICESCR). Making emergency medical services more widely available makes other human rights beyond the right to health more accessible as well (Ryan, Beattie, & Young, 2014, p. S72).

The realisation of economic, social and cultural rights, which are not as distinct from civil and political rights (including the right to life, Yliruusi, 2014, p. 16) as classical human rights teaching might indicate (Yliruusi, 2014, p. 15), is meant to occur over time (Yliruusi, 2014, p. 14) and it is dependent on state resources (Yliruusi, 2014, p. 14). The right to communicate, including the right to internet access, is not just a stand-alone right, as it has to be understood in light of the rights the enjoyment of which is made possible or at least easier due to mobile internet access. The ability to communicate in an emergency situation, particularly in a remote area, has to be complemented by the provision of services, such as police or emergency medical services. This will also require substantial resources, although it is something that the state already has an obligation to do under existing international human rights obligations. This requires the creation and maintenance of infrastructures. The closure of police stations and the reduction in emergency medical services in rural areas in Finland (see Yliruusi, 2014, p. 32) poses a serious threat to the well-being of residents and visitors, and it could lead to violations of fundamental human rights (cf. Yliruusi, 2014, p. 38), such as the right to life and the right to emergency medical health care, due to the state's failure to undertake the required positive actions and to maintain the infrastructures in question. Accordingly, it has been found that changes made to the rules applicable to hospital districts in Finland violate human rights (Yliruusi, 2014, p. 37).

The right to life, which is protected by numerous human rights norms, including Article 2 of the ECHR, not only requires states to refrain from harming individuals but also imposes a duty on states to take positive action to protect human life (Marochini, 2013, p. 734). This duty "not only to refrain from the intentional or unlawful taking of life, but also to take all appropriate steps to safeguard lives of those within their jurisdiction" (Marochini, 2013, p. 735) can include "a duty to provide medical services" (Marochini, 2013, p. 735). A duty to provide healthcare services might also follow from the right to private life enshrined within Article 8 of the ECHR (Marochini, 2013, p. 739 et seq.) or from the European Social Charter (ESC), which contains several health-related provisions (Marochini, 2013, p. 749) and which complements the ECHR similarly to how the ICESCR and the ICCPR have to be seen as two parts of one whole (cf. Marochini, 2013, p. 748). In order to be both effective (an important concept when it comes to the implementation of the ECHR at the domestic level) and useful for individuals, such services have to be accessible too.

In terms of the right to access mobile phone and mobile internet services, it is clear that the increasing dependency on internet-based technologies will only increase the need for connectivity without discrimination.

6.4.3 *A Legal Basis for the Right to Mobile Internet Access*

The identification of the legal basis for a right to internet access in the existing law can prove challenging because the relevant international treaties were created long before the widespread adoption of the internet (Jørgensen & Marzouki, 2015, p. 13).

Communication is certainly a human right (Jørgensen & Marzouki, 2015, p. 1), as is the right to access information (see Arai, 2014, p. 620 et seq.). Additionally, Article 19 of the Universal Declaration of Human Rights (UDHR) protects the freedom to communicate (Shackelford, 2017). This right also applies in cyberspace (Howell & West, 2016). The right to communicate, which is enshrined in, for example, Article 8 of the ECHR, is usually considered from the perspective of government interference with communication, for example, through surveillance (cf. Heringa & Zwaak, 2006, p. 730 et seq.; Leach, 2005, p. 289 et seq.; Reid, 2007, p. 560 et seq.). It is not immediately clear from the wording of Article 8 of the ECHR that this right should include an obligation on the part of the state to provide the infrastructure necessary for communication. This is unsurprising, given that the ECHR is concerned with civil and political rights, rather than with social, fundamental and human rights. It should be noted, however, that further development is not impossible. Not only is there an understanding that the ECHR reflects "the shared political and cultural values of the Western democracies" (Guerra Martins, 2013, p. 193; translation by the author of this chapter), part of which is an increasing awareness of the importance of fair access to the internet, there is also a built-in ability on the part of the European human rights system to evolve (Koivu, 2015, p. 382; Letsas, 2009, p. 58 et seq.). Against this background, the following paragraphs will show how the ECHR might be utilised to argue in favour of the existence of a right to mobile internet access as part of the right to free speech, which is protected by Article 10 of the ECHR.

Article 10 of the ECHR is not currently understood as providing a freedom of information-type of right (Harris, O'Boyle, Bates, & Buckley, 2014, p. 620). While the parallel interpretation of the respective norms contained within the European Union's (EU's) Charter of Fundamental Rights and in the ECHR, which has been emphasised by the presidents of the European Court of Justice and the European Court of Human Rights (European Court of Human Rights, 2011), might allow for such an interpretation (cf. Harris et al., 2014, p. 620), such a right would refer to information held by the state (Harris et al., 2014, p. 620) and, therefore, primarily to communication between the individual and the state (see also Harris et al., 2014, p. 620). While Article 10 of the ECHR does not provide for universal access to state-held information (Reid, 2007, p. 364), this is usually not what is at stake when an individual wants to access the internet and "Article 10 [of the ECHR] cannot be used to derive a general right to access to information" (Reid, 2007, p. 364) held by the state. In the 1998 case of *Guerra and Others v. Italy*, the Court repeated *verbatim* the long-standing view, which can be traced back to the 1987 judgment in *Leander v. Sweden* (European Court of Human Rights, *Leander v. Sweden*, Judgment of 26 March 1987, para. 74), that Article 10 (2) of the ECHR "basically prohibits a government from restricting a person from receiving information that others wish or may be willing to impart to him" (European Court of Human Rights, *Guerra and Others v. Italy*, Application No. 14967/89, Judgment of 19 February 1998, para. 53), as it also did in *Gaskin v. United Kingdom* (European Court of Human Rights, *Gaskin v. United Kingdom*, Application No. 10454/83, Judgment of 7 July 1989, para. 52), *Sîrbu and Others v. Moldova* (European Court of Human Rights, *Sîrbu and Others v. Moldova*, Applications Nos. 73562/01, 73565/01, 73712/01, 73744/01, 73972/01and 73973/01, Judgment of 15 June 2004, para. 18) and *Gillberg v. Sweden* (European Court of Human Rights, *Gillberg v. Sweden*, Application No. 41723/06, Judgment of 3 April 2012, para. 83). The point of departure for the interpretation of the freedom to receive information is, therefore, the idea that there is a "negative duty of the government not to interfere with communication of information among individuals *inter se*" (Harris et al., 2014, p. 621). The freedom of expression is usually seen as obliging the state to refrain from interference (*status negativus* obligations). This follows from the way Article 10

(2) of the ECHR is phrased: "The restrictions allowed under the second paragraph of Art. 10 must be narrowly interpreted and the necessity for restrictions convincingly established, in addition to being lawful and pursuing legitimate aims" (Reid, 2007, p. 343).

The importance of communication as a human right indicates that it is necessary to question whether (and, if so, on which legal basis) states have positive obligations to enable communication through the creation of legal frameworks that facilitate the construction and maintenance of mobile communication infrastructures. Actively limiting internet connections, for example, through internet shutdowns (Howell & West, 2016), constitutes a violation of human rights (Boyle, 2016; Kravets, 2011). Hence, there is a *status negativus* obligation—but is there also a positive obligation on the part of states to make internet access possible? The right to internet access can be based on a number of other rights, such as the right to a private life, the enjoyment of which requires right holders to be able to communicate effectively. Often, this is not possible in the absence of sufficiently fast mobile internet connections. Even in developed countries, access to sufficiently fast mobile internet services in rural regions is often still lacking or spotty at best. While human rights can have negative dimensions (meaning the prohibition of interference with the right in question), they can also have a positive dimension (corresponding to a duty to protect) and even horizontal effects ("horizontal" here refers to the legal relationship between private actors, including the legal relationship between customers and corporations that provide mobile phone services, although the de facto dependency of customers will usually mean that they do not see their relationship with corporations as one among equals). In certain special circumstances, the effective protection of human rights will mean that rights have to evolve into participatory rights (which require legal institutions, for example, in the case of the right to vote) or even into infrastructure rights. It is the latter dimension that is relevant here. It would have to be the case that a failure on the part of the state to take action to change living conditions would amount to a violation of human rights. Classical examples of this include the duty of the state to ensure that there are sufficient hospitals, police officers, fire brigades, etc., in order to protect the right to life. It is, however, far from clear if the material scope of, for example, the right to a private life actually includes access to mobile internet services. The drafters of the ECHR did not refer to infrastructure rights as such, although the concept is visible in the text of the Convention because the entire ECHR is based on the premise that it applies in a democratic society (see Zand, 2017, p. 197 et seq.). Access to services is often seen as falling within the realm of social rather than civil or political rights. It would exceed the scope of this chapter to investigate all kinds of rights. Therefore, in the following, the focus will be on the right to communicate, a human right that is relevant both as a civil right and as a social right.

The question then becomes whether the aforementioned negative obligation on the part of the state is accompanied by a parallel positive obligation. In principle, there are "[p]ositive obligations to protect the exercise of the freedom of expression" (Reid, 2007, p. 365). According to Article 10 (2) of the ECHR, the state may only interfere under specific conditions and it has to create an environment in which free speech is possible, although normally it would not have to create the forum in which such free speech is exercised or the tools with which opinions are spread. To use a classical example, the state has to accept the existence of a print media market in which different opinions are expressed, although it does not have to provide newspapers in order to allow the individual to express his or her opinion. The duty of the state to require non-state actors to provide third parties with infrastructure for communication desired by those third parties is limited (Reid, 2007, p. 365): "Article 10 [of the ECHR] does not bestow freedom of forum, though the Court did not exclude that where the bar on access to private property effectively destroyed the essence of the right [...], a positive obligation might arise for the state to regulate property rights in favor of exercise of freedom of expression" (Reid, 2007, p. 365). This logic, which was employed in *Appleby v. United Kingdom* (European Court of Human Rights, *Appleby v. United Kingdom*, Application No. 44306/98, Judgment of 6 May 2003, para. 47), could be developed further and thus serve as a starting point for a discussion regarding the justification for the interference of the state with the rights of mobile internet service providers. Such an interference, for example, by requiring universal coverage, could be justified by the duty of the state to facilitate access to other human rights, as the right to mobile internet access goes beyond the kinds of rights one might initially associate with Article 10 of the ECHR. The duty of the state to create a regulatory environment that is conductive to free communication can be evidenced by an old example that involves a denser regulatory web than print media, namely radio. It has long been accepted that radio broadcasts require licences issued by the state. Through the establishment of such a broadcasting licensing system, the state creates a regulatory environment that facilitates the use of the freedom protected under Article 10 of the ECHR (Reid, 2007, p. 354). The regulations that apply to such a communicative environment, such as the conditions attached to licences for radio broadcasts or for mobile internet services, have to comply with the limits imposed on the state by paragraph 2 of Article 10 of the ECHR (Reid, 2007, p. 354). The de facto availability of mobile internet services has a similarly limiting effect. The ideas contained within Article 10 (2) of the ECHR thus apply *mutatis mutandis* to the role that the state plays in the provision of mobile internet services. While such services are usually provided by private corporations, it is the state that creates the necessary legal framework, including licensing conditions such as universal equal coverage requirements. In practice, this means that in order to make a profit from customers in urban centres, such as Helsinki, mobile internet providers also have to service remote and sparsely populated parts of the county, even if they do so at a loss. Such a holistic and egalitarian perspective is characteristic of the Nordic countries, especially when contrasted with the largely liberalised telecommunications markets seen in the richer countries of Western and Central Europe.

While the word "restrict", which has frequently been used by the Court since the aforementioned case of *Leander v. Sweden*, may primarily be understood to refer to *status negativus* obligations, the notion of effectiveness (which permeates the Convention) allows for the suggestion that any action or omission by the state that makes communication impossible may constitute an infringement of the right protected under Article 10 (1) of the ECHR. Although this might not necessarily require the state to actively accommodate every new technology, it can be argued that the common modes of communication have to be secured in order to ensure as effective a protection of the right as possible. A few generations ago, these common modes of communication might have included mail and telephone, while today the effective realisation of the right to communicate must also cover mobile internet services. The inclusion of modern forms of communication is not simply based on the desire of users; in fact, it is due to the principle of the effectiveness of the protection of human rights, which permeates the Convention. This practical need influences the interpretation of the Convention. It would behove the European Court of Human Rights, especially in light of its own interpretation of the ECHR as a "living instrument" (European Court of Human Rights, *Tyrer v. United Kingdom*, Application No.

5856/72, Judgment of 25 April 1978, para. 31), to provide clarification regarding the positive infrastructure-related obligations of states under Article 10 (1) of the ECHR, if it is ever given the opportunity to do so. Due to the high degree of mobile internet coverage in the European High North, it appears unlikely that a case which would give the Court this opportunity would originate in Northern Europe.

The importance of mobile internet access raises the question of whether there is already a norm of customary international law to the effect that access to the internet is indeed a human right. The emergence of such a norm would require a practice of states and a corresponding *opinio juris*, that is, a belief that the practice of states is based on an obligation that stems from international law. While a detailed analysis of this issue would be beyond the scope of the present chapter, a few observations can be made.

Customary international law requires a consistent practice of states, which is based on the conviction (*opinio juris*) that this practice is required by international law. The paradox inherent in the role of *opinio juris* in the genesis of customary international law (see, e.g. Meguro, 2017) is that for a legal obligation to come into existence, it is necessary for states to already believe in its existence. This is not merely an academic problem, as it seems fair to assume that currently the international community, at least in the European High North, has reached a stage where a norm of (at least regional) customary international law is emerging to the effect that there is, or soon will be, a right to mobile internet access under customary international law.

There is already a significant amount of state practice from a range of state actors that indicates internet access to be perceived as a human right: "For reasons of economic growth, geographic cohesion, social equity and to support the achievement of other rights governments, regulators, parliaments and the courts have pressed operators to extend their services geographically to remote areas and socially to poorer citizens and to the disabled. To boost economic growth and improve productivity a range of legislative, policy and regulatory tools have been devised to facilitate the introduction and adoption of new technologies" (Sutherland, 2018, p. 10). The right to internet access has been discussed in a number of international forums, such as the General Assembly of the United Nations and the UN's Human Rights Council (Jørgensen & Marzouki, 2015, p. 1), as well as at the 2015 World Social Forum (Jørgensen & Marzouki, 2015, p. 3). This interest reflects the idea that the internet is meant to serve everybody (cf. Jørgensen & Marzouki, 2015, p. 3). It also reflects the call made in 2011 by Frank LaRue, then UN Special Rapporteur on the promotion and protection of the right to freedom of opinion and expression, "to make universal internet access a priority for all states" (Jørgensen & Marzouki, 2015, p. 9, with further references; see also Canazza, 2018, p. 4). At that time, it was already clear that this would not happen overnight (Jørgensen & Marzouki, 2015, p. 9), although a number of states have taken action accordingly (Jørgensen & Marzouki, 2015, p. 9): "several countries - especially within Europe - have already stipulated universal internet access in national legislation" (Jørgensen & Marzouki, 2015, p. 10), including "the Estonian Telecommunication Act (2000), a ruling by the French Constitutional Court (2009), [an] Amendment to the Finnish Communication Market Act (2010), a ruling by the Costa Rican Constitutional Court (2010), and an amendment to the Spanish Act on sustainable economy (2011)" (Jørgensen & Marzouki, 2015, p. 10, fn. 22; see also Canazza, 2018, p. 4). In 2014, the Council of Europe adopted Resolution 1987, in which it called for the recognition of the human right to internet access (Jørgensen & Marzouki, 2015, p. 11), while "Brazil [...] approved Law 12965 [, which] ensures the right to Internet access for all" (Canazza, 2018, p. 4). While these measures reflect the actions of only a limited number of states and organisations and, therefore, might not yet be sufficient "evidence of a general practice" (Article 38 1. b. Statute of the International Court of Justice) to assume the existence of sufficient state practice, they nevertheless indicate the existence of a strong trend in international law towards the emergence of a human right to internet access.

The motives behind the above-mentioned measures differ from state to state. In order to be relevant to the creation of a norm of customary international law, such state practice would have to be accompanied by a corresponding *opinio juris*, meaning that the state has acted in a certain way due to a conviction that the particular course of action was required by existing international law (cf. Article 38 1. b. Statute of the International Court of Justice). States may promote the development of internet accessibility in order to promote national economic development: "The social and economic attributes of the Internet as a [Global Public Good] produce two consequences that underpin the policymaking role of governments and multilateral organisations (MOs) in Internet governance. First, they create a strategic *incentive* for governments and

MOs, sparked by their respective constituencies, to maximise Internet development, provide Internet access as a public good, and bridge the digital divide within and among countries. Second, they reveal a strong *need* for policymaking and multilateral cooperation to address the Internet's nonrivalry and nonexcludability; to correct market failures; to enhance (mitigate) the Internet's positive (negative) externalities; and to achieve a welfare-maximising global supply of the Internet" (Canazza, 2018, p. 2, italics in the original). States also have a vested interest in a functioning mobile communication infrastructure, not only for their own purposes but also in the interests of maintaining a functioning economy and ensuring the well-being of all individuals and communities within their respective borders. In this case, the interests of individuals, communities and public authorities align. The privatisation of telecommunications infrastructures and services, however, meant that states had to give up part of the practical power necessary to ensure the provision of a mobile internet communication infrastructure. When privatising telecommunications services, many states de facto laid the groundwork for an approach to mobile communication services that is based on the generation of income for private companies rather than on the provision of public services to all residents. In many countries, for example, Germany, this led to a lack of mobile phone (let alone mobile internet) coverage in sparsely populated areas. Giving up decision-making power concerning the provision of mobile internet services in the rural European High North in this way meant that sufficient coverage in remote regions could only be achieved through regulatory measures, such as the requirement in Finnish law to provide mobile phone (including mobile internet) coverage equally across the country. While this difference in attitude towards mobile coverage in rural areas between Finland and central European countries might be explained in political terms by the way the concept of equality permeates Finnish (legal) culture, the question of the equal treatment of urban and rural residents is more than a question of respect (cf. also Seelmann, 2008). Such an interest does not necessarily mean that there is no *opinio juris* to the effect that the state considers itself bound by a norm of customary international law, although it might make it harder to prove that this is the case because such an *opinion juris* might not be sufficiently documented. As with the state practice, the corresponding conviction of having acted based on a legal duty requires development before a norm of customary law can be deemed to have come into effect.

From an international law perspective, it remains necessary to question whether there is a legal basis in international law for the assumption that access to mobile internet services is a legally protected human right. At the moment, such a norm of customary international law to the effect that there exists a right to mobile internet access can be deemed to be in the process of emerging. In light of the economic considerations outlined above, it seems reasonable to categorise such a right primarily as a social right, even though it would help with the realisation of civil and political rights as well. The fact that such a right is currently in the process of emerging as a rule of customary international law does not mean that a long process would be necessary to prove its existence. Indeed, such a right could exist today because customary law can develop quickly when dramatic technological changes are involved. The development of customary international law norms concerning the use of outer space during the 1950s and 1960s is a case in point. Back then, customary law norms such as state responsibility for spacecraft and the duty to render assistance to astronauts or cosmonauts in emergency situations developed practically in parallel to the first steps taken into outer space, even though at the time only two states, the United States and the USSR, were actually operating spacecraft. Given the global relevance of mobile internet services, it could be argued that there is already such a right or that it is currently emerging. As this right is emerging, its enforcement as a subjective right might be difficult. The next part of this chapter will look at the way international law might provide a framework for the practical implementation of a right to mobile internet access once it has been hardened into a legally binding objective obligation that corresponds to a subject right of individuals and communities.

Even if the right to mobile internet access is not included in human rights treaties (Cradduck, 2017; Jørgensen & Marzouki, 2015, p. 12) *expressis verbis*, states have a duty to refrain from interfering with the right to communicate (Jørgensen & Marzouki, 2015, p. 12). Further, it appears likely that there is an emerging norm of customary international law to the effect that the obligation of states is not only negative but that it has a positive dimension that translates into a human right to mobile internet access, if not as a right on its own then at least as a necessary corollary of the enjoyment of other human rights. The case law of the European Court of Human Rights shows that it is possible to interpret the ECHR so as to provide for an obligation on the part of the state to ensure access to mobile internet services for rural residents.

6.5 Realising the Right to Mobile Internet Access in the European High North

6.5.1 Public Goods, Corporations and Social Rights

The internet has rightly been described as "a global public good" (Canazza, 2018, p. 2) and "as an enabler of other public goods" (Canazza, 2018, p. 3). Global public goods "require government intervention and international cooperation to overcome market failure and achieve efficiency and equity in their allocation and provision" (Canazza, 2018, p. 2). This is exactly what happened in Finland. Here, the state intervened to make mobile internet access a realistic possibility for everybody. When it comes to social rights, the realisation of which is cost-intensive, it is often necessary to strike a balance between the needs of individuals and the capacities of public authorities. Yet, it can be argued that the widespread availability of sufficiently fast mobile internet is in the public interest. Mobile internet has become so pervasive, as well as so important for personal safety and for the conducting of business, that its absence can be felt acutely. Providing sufficient communication infrastructure requires funding. In countries in which the telecommunications sector has been completely privatised, this means that there has to be at least a reasonable expectation of achieving a profit for a mobile phone provider to choose to install a mobile phone antenna in a specific location. Due to sharply falling rates for calls and consumer expectations of fast mobile internet at flat rates, mobile phone service providers' profit margins are likely shrinking, making it even less profitable for companies to invest in the installation of new infrastructure in rural areas with very low population densities, let alone in areas that are only used infrequently, for example, by nomadic herders. This would normally constitute a substantial barrier to universal mobile phone service accessibility in remote regions. The lack of profitability is also the key reason why universal mobile phone coverage (even basic coverage that is sufficient for only phone calls and short message service [SMS] rather than for mobile internet) has not been achieved in some relatively rich countries, such as Germany, where the mobile phone service sector is completely privatised. The profitability hurdle can be overcome by creating a legal requirement, based on the notion of equality, that services have to be provided without discrimination, including without discrimination based on place of residence.

6.5.2 *The Need for Infrastructure and Land-Use Conflicts in the European High North*

Even if mobile phone service providers are forced to provide equal coverage across a country, the provision of the necessary infrastructure requires land, for example, for the construction of mobile phone antenna towers. Although existing infrastructure, such as pre-existing buildings, can often be used for this purpose, in many cases, especially in remote regions, the construction of new infrastructure will be inevitable. This can lead to land-use conflicts. In the European High North, land-use conflicts involving indigenous communities are particularly relevant. While international law recognises some land rights of indigenous peoples, those rights do not usually amount to full ownership, especially in regions such as Finnmark in Norway or Finnish Lapland, where large parts of the land are owned by the state, either directly or indirectly. Indigenous land ownership rights remain a much discussed (cf. Bankes, 2013; Webber, 2013) issue in the international legal discourse. The lack of recognition of indigenous ownership rights in the European High North (albeit not exclusively there [cf. Herrera Sarmieto, 2009]) is not only due to different ideas regarding ownership as a concept (cf. Xanthaki, 2017, p. 4) or to the lack of recognition of the fact that indigenous legal concepts might differ from those of the dominant society (cf. Magga, 2017). The absence of indigenous land rights in the European High North is also a result of the outdated legal idea prevalent in the European High North for centuries that land for which no ownership could be proven belonged to the state (Ravna, 2013, p. 180). In essence, this is a long-term effect of the concept of *terra nullius* (cf. Uimonen, 2014, p. 67; on the historical development in Sápmi, see Heikkilä, 2006, p. 97; Lehtola, 2010, p. 30 et seq.). In the European High North, the disregard for indigenous legal systems and concepts of usage rights, which did not conform to the expectations of ownership concepts held by the dominant societies, meant that in practice the ownership rights of nomadic communities were denied (Ravna, 2013, p. 181). Even during the twentieth century, this line of thinking was used to deny indigenous land ownership rights (Ravna, 2013, p. 183) and only recently have there been a few efforts in some parts of the European High North, particularly in Norway's Finnmark, to examine indigenous land rights based on land usage (Ravna, 2013, p. 192 et seq.). This novel (Ravna, 2013, p. 189) legal development in Norway is due in large part to

Norway's ratification of the International Labour Organisation's (ILO's) Convention No. 169 (Ravna, 2013, p. 183; on the importance of ILO 169 for indigenous land ownership rights, see Joona, 2012, p. 95 et seq.). There have been no parallel developments in other countries of the European High North, as Norway is the only state in the region to have actually ratified this international treaty, which remains a cornerstone of indigenous rights protection in international law (cf. Anaya, 2009, p. 133 et seq.). International human rights obligations towards indigenous communities are not yet fully realised, neither the obligations of states nor the obligations of non-state actors (cf. Westra, 2013, p. 228 et seq.). This is also the case in the European High North. Realising the right to mobile internet communication could clash with the land rights of indigenous communities, which have called these remote and sparsely populated areas their home for thousands of years. In order to prevent violations of one human right (the land rights of indigenous communities) while trying to implement a different right (the right to mobile internet access), both rights, ancient and modern, have to be balanced in a reasonable and effective manner. In addition to land rights, which remain an emerging issue (cf. Koivurova, 2014, p. 95), indigenous peoples have a right to be involved in issues that concern them (Heinämäki, 2013, p. 141; on the role of ILO 169 in this context, cf. Xanthaki, 2007, p. 76 et seq.), as well as to be involved in decision-making processes at the national level which affect them (Tomaselli, 2016, p. 206 et seq.), in an "effective" (Rombouts, 2014, p. 79) manner. While the concept of Free, Prior and Informed Consent (FPIC) (Rombouts, 2014) does not necessarily have to result in a veto power on the part of indigenous communities, at the very least "a balance between effective decision-making and development goals on one hand and fair inclusion of culturally distinct groups in the decision-making on the other" (Rombouts, 2014, p. 415) must be struck.

6.6 Conclusions and Outlook

A common problem in international law concerns the implementation and enforcement of existing obligations. As discussed in this chapter, this is particularly true in relation to social rights that are to be implemented over time and even more so in relation to rights that are not safeguarded through an institutional system that would allow, for example, for human rights litigation. As the right to mobile internet access is a social right that is currently still in the process of emerging in customary international law but that has so far received limited codification in international treaty law, implementing it appears challenging. When it comes to telephones, there has long been an expectation of "universal services" (Cerf, 2012). Today, the same applies to mobile internet services. Access to the internet not only makes it easier to enjoy human rights and to participate in democracy (Wang, 2016, p. 108), it also contributes to the realisation of the social development goals that have been agreed upon by the member states of the United Nations (cf. Howell & West, 2016). Additionally, in countries that have not recognised a right to internet access outright, there is an increasing sense that internet access ought to be a human right (cf. White, 2015). It can be argued that there is a right to mobile internet access under existing international treaties, particularly the ECHR, and that there might already be a customary international law right to internet access. At the very least, such a right is in the process of emerging. While the implementation of a right to mobile internet access would not be without challenges, as a social right it would (once it was actually part of an international treaty or customary international law, which might soon be the case) have to be implemented over time. In the long run, what is currently taken for granted will turn into a legally protected right. This will eventually also apply to discrimination-free mobile internet access. It is clear that the European High North in general and Finland in particular have played an important role in making this development possible.

References

Anaya, S. J. (2009). *International human rights and indigenous peoples* (1st ed.). New York, NY: Aspen Publishers.

Arai, Y. (2014). Article 10: Freedom of expression. In D. Harris, M. O'Boyle, E. Bates, & C. Buckley (Eds.), *Law of the European Convention on Human Rights* (3rd ed., pp. 613–709). Oxford: Oxford University Press.

Bankes, N. (2013). Recognising the property interests of indigenous peoples within settler societies: Some different conceptual approaches. In N. Bankes & T. Koivurova (Eds.), *The proposed Nordic Saami convention: National and international dimensions of indigenous property rights* (1st ed., pp. 19–43). Portland, OR: Hart Publishing.

Bederman, D. J. (2008). *Globalization and international law* (1st ed.). New York, NY: Palgrave Macmillan.

Boyle, E. (2016, July 5). UN declares online freedom to be a human right that must be protected. *The Independent*. Retrieved from https://www.independent.co.uk/life-style/gadgets-and-tech/un-declares-online-freedom-to-be-a-human-right-that-must-be-protected-a7120186.html.

Brown, S., & Mickelson, A. (2018). A decision framework for choosing telecommunication technologies in limited-resource settings. *Future Internet, 10*(8). https://doi.org/10.3390/fi10010008.

Canazza, M. R. (2018). The Internet as a global public good and the role of governments and multilateral organizations in global internet governance. *Meridiano 47—Journal of Global Studies, 19*. http://dx.doi.org/10.20889/M47e19007.

Cerf, V. G. (2012, January 4). Internet access is not a human right. *New York Times*. Retrieved from https://www.nytimes.com/2012/01/05/opinion/internet-access-is-not-a-human-right.html.

Charter of Fundamental Rights of the European Union. (2012). *Official Journal C 326*. Luxemburg: European Union (Publications Office).

Clare, H. (2017). *Icebreaker—A voyage far North* (1st ed.). London: Chatto & Windus.

Council of Europe. (1953). *European Convention on Human Rights (ECHR)*. Retrieved from https://www.echr.coe.int/Documents/Convention_ENG.pdf.

Cradduck, L. M. (2017). Legislating for Internet "access"-ability. In J. Hunsinger, L. Klastrup, & M. Allen (Eds.), *Second international handbook of Internet research* (1st ed.). Dordrecht: Springer. https://doi.org/10.1007/978-94-024-1202-4_14-1.

Electronic Communication Services Act. (2014). *Laki sähköisen viestinnän palveluista* [Electronic Communications Services Act]. Retrieved from https://www.finlex.fi/fi/laki/ajantasa/2014/20140917.

European Court of Human Rights, *Tyrer v. United Kingdom*, Application No. 5856/72, Judgment of 25 April 1978.

European Court of Human Rights, *Leander v. Sweden*, Judgment of 26 March 1987.

European Court of Human Rights, *Gaskin v. United Kingdom*, Application No. 10454/83, Judgment of 7 July 1989.

European Court of Human Rights, *Guerra and Others v. Italy*, Application No. 14967/89, Judgment of 19 February 1998.

European Court of Human Rights, *Appleby v. United Kingdom*, Application No. 44306/98, Judgment of 6 May 2003.

European Court of Human Rights, *Times Newspapers Ltd v. the United Kingdom (nos. 1 and 2)*, Applications Nos. 3002/03 and 23676/03, Judgment of 10 March 2003.

European Court of Human Rights, *Sîrbu and Others v. Moldova*, Applications Nos. 73562/01, 73565/01, 73712/01, 73744/01, 73972/01and 73973/01, Judgment of 15 June 2004.

European Court of Human Rights. (2011). *Joint communication from Presidents Costa and Skouris*. Retrieved from https://www.echr.coe.int/Documents/UE_Communication_Costa_Skouris_ENG.pdf.

European Court of Human Rights, *Gillberg v. Sweden*, Application No. 41723/06, Judgment of 3 April 2012.

European Court of Human Rights, *Cengiz and Others v. Turkey*, Applications Nos. 48226/10 and 14027/11, Judgment of 1 December 2015.

European Court of Human Rights, *Delfi AS v. Estonia*, Application No. 64569/09, Judgment of 16 June 2015.

Eurostat. (2018). *Population density by NUTS 2 region*. Retrieved from https://ec.europa.eu/eurostat/tgm/table.do?tab=table&init=1&language=en&pcode=tgs00024&plugin=1.

Feldstein, S. (2017, June 1). Why Internet access is a human right. *Foreign Affairs*. Retrieved from http://carnegieendowment.org/2017/06/01/why-internet-access-is-a-human-right.

Global Scan. (2010). *Four in five regard internet access as a fundamental right: Global poll*. Survey for the BBC World Service. Retrieved from http://www.globescan.com/news_archives/bbc2010_internet/BBC_Internet_Poll.pdf.

Goodman, M. (2016). *Future crimes—Inside the digital underground and the battle for our connected world* (2nd ed.). London: Corgi Books.

Grabenwarter, C. (2008). *Europäische Menschenrechtskonvention* [European Convention on Human Rights] (3rd ed.). Munich: C. H. Beck.

Grabenwarter, C. (2014). *European Convention on Human Rights—Commentary* (1st ed.). Munich: C. H. Beck.

Guerra Martins, A. M. (2013). *Direito Internacional dos Direitos Humanos – Relatório – Programa, Conteúdos e Métodos de Ensino Teórico e Prático* [International human rights law—Report—Program, content and methods of theoretical and practical training] (1st ed.). Coimbra: Almedina.

Harris, D., O'Boyle, M., Bates, E., & Buckley, C. (2014). *Law of the European Convention on Human Rights* (3rd ed.). Oxford: Oxford University Press.

Hawes, J. (2017). *The shortest history of Germany* (1st ed.). London: Old Street Publishing.

Heikkilä, L. (2006). *Reindeer Talk—Sámi Reindeer Herding and nature management* (1st ed.). Rovaniemi: Lapland University Press.

Heinämäki, L. (2013). The Nordic Saami convention: The right of a people to control issues of importance to them. In N. Bankes & T. Koivurova (Eds.), *The Proposed Nordic Saami Convention: National and international dimensions of indigenous property rights* (1st ed., pp. 125–147). Portland, OR: Hart Publishing.

Heringa, A. W., & Zwaak, L. (2006). Right to respect for privacy (Article 8). In P. van Dijk, F. van Hoof, A. van Rijn, & L. Zwaak (Eds.), *Theory and practice of the European Convention on Human Rights* (4th ed., pp. 663–750). Antwerp: Intersentia.

Herrera Sarmieto, E. (2009). Derechos territoriales indígenas en la Amazonía boliviana y creación de lo<<tacana>>[Territorial rights of indigenous persons in the Bolivian Amazon and the creation of the "tacana"]. In V. Robin Azecedo & C. Salazar-Soler (Eds.), *El regreso de la indígena – Retos, problemas y perspectivas* [The return of the indigenous—Challenges, problems and perspectives] (1st ed.) (pp. 159–185). Lima: Instituto Francés de Estudios Andinos.

Hilton, S. (2016). *More human—Designing a world where people come first* (2nd ed.). London: WH Allen.

Hosting Facts. (2018). *Internet Stats & Facts for 2019*. Retrieved from https://hostingfacts.com/internet-facts-stats/.

Howell, C., & West, D. M. (2016, November 7). The internet as a human right. *Techtank*. Retrieved from https://www.brookings.edu/blog/techtank/2016/11/07/the-internet-as-a-human-right/.

International Court of Justice. (1945). *Statute of the international court of justice*. Retrieved from http://legal.un.org/avl/pdf/ha/sicj/icj_statute_e.pdf.

International Labour Organization. (1989). *Indigenous and Tribal Peoples Convention (ILO 169)*. Retrieved from https://www.ilo.org/dyn/normlex/en/f?p=NORMLEXPUB:12100:0::NO::P12100_ILO_CODE:C169.

International Covenant on Civil and Political Rights (ICCPR) Internet World States. (2019). *Internet in Europe Stats, Internet User Statistics & 2019 Population for the 53 European countries and regions*. Retrieved from https://www.internetworldstats.com/stats4.htm.

Internet.org. (2015a). *Free basics by Facebook*. Retrieved from https://info.internet.org/en/story/free-basics-from-internet-org/.

Internet.org. (2015b). *Free basics platform*. Retrieved from https://info.internet.org/en/story/platform/.

Internet.org. (2015c). *Participation guidelines*. Retrieved from https://developers.facebook.com/docs/internet-org/participation-guidelines.

Jaeger, P. T., Wentz, B., & Bertot, J. C. (2015). Libraries and the future of equal access for people with disabilities: Legal frameworks, human rights, and social justice. In B. Wentz, P. T. Jaeger, & J. C. Bertot (Eds.), *Accessibility for persons with disabilities and the inclusive future of libraries, 40 advances in librarianship* (1st ed., pp. 237–253). Bingley: Emerald Publishing.

Joona, T. (2012). *ILO Convention No. 169 in a Nordic context with comparative analysis: An interdisciplinary approach* (1st ed.). Rovaniemi: Lapland University Press.

Jørgensen, R. F., & Marzouki, M. (2015). *Internet governance and the reshaping of global human rights legacy at WSIS+10*. Paper prepared for the 10th GigaNet Annual Symposium, 9 November 2015, João Pessoa, Brazil.

Katyal, S. (2016). *Infrastructural entitlements and the civil right to technology* (UC Berkeley Public Law Research Paper No. 2716368). Retrieved from https://papers.ssrn.com/sol3/cf_dev/AbsByAuth.cfm?per_id=115375.

Koivu, V. (2015). *European Convention on Human Rights and transition of the legal culture—Reception of the argumentation of the European Court of Human Rights by the Finnish supreme jurisdictions* (1st ed.). Rovaniemi: Lapland University Press.

Koivurova, T. (2014). *Introduction to international environmental law* (1st ed.). London: Routledge.

Kravets, D. (2011, June 3). U.N. report declares internet access a human right. *Wired*. Retrieved from http://www.wired.com/2011/06/internet-a-human-right.

Leach, P. (2005). *Taking a case to the European court of human rights* (2nd ed.). Oxford: Oxford University Press.

Lehtola, V.-P. (2010). *The Sámi People—Traditions in transition* (4th ed.). Inari: Kustannus-Puntsi Publisher.

Letsas, G. (2009). *A theory of interpretation of the European Convention on Human Rights* (1st ed.). Oxford: Oxford University Press.

Livingstone, S., Carr, J., & Byrne, J. (2015). *One in three: Internet governance and children's rights* (Global Commission on Internet Governance Paper Series No. 22). London: Global Commission on Internet Governance.

Magga, A.-M. (2017). Indigenous peoples' right to own legal orders and governance systems in the international human rights regime. In A. Xanthaki, S. Valkonen, L. Heinämäki, & P. Nuorgam (Eds.), *Indigenous peoples' cultural heritage: Rights, debates and challenges* (1st ed., pp. 273–293). Leiden: Brill Nijhoff.

Marochini, M. (2013). Council of Europe and the right to healthcare—Is the European Convention on Human Rights appropriate instrument for protecting the right to healthcare? *Zbornik Pravnog Fakulteta Sveucilista u Rijeci, 34*(2), 729–760.

Meguro, M. (2017). Distinguishing the legal bindingness and normative content of customary international law. *ESIL Reflections, 6*(11). https://esil-sedi.eu/wp-content/uploads/2017/12/ESIL-Reflection-Meguro.pdf.

Meyer-Ladewig, J., & Nettesheim, M. (2017). Artikel 8 Recht auf Achtung des Privat- und Familienlebens [Article 8 right to respect of private and family life]. In J. Meyer-Ladewig, M. Nettesheim & S. von Raumer (Eds.), *Europäische*

Menschenrechtskonvention - Handkommentar [European Convention on Human Rights—Commentary-manual] (4th ed.) (pp. 317–366). Baden-Baden: Nomos.

Nissen, O. (2017). A right to access to emergency health care: The European Court of Human Rights pushes the envelope. *Medical Law Review, 25*. https://doi.org/10.1093/medlaw/fwx059.

Office of the United Nations High Commissioner for Human Rights/World Health Organisation. (2008). *The right to health* (Fact Sheet No. 31). Geneva: Office of the United Nations High Commissioner for Human Rights.

Orgad, L. (2018). The future of citizenship: Global and digital—A rejoinder. In R. Bauböck (Ed.), *Debating transformations of national citizenship* (1st ed.) (pp. 353–358). Cham: Springer. https://doi.org/10.1007/978-3-319-92719-0_61.

Petri, C. (2018). Rural libraries and the human right to Internet access. In B. Real (Ed.), *Rural and small public libraries: Challenges and opportunities, 43 advances in librarianship* (1st ed., pp. 13–35). Bingley: Emerald Publishing.

Rainey, B., Wicks, E., & Ovey, C. (2014). *The European Convention on Human Rights* (6th ed.). Oxford: Oxford University Press.

Ravna, Ø. (2013). The draft Nordic Saami convention and the assessment of evidence of Saami use of land. In N. Bankes & T. Koivurova (Eds.), *The proposed Nordic Saami convention: National and international dimensions of indigenous property rights* (1st ed., pp. 177–205). Portland, OR: Hart Publishing.

Rehbein, B., & Schwengel, H. (2008). *Theorien der Globalisierung* [Theories of globalisation] (1st ed.). Konstanz: UVK Verlagsgesellschaft.

Reid, K. (2007). *A practitioner's guide to the European Convention on Human Rights* (3rd ed.). London: Sweet & Maxwell.

Robitzski, D. (2018, January 22). Finland is the mobile data capital of the world. *IEEE Spectrum*. Retrieved from https://spectrum.ieee.org/tech-talk/telecom/wireless/what-the-finnish-obsession-with-mobile-data-says-about-5g.

Rombouts, S. J. (2014). *Having a say—Indigenous peoples, international law and free, prior and informed consent* (1st ed.). Oisterwijk: Wolf Legal Publishers.

Ryan, M., Beattie, L., & Young, H. (2014). The expanding role of emergency medicine in safeguarding human rights. *Annals of Emergency Medicine, 64*(4S), S72.

Seelmann, K. (2008). Respekt als Rechtspflicht? [Respect as legal duty?] In W. Brugger, U. Neumann, & S. Kirste (Eds.), *Rechtsphilosophie im 21. Jahrhundert* [Legal philosophy in the 21st century] (1st ed.) (pp. 418–439). Frankfurt am Main: Suhrkamp.

Sergeev, A. (2017). The right to internet access: Assessing the impact and the merits of compliance through an example of modern China. *International Journal of Law and Information Technology, 25*(4), 309–335. https://doi.org/10.1093/ijlit/eax018.

Shackelford, S. (2017, February 14). Should cybersecurity be a human right? *The Conversation*. Retrieved from http://theconversation.com/should-cybersecurity-be-a-human-right-72342.

Smith, R. K. M. (2014). *Textbook on international human rights* (6th ed.). Oxford: Oxford University Press.

Sutherland, E. (2018). *The Internet and human rights: Access, shutdowns, and surveillance*. Paper presented at the WG Hart Legal Workshop 2018—"Building a 21st Century Bill of Rights"—Held at the Institute for Advanced Legal Studies, University of London, 11th and 12th June 2018. Retrieved from https://papers.ssrn.com/sol3/papers.cfm?abstract_id=3203883.

Teubner, G. (2008). Die anonyme Matrix: Zu Menschenrechtsverletzungen durch≫private<< transnationale Akteure [The anonymous matrix: On human rights violations by "private" transnational actors]. In W. Brugger, U. Neumann & S. Kirste (Eds.), *Rechtsphilosophie im 21. Jahrhundert* [Legal philosophy in the 21st century] (1st ed.) (pp. 440–472). Frankfurt am Main: Suhrkamp.

Tomaselli, A. (2016). *Indigenous peoples and their right to political participation* (1st ed.). Baden-Baden: Nomos.

Tully, S. (2014). A human right to access the Internet? Problems and prospects. *Human Rights Law Review, 14*(2), 174–195. https://doi.org/10.1093/hrlr/ngu011.

Uimonen, J. (2014). *From unitary state to plural, asymmetric state: Indigenous quest in France, New Zealand and Canada* (1st ed.). Rovaniemi: Lapland University Press.

United Nations. (1948). *Universal Declaration of Human Rights (UDHR)*. Retrieved from http://www.un.org/en/udhrbook/pdf/udhr_booklet_en_web.pdf.

United Nations. (1976a). *International Covenant on Civil and Political Rights (ICCPR)*. Retrieved from https://treaties.un.org/doc/publication/unts/volume%20999/volume-999-i-14668-english.pdf.

United Nations. (1976b). *International Covenant on Economic, Social and Cultural Rights (ICESCR)*. Retrieved from https://www.ohchr.org/en/professionalinterest/pages/cescr.aspx.

United Nations General Assembly. (2016). *Human Rights Council, Thirty-second session, agenda item 3, Promotion and protection of all human rights, civil, political, economic, social and cultural rights, including the right to development*. UN Doc. A/HRC/32/L.20, 27 June 2016.

von Lutterotti, N. (2018, August 26). Wie sicher ist das kranke Herz vor Cyber-Attacken? [How secure is the sick heart against cyber attacks?] *Frankfurter Allgemeine Zeitung*. Retrieved from http://www.faz.net/aktuell/wissen/medizin-ernaehrung/hackerangriffe-auf-herzschrittmacher-und-defibrillatoren-15747706.html.

Walzer, M. (1995). Was heißt zivile Gesellschaft? [What does civil society mean?] In B. van den Brink & W. van Reijen (Eds.). *Bürgergesellschaft, Recht und Demokratie* [Civil society, law and democracy] (1st ed.) (pp. 44–70). Frankfurt am Main: Suhrkamp.

Wang, X. (2016). *Human Rights and Internet Access—A philosophical investigation* (Unpublished doctoral dissertation). Universiteit Utrecht, Utrecht, Netherland. Retrieved from https://dspace.library.uu.nl/bitstream/handle/1874/334446/wang.pdf?sequence=1.

Webber, J. (2013). The public law dimension of indigenous property rights. In N. Bankes & T. Koivurova (Eds.), *The proposed Nordic Saami convention: National and international dimensions of indigenous property rights* (1st ed., pp. 79–102). Portland, OR: Hart Publishing.

Westra, L. (2013). *Environmental justice and the rights of indigenous peoples—International and domestic legal perspectives* (1st ed.). Abingdon: Earthscan.

White, A. (2015). *Judicial infringement of the right to Internet access by the imposition of special sentencing conditions.* Victoria University of Wellington Legal Research Paper Series (Student/Alumni Paper No. 22). Retrieved from https://papers.ssrn.com/sol3/cf_dev/AbsByAuth.cfm?per_id=1789295.

Xanthaki, A. (2007). *Indigenous rights and United Nations standards—Self-determination, culture and land* (1st ed.). Cambridge: Cambridge University Press.

Xanthaki, A. (2017). International instruments of cultural heritage: Tales of fragmentation. In A. Xanthaki, S. Valkonen, Sanna L. Heinämäki, & P. Nuorgam (Eds.), *Indigenous peoples' cultural heritage: Rights, debates and challenges* (1st ed.) (pp. 1–19). Leiden: Brill Nijhoff.

Yliruusi, J. (2014). *Access to emergency medical services in sparsely populated areas: Perspective of international human rights law* (1st ed.). Tallinn: Tallinn University of Technology.

Zand, J. (2017). The concept of democracy and the European Convention on Human Rights. *University of Baltimore Journal of International Law, 5,* 195–227.

CHAPTER 7

The Legal Regime Governing Submarine Telecommunications Cables in the Arctic: Present State and Challenges

Daria Shvets

7.1 Introduction

The twenty-first century is an era characterised by the advent of advanced technologies, especially in relation to the field of telecommunications. Such technologies play an essential role in all modern societies, although they are of particular importance for northern territories.

In the severe conditions of the Arctic, telecommunications link people together and connect them to the rest of the world. Individuals, private entities, small communities and governments located in different parts of the north can engage in business with each other despite residing in geographically diverse locations. The use of the internet, for instance, is crucial for meeting local needs and, therefore, it is of great importance in the north. The material comforts available in densely populated territories are considered precious in the north due to their limited availability. Telecommunications help to reduce this gap by providing, for example,

D. Shvets (✉)
Universitat Pompeu Fabra, Barcelona, Spain
e-mail: daria.shvets01@estudiant.upf.edu

M. Salminen et al. (eds.),
Digitalisation and Human Security, New Security Challenges,
https://doi.org/10.1007/978-3-030-48070-7_7

access to high-quality education through online platforms and courses, facilitating medical treatment via telemedicine, providing online employment for the local population and supporting domestic business. The internet also facilitates life in the north by incorporating new technologies into the traditional way of life. Currently, the high-quality internet connection used for audio and video conferences helps to reduce carbon dioxide (CO_2) emissions by eliminating the need to hold in-person meetings, which contributes to the sustainable development of the region (Burnett & Carter, 2017). The use of new technologies, including the internet, can be considered within the framework of a human security concept. People who reside in the Arctic do not feel isolated, as telecommunications and online services mean that they can be involved in the social and cultural life of their community and, further, that they can contribute to their own economic well-being. Therefore, connectivity and access to the internet arguably matter more in the Arctic than in any other region.

The internet, in addition to other data running daily to and from the northern regions, requires a reliable infrastructure that is ready to transmit information at high speed and without losing quality. In the Arctic, the percentage of telecommunications transferred by submarine cables is constantly growing, as there are increasing opportunities to create more connections and to extend the submarine cables network (Arctic Economic Council, 2016). This chapter will focus on submarine telecommunications cables in the Arctic (paying special attention to the European High North), which are also referred to by general collocation "submarine cables".

The primary purpose of the submarine cables in the Arctic is to transmit telecommunications and to ensure that data are transferred without any interruptions. It is important to recognise that one of the key aspects of human security is stable communications being enjoyed by the population on a daily basis. Once the local population is connected to the internet and high-speed telecommunications services, it relies heavily on this infrastructure to meet its needs. A lack of access to the digital environment can thus be perceived as a threat to the realisation of human rights (Skepys, 2012).

Special attention should be paid to plans to lay new submarine telecommunications cables through the Arctic due to their importance for local communities. However, several disadvantages associated with the legal regime governing submarine telecommunications cables could complicate the implementation of such cable projects in the Arctic. This could, in turn, affect the digitalisation of the region and, as a consequence, the level of human security.

In the context of the legal regime governing submarine telecommunications cables in the Arctic, the following aspects are of particular interest. First, the general legal regime governing submarine cables under international law. Submarine cables not only constitute a substantial infrastructure but are also considered as an object of legal regulation under international law. Unsurprisingly, the legal perspective concerning activities related to submarine cables has attracted the attention of legal scholars (Burnett, 2010; Davenport, 2015; Ekwere, 2016; Van Logchem, 2014). Second, the legislation of the Arctic states regarding submarine cables, with an emphasis on the legislation of states in the European High North. Third, how the general legal regime applies to the realities of the Arctic in light of current and upcoming submarine cables projects as well as which challenges associated with the present legal regime could affect the development of the submarine cable network in the Arctic.

7.2 The Legal Regime Governing Submarine Cables Under International Law

The legal regime governing submarine cables is a complex regime characterised by the presence of multiple actors and several levels of legal norms. Most of the norms concerning submarine cables at the international law level are contained within the United Nations Convention on the Law of the Sea (UNCLOS) (United Nations [UN], 1982), the so-called "constitution of the oceans" (Koh, 1982). The UNCLOS is an authoritative international agreement aimed at collecting various issues regarding the law of the sea together in a single document. Currently, 168 states are members of the convention (UN, 2019). Bilateral intergovernmental treaties in the area of submarine cables are rare. Cooperation between states is mainly expressed through mutual negotiations, declarations and reports in which states express the intention to cooperate in laying a cable. This can be explained by the fact that contracts for the installation of submarine cables are concluded between private companies, most commonly without state participation, and regulated by private international law. Thus, it can be argued that terms of the UNCLOS shall apply directly to private parties (Ekwere, 2016). Therefore, the UNCLOS remains the main regulator in international law.

The UNCLOS does not contain a separate chapter or a separate section devoted to submarine cables. The provisions relating to various aspects of submarine cables regulation are spread throughout different chapters of the convention. These provisions are distributed according to the different maritime areas that comprise the territory of the World Ocean. The Arctic states are considered to be coastal states and, with regards to the submarine cables laid in their maritime zones, the UNCLOS currently establishes the following legal regime. There are maritime zones that are under the sovereignty of coastal states (territorial sea), maritime zones that are not under the sovereignty of states but where they have rights to exercise jurisdiction (exclusive economic zone and continental shelf) and maritime zones that are not subject to the jurisdiction of states (high seas and international seabed area).

7.2.1 *Territorial Sea*

The UNCLOS awards coastal states with sovereignty over their territorial sea, which is a maritime area of up to 12 nautical miles, as measured from a given coastal state's baselines. All the Arctic states have established territorial seas of 12 nautical miles. The only exception is the territorial sea surrounding Greenland, which is only three nautical miles (UN, 1963). Within its territorial sea, a coastal state has the right to regulate all activities on the part of all entities, including activities related to the laying, operating and repairing of submarine cables. This includes provisions intended to protect submarine cables from damage by other maritime users that are exercising the right of innocent passage (UN, 1982).

However, the freedom of a coastal state to adopt laws and regulations in its territorial sea is not unlimited. Such laws and regulations must be in accordance with the rules of international law. In particular, they must comply with other provisions of the UNCLOS, and they be interpreted in light of the convention in general as well as other international law agreements.

7.2.2 *Exclusive Economic Zone and Continental Shelf*

An exclusive economic zone is a maritime area extending up to 200 nautical miles from the same baselines of a coastal state from which a territorial sea is measured. The regime covers the water column, while the subsoil is regulated under the legal regime for the continental shelf. In an exclusive economic zone, coastal states have sovereign rights and jurisdiction, including several rights concerning submarine cables. They enjoy the same rights as in the high seas, albeit with several exceptions. One such exception is that states must respect the rights and obligations of the coastal state and comply with its laws and regulations (UN, 1982).

A continental shelf is a maritime area beyond the limits of a territorial sea, which, as a general rule, is up to 200 nautical miles from the baselines. Article 79 of the UNCLOS provides that all states have the right to lay submarine cables on the continental shelf. A coastal state, when exercising its rights with respect to the continental shelf, must not prejudice other activities (e.g. navigation, fishing) or other freedoms of other states provided for by the UNCLOS. Additionally, it must not lead to any unjustifiable interference with their implementation. This also applies to submarine cables.

Paragraph 3 of Article 79 of the UNCLOS is of particular interest here. It contains the following rule: "The delineation of the course for the laying of such pipelines on the continental shelf is subject to the consent of the coastal State". Thus, the delineation of a submarine cable route is not subject to the consent of a coastal state and is instead determined freely, based on scientific evidence. Submarine cables and pipelines already in position on the seabed must also be taken into account when laying a new cable on the continental shelf.

7.2.3 "The Area" and the High Seas

Articles 87 and 112 of the UNCLOS recognise the right of all states to lay submarine cables on the bottom of the high seas, beyond the continental shelf. This freedom refers to the legal regime of "the Area", a newly established maritime zone under the UNCLOS. Although submarine cables are not directly mentioned in Part XI "The Area", the relevant regulation is provided by the regime applicable to the high seas since these two maritime areas occupy the same geographical space, with the only difference being that the regime of the high seas applies to the water column, while the regime of the Area applies to the seabed and the subsoil thereof.

The right of all states to lay submarine cables is restricted by the freedoms of other users of the seabed. A state that is planning to lay submarine cables in the Area must pay due attention to the already existing submarine cables and not interfere with the repair possibilities of those cables. The coastal state must also take into account the interests of other states exercising their rights and freedoms as provided by the UNCLOS with respect to activities in the Area (UN, 1982).

7.3 National Legislation of the Arctic States Concerning Submarine Cables

The UNCLOS applies to the Arctic Ocean on the same basis as it does to any other ocean around the world. As all the Arctic states except for the United States are parties to the UNCLOS, they (including the United States bound by the provisions of the UNCLOS on customary law basis) should comply with its provisions. As the convention establishes only a general framework that requires further implementation, the Arctic states have developed their national legislation concerning submarine cables during the course of implementing the convention.

This section aims to explore how successful the Arctic states have been in terms of implementing the provisions of the UNCLOS through their national legislation as well as how well they have extended its framework in relation to submarine cables. The initial focus will be on the national legislation of Finland, Sweden and Norway, as states representing the European High North, before switching to the legislation of other Arctic states in order to complete the picture of the legal regime governing submarine cables in the Arctic.

7.3.1 Finland

The construction of submarine cables in Finland's territorial sea is subject to general rules regarding network construction and maintenance. As an infrastructure that provides public communications services, the installation of a new submarine cable would require both an environmental assessment and a process of public hearings. The UNCLOS does not require the assessment of the environmental impact of submarine cables because submarine cables are considered to not cause pollution to the marine environment since pollution is only mentioned in the context of pipelines in Article 79, paragraph 2. However, in its territorial sea, a coastal state has the right to establish laws and regulations extending the framework of the UNCLOS.

During the process of constructing a new cable project, the local rules and statutes of all the territories in which cables are planned to be laid must be obeyed, with the relevant maritime authorities subsequently being provided with maps reflecting the locations of any new submarine cables. Finland has specific liability rules concerning damage to cables caused by different sources, although the main problem has always been damage caused by humans, particularly by the shipping industry (Carter & Burnet, 2011). Environmental permits are required in relation to submarine cables that affect the seabed, while appropriation rules apply with respect to private property (Global Legal Group, Finland, 2010a).

The Act on the Protection of Certain Submarine Cables and Pipelines 145/1965 is currently in force in Finland. Section 3 of the Act provides that if, during the laying, constructing or repairing of a submarine cable, damage is caused to another cable or pipeline, then the owner of the newly constructed cable (even if she/he cannot be punished for causing injury) must compensate the owner of the damaged cable or pipeline for the cost of repairing the damage. Section 3 of the Act on the Exclusive Economic Zone of Finland, as of 26 November 2004, provides that it shall be in force in the Finnish exclusive economic zone. Outside Finnish territorial waters and the exclusive economic zone, this Act applies to vessels under the Finnish flag and to subjects under Finnish jurisdiction. It complies with the obligation on coastal states under Article 113 of the UNCLOS to establish liability for causing damage to submarine cables. Thus, in its national legislation, Finland has successfully addressed those provisions left by the UNCLOS to coastal states for further regulation.

7.3.2 Sweden

In Sweden, the construction of submarine cables involves several legal documents, the participation of several authorities and the necessity to obtain various permits (ICPC Members' Database, 2018). According to the Swedish Public Water Areas (Boundaries) Act 1950:595, in order to agree the route of a new submarine cable in public waters (all waters in Sweden are divided into private and public), the permission of the Legal, Financial and Administrative Services Agency is required. In any case, it is necessary to obtain the permission of several parties for installation and further operations. The environmental assessment of a project must take place in accordance with the Swedish Environmental Code 1998:808, which can be a time-consuming process. The County

Administrative Boards ("Länsstyrelse") and the Regional Environmental Court ("Miljödomstol") are responsible for these procedures. If a cable is planned to be laid outside the Swedish territorial sea, a permit will be required under the Swedish Continental Shelf Act 1966:314.

In terms of damage caused to submarine cables in the Swedish exclusive economic zone or on the high seas (by Swedish nationals or ships flying the Swedish flag), there is the Swedish Act on Liability to Pay Compensation for Damage Caused to Existing Submarine Cables and Pipes 1996:518. This Act repeats the relevant provision of the UNCLOS and establishes responsibility to compensate the owner of a submarine cable if his cable is damaged. There also exists the right of a shipowner to be compensated if he can prove that he sacrificed the ship's anchor in order to not cause damage to a submarine cable upon taking all the precautionary measures set out in Article 115 of the UNCLOS. Hence, Sweden has fulfilled its obligation under the UNCLOS to ensure national legislation is implemented in relation to matters concerning damage to submarine cables.

The Planning and Building Act 1987:10 and the Utility Easements Act 1973:1144 are among the pieces of general legislation relating to the installation of cable lines in the territory of Sweden. In the exclusive economic zone, the construction of a new submarine cable is subject to permits being granted by the Swedish government. A permit is also required from the local Water Authority. Other permits, for instance, building permits, are granted by the County Administrative Board and the Municipal Board (Global Legal Group, Sweden, 2010b).

Generally speaking, the Swedish system of laws relating to submarine cables includes several legislative acts providing for the regulation of the laying of submarine cables in different maritime areas and addressing the main issues left by the UNCLOS to the realm of domestic law.

7.3.3 Norway

Norway has issued several documents addressing the issue of digitalisation, the development of telecommunications and the laying of submarine cables. Thus, the government of Norway recognises that "environmental and economic considerations are often important when establishing an infrastructure, for example, several companies using the same route for installing cables, many companies sharing a communications mast, and the colocation of technical equipment for several companies" (Norwegian Ministers, 2012, p. 18).

Norway has a specific legal document devoted to submarine cables, namely the Act on the Protection of Submarine Cables and Pipelines. It focuses on the protection of submarine cables from different kinds of damage. In general, it repeats the provisions of the UNCLOS relating to damage caused to submarine cables and the possibility of compensation being due to a vessel's owner. In addition to the regulations contained within the Electronic Communications Act No. 83 (in which electronic communications are defined as being transmitted by submarine cables), the fixed installations to which submarine cables are attached are also regulated by the Planning and Building Act No. 71, 2008. It must be noted that the government of Norway may issue a decision or give consent to undertake the compulsory purchase of the title to or right of use of real property for the installation of electronic communications networks and equipment for electronic communications (Opperud, 2010). This can be explained by the importance of the submarine cables infrastructure for the whole country and, therefore, its precedence over individual interests. The Norwegian Post and Telecommunications Authority, an autonomous agency of the Ministry of Transport and Communications, is responsible for supervising telecommunications services, including those provided by submarine cables.

Although the Norwegian legal regime governing submarine cables is not as diversified as the Swedish legal regime, it has the advantage of having a separate authority to deal with submarine cables at the national level.

7.3.4 *Other Arctic States*

All the Arctic states have adopted national legislation concerning submarine cables, albeit with an emphasis on different aspects of regulation.

The Canadian legislation is represented firstly by the Telecommunications Act 1993. It defines the notion of international submarine cable and makes the laying of submarine cables subject to a licensing procedure. The Canadian Telecommunications Act prohibits the laying of submarine cables except for in accordance with licences issued for a period of up to ten years. Applications for such licences need to include documentation that complies with the requirements of the Canadian Environmental Assessment Act 2012.

The Danish priority with regards to submarine cables is their security. There are specially designated 200-metre wide protection zones along each side of the cable. In such zones, due to the Danish Order on the Protection of Submarine Cables and Pipelines No. 939, anchoring is prohibited because it can be extremely dangerous if anchors come into contact with, cut or damage submarine cables. To promote the significance of submarine cables and to ensure their security, the Minister for Business and Growth, in accordance with the Act on Safety at Sea No. 72, may set out regulations, take measures and issue general and specific prohibitions with regards to submarine cables.

The Icelandic legislation foresees the involvement of several governmental authorities in the process of laying submarine cables in Icelandic waters. The Minister for the Environment, in consultation with several other ministers and the Union of Local Authorities in Iceland, sets general provisions regulating the laying of submarine cables. In any case, doing so is subject to the approval of the Environment and Food Agency following the promulgation of the Act on Marine and Coastal Antipollution Measures No. 33. The Regulation on Authorization for the Laying of Submarine Cables and Underwater Pipelines No. 600 further extends the protection of the sea, its resources and its ecosystem by ensuring that subjects authorised to lay submarine cables in Icelandic waters fulfil all the requirements to do so. Other vessels that see or should be able to see signals from a cable vessel engaged in cable repair operations should keep to a distance of no less than one-quarter mile from the cable vessel (Ministry of the Interior of Iceland, 2003).

The Russian Federation has enacted a variety of laws addressing this issue. The competent authorities have the right to determine the route and conditions for laying submarine cables (State Duma of the Russian Federation, 1995). However, according to the UNCLOS, this determination is not necessary since the convention only mentions pipelines in this context. In particular, permission from the Federal Service for the Supervision of Natural Resources, following consultation with numerous other authorities is required. Therefore, the process for obtaining permission is a long one. Submarine cable projects in the exclusive economic zone and on the continental shelf are the objects of state ecological expertise according to the Federal Law on the Exclusive Economic Zone of the Russian Federation No. 191-FL, which increases the application processing time even more. Foreign cable vessels are subject to specific requirements. The operators of such vessels shall, for example, ensure the participation of up to ten Russian specialists in cable repairing operations, provide daily reports to border authorities and notify the competent authorities about changes in the cable vessel's operation areas.

The United States focuses on damage to submarine cables by providing that someone who wilfully and wrongfully breaks or injures, or attempts to break or injure, submarine cables will be liable to imprisonment for a term not exceeding two years, or to a fine not exceeding 5000 United States dollars, or to both a fine and imprisonment, at the discretion of the court. Various types of damage caused in different circumstances are further specified. This is followed by rules concerning vessels laying submarine cables, rules on safety (i.e. keeping fishing nets away from submarine cables) and rules on licensing procedures (United States, 2019). All the regulations are designed to ensure the protection of submarine cables and their stable functioning.

7.3.5 Conformity of the National Legislation of the Arctic States with International Law

Among the policies and laws of the various states of the European High North, different directions and aspects are prioritised regarding the development of telecommunications, particularly with regards to submarine cables. The Norwegian government's aim is to "facilitate economically profitable laying of fibre-optic cable to other countries to strengthen the basis for establishing large data centres and other data-based industry in Norway" (Norwegian Ministry of Local Government and Modernisation, 2015–2016, p. 32). The Swedish government highlights the importance of broadband infrastructure that is open to everyone. It specifies that fibre-optic cables and equipment for transferring information inspire the "competition of services, development of services and innovation" (Ministry of Enterprise and Innovation of Sweden, 2016, p. 15). The Finnish authorities aim to help businesses and citizens to improve their information security (Ministry of Transport and Communications of Finland, 2016).

The states of the European High North and all the other Arctic states have generally complied with their obligations under the UNCLOS to implement the legal regime of submarine cables within their national legislation. Indeed, a relatively well-developed system of national acts exists in each country covering different aspects of the submarine cable legal regime.

The development of activities in the Arctic has always been a priority of the Arctic states, although now with the possibility to "discover" the Arctic in new ways, such as the laying of submarine cables, the movement of states in this direction has further increased. Additionally, the Arctic Council has now also recognised the importance of telecommunications as a factor for sustainable development in the Arctic (Arctic Council, 2017).

7.4 Submarine Telecommunications Cables in the Arctic

The construction of submarine cables is a new activity in the Arctic. In comparison to fishing, navigation, marine scientific research and other traditional activities that have been conducted in the Arctic for a long time, telecommunications reached the Arctic relatively recently. To date, several submarine cables projects have been successfully implemented in the Arctic. For instance, the "Greenland Connect" cable system is a 4500 km submarine cable that connects Canada, Greenland and Iceland. Another example is the "Svalbard Undersea Cable System", which provides a broadband connection between Svalbard and mainland Norway (Arctic Economic Council, 2016). Currently, more projects involving the construction of submarine cables in the Arctic region are underway. Global warming has increased the possibility of laying submarine cables in the Arctic waters. Additionally, the presence of ice-free areas and navigation routes that are open for a longer period has also facilitated development and construction of submarine cables projects (Hsu, 2016).

The "Telecommunications Infrastructure in the Arctic" report by the Arctic Council specifies that "fiber-optic cable is the primary way Internet and other traffic is transmitted between continents or countries separated by the high seas" and, further, that in the Arctic "providers need to select their submarine fiber routes carefully, given the risk of ice scour in some areas, and to ensure reliable service backup plans to carry end-users' traffic if a submarine cable breaks during the winter months when weather conditions can make repair infeasible" (Arctic Council, 2017).

Taking into account the important role played by telecommunications, which can be brought to the northern territories by submarine cables, possibilities for constructing new submarine cables are being considered in the Arctic community. New submarine cables are planned to extend the existing network and to cover more territories. Nowadays, there are several projects involving the laying of submarine cables in the Arctic that are worthy of attention.

7.4.1 ROTACS

The ROTACS (Russian Optical Transarctic Cable System) project initiated by Polarnet Project, a Russian company, aims to lay cables along the Russian Arctic coast and to connect Tokyo and London. Aside from the final destinations, the submarine cables will create a network of additional cables landing stations on northern territories, thereby providing northern communities with a high-speed and reliable internet connection (Anthony, 2012).

The ROTACS project has existed since the early 2000s. The negotiations are ongoing and some exploratory work has been conducted, but the construction schedule has been repeatedly postponed (Titarenko, 2017).

In 2012, Polarnet Project became the first and only company to have completed extensive marine survey operations intended to develop the route for submarine cables in the Arctic region. After the work was completed, it became clear that a submarine cables system could be constructed in the icy conditions and Arctic climate. Although preparatory works have been performed, the project has still not entered the active construction phase and, therefore, it has gradually been overtaken by other cable projects. However, the project deserves special attention because it was pioneering in terms of large-scale submarine cables in the Arctic, with the initial work needed to determine the possibility of laying submarine cables in the Arctic having been conducted within the project's framework.

7.4.2 Alternative Cable Projects in the Arctic

The shortest way to connect Asia to Europe is through the Arctic. Doing so would reduce the time and cost associated with the construction of the cable and ensure the faster transmission of data once the cable was installed. In addition to the Russian project discussed above, proposals for laying submarine cables along a similar route in the

Arctic have been expressed by China and Finland (Pro-Arctic, 2016). Of particular note here is the Arctic Connect Project cable system initiated by Cinia after being commissioned by the Finnish Ministry of Transport and Communications to construct a fibre-optic cable along the Russian Arctic coast. Cinia Ltd is a young connectivity company based in Finland. It is owned by the state (Ministry of Transport and Communications) as the main shareholder and the company's primary goals include providing intelligent networking, cloud connectivity and software services. Cinia currently plays a key role in enabling digital business operations for both Finnish and international organisations (Cinia, 2019). Being part of an international submarine cables project such as the Arctic Connect Project is of strategic importance and ensures the presence of Finland in the global connectivity and telecommunications markets.

The Arctic Connect Project is considered necessary and useful not only by Finland but also by other states. It is interesting for China in light of Beijing's Belt and Road Initiative (BRI), a large-scale project intended to create connection infrastructure to be adopted in China (Deloitte Insights, 2018). The Arctic part of this initiative was discussed between Russia, China and Finland, with all the parties expressing an interest in completing the project. However, the project remains more of an initiative than a real project. Ari-Jussi Knaapila, president and chief executive officer (CEO) of Cinia, recently commented: "I would rather call it an initiative because there is a lot of interest. A lot of pre-planning has been done, but there have not been any decisions made for the final investment or even the first phase" (Bannerman, 2018, p. 37).

There are also plans for several other cables to be laid in the Arctic. One alternative is the Eastern Arctic Undersea Fibre Optic Network (EAUFON) (Delaunay, 2017). This cable is not yet in use, although the plan is to have landing points on the Canadian coast so as to connect multiple northern communities in Canada's Nunavik region. In February 2018, WFN Strategies, the company planning to lay the EAUFON cable, received a contract from the Kativik Regional Government (Canada) to conduct a cable route survey for the proposed submarine cables system (*Subsea World News*, 2018). The preliminary date for the EAUFON to enter into service is 2020.

There are a number of other examples of submarine cables projects in the Arctic. For instance, the Mackenzie Valley and Dempster Highway Fibre project, which is planned to extend along the Mackenzie Valley and Beaufort Delta. Midgardsormen is a Norwegian project seeking to construct a Norway-centric transatlantic cable system to connect Norway and Sweden to the East Coast of the United States. The Norilsk Fiber Optic Broadband project, which was initiated by the Russian state-owned company Nornickel, and Nxtvn's Oulu Nordic Express in the Nordic regions of Finland and Sweden (with an extension of the cable network to Norway) are also examples of the development of submarine cable networks in the Arctic (Arctic Economic Council, 2016).

Aside from the cables mentioned above, an intention to lay a submarine cable under the auspices of the Ivaluk Network project was expressed by the Canadian company Nuvitik Communication in 2014 (Delaunay, 2014). Thus far, this cable has not been laid and its route has not yet appeared on the TeleGeography submarine cables map (TeleGeography, 2019), an authoritative online map reflecting the positions of telecommunications cables laid worldwide. The project should have had three parts: Nunavik, Southeast Nunavut and Northwest Nunavut. Each part was planned to connect every coastal community on its route while travelling through the Arctic territories. The installation was designed to include two phases, with the Nunavik and Southeast Nunavut network routes intended to be operational by the fall of 2016 and the Northwest installation scheduled for 2017. However, the project was brought to a standstill by corporate changes. Arctic Fibre, a telecommunications company, had begun to implement it, but in May 2016, it announced that Quintillion Submarine Holdings has acquired its assets (Arctic Economic Council, 2016). Hence, further development of the cable will be performed by Quintillion.

7.4.3 Quintillion Submarine Cable System

The Quintillion Submarine Cable System is an alternative submarine cable project in the Arctic. This project should be examined separately from the above-mentioned projects because, in contrast to the ROTACS project, which is currently stalled, the EAUFON cable project, which is not yet in service, and all the other discussed submarine cables, which are at various stages of development, the construction of this submarine cable is currently actively in progress (Quintillion, 2016).

The project consists of three stages. The first stage runs along the coast of Alaska, the second is the Pacific stage (connecting Alaska to Japan) and the third is the stage connecting Alaska to Europe via the

Canadian Arctic, running through the Northwest Passage to the United Kingdom (Arctic Economic Council, 2016). The first stage of the submarine cable ("Alaska"), a 1200-mile fibre-optic line off the coast of Alaska, is already in use. In other words, this section of the cable has already begun transmitting signals and data, thereby bringing telecommunications to the north. Initially, the introduction of the first stage was planned for December 2017, but the cable has been running in a test mode since November 2016 (Pierce, 2017). The second part of the cable ("Asia") is planned to be laid in Western Asia and it will connect Asia to the United States. The start of the installation of the second part is scheduled for 2019, with the preliminary installation being due to be completed in 2020 (Bannerman, 2018). The third part ("Canada-United Kingdom") is designed to further expand the cable system network and to enable the northern territories of Canada to be connected to it. The cable is already laid and it will be owned by Quintillion, a private company registered in the United States (Quintillion, 2016). However, the overall route is intended to pass through the maritime zones of several states, including Japan, the United States, Canada and the United Kingdom. The company is even considering the possibility of building its own data centre as the project progresses (Bannerman, 2018).

Quintillion is an innovator in the submarine cable industry in the Arctic and it "has become something of an expert at deploying networks in hostile Arctic conditions, but that doesn't mean the project hasn't come with its share of challenges" (Bannerman, 2018, p. 36). Those challenges concern the legal environment that the Quintillion project might face, as would any other projects planned in the Arctic, which will be discussed in the next section.

7.5 Challenges Concerning Submarine Cables in the Arctic

The possibility of laying submarine cables in the Arctic is the subject of increasing attention as such cables have now become a critical infrastructure. They can also be considered one of the key elements associated with further Arctic development. Disruptions to telecommunications would negatively affect local inhabitants due to cutting off their means of communication with the outside world.

Although laid on the seabed, far from human eyes, submarine cables are vulnerable to various risks, which could affect the stability of communications. These risks stem from human beings as well as from natural factors. There are certain environmental challenges associated with submarine cables in the Arctic that can prejudice their installation and stable performance. As a consequence, human security in the Arctic could be affected by such risks because it depends on stable communications facilitated by submarine cables. Thus, the following sub-section is devoted to that aspect of human security that specifically relates to environmental security. Several factors that can influence the successful development of telecommunications and the digitalisation of the Arctic in general will be examined in more detail.

7.5.1 Environmental Dimension

The installation of submarine cables in the Arctic is an extremely challenging task given the unique climate conditions of the region and the lack of established practices. Submarine cables have never been laid there before, and there is a lack of scientific knowledge regarding how exactly the Arctic environment will affect them in the short- and long-term. The laying of a submarine cable in the Arctic waters is not the same as the laying of a cable in normal seabed conditions and normal temperatures, for which sufficient experience exists and enough research has been done. Submarine cables in the Arctic may be affected by ice, which is capable of breaking them as they are being shifted and moved. Icebergs migrate on the open sea and can exert pressure on ridges on the seabed, causing gouges or scours (Wilson, 2013). If a cable was damaged during the non-navigable season, its repair would be impossible until the re-opening of the route. The same applies to cable facilities, such as onshore stations, which could also be damaged by moving ice. The air and sea temperatures, combined with the higher energy of the waves, could be considered additional potential threats to submarine cables in the Arctic (Wilson, 2013).

The process of laying a cable in the Arctic climate is characterised by a short navigation season during the period of open ice when a cable ship can perform its work. Thus, the ship is forced to perform its continuing operations in this short period, making the process constant and increasing the stress on the environment. As a result, more disruptive effects on the environment may occur. The Arctic waters also provide only limited opportunities for the supply of special mechanisms, including the remotely operated vehicles that facilitate the laying of cables on the seabed. Additionally, there is a need to install further protection for the buried submarine cable so as to prevent damage by ice, which means an extended presence on the seabed when compared to the usual conditions for the installation of cables. Thus, submarine cables are vulnerable to damage from the Arctic environment.

At the same time, submarine cables could affect the Arctic marine environment. Yet, in the scientific literature, it is widely recognised that submarine cables do not cause serious harm to the environment (Heezen & Johnson, 1969; Perez-Alvaro, 2013). This issue was explicitly addressed in the joint report by the United Nations Environment Programme (UNEP) and the International Cable Protection Committee, in which they emphasised that "the small physical size of a telecommunications cable implies that its environmental footprint is likely to be small and local [...] a suggestion that is borne out by several studies" (Carter et al., 2009, p. 9). The founders of the UNCLOS adopted the same position, as evidenced by the wording of the relevant provisions. Article 79, paragraph 2, says that a coastal state may "take reasonable measures for the exploration of the continental shelf, the exploitation of its natural resources and the prevention, reduction, and control of pollution from pipelines". Even though the article addresses both pipelines and submarine cables, the wording clearly indicates that pollution is only mentioned in the context of pipelines. However, the norms of the UNCLOS provide a general regime for laying and using submarine cables. They do not distinguish the Arctic from other parts of the World Ocean.

The Arctic has a unique and vulnerable environment, which should be protected in a specific way. Some authors note that submarine cables may have a negative impact on the marine environment (Worzyk, 2009). In 2009, the OSPAR Commission published an environmental impact assessment concerning submarine cables (OSPAR Commission, 2009). The commission consists of representatives of contracting parties to the Convention for the Protection of the Marine Environment of the North-East Atlantic, who are working to implement the convention and its strategies, recommendations and agreements relating to the threats to the marine environment (OSPAR Commission, 2019).

According to this assessment, there exists uncertainty as to the impact of electromagnetic fields and thermal radiation on organisms that are sensitive to them. Therefore, the possible significant negative impact of submarine cables cannot be excluded. Although the impact is reduced and limited, it is still an influence that could be of concern in particularly vulnerable areas such as the Arctic.

The environmental dimension of submarine cables poses risks from the human security perspective as well. The laying of submarine cables is a relatively long process. The operations of a cable ship might affect the traditional fishing practices of locals in certain areas since they prohibit all other vessels from being in the vicinity of the cable installation site. However, the most dangerous factor is that the local population, who are not used to the rules of behaviour relevant to cable laying areas, could come too close to the cable ship or other marine vehicles, which could pose a danger to life and equipment.

In addition, a cable ship and any other vessels navigating in the Arctic waters are subject to the potential risk of collision with icebergs or other vessels. During construction activities, spillages of hazardous materials (oil, lubricants or brake fluid) might occur, which would negatively impact local inhabitants by contaminating adjacent waters, polluting the air via greenhouse gas emissions as well as poisoning fish stocks and other living resources utilised by them.

The cable company's staff members are also subject to life- and health-related risks while performing work on board the cable ship. Cable installation companies usually adopt health and safety policies to instruct their staff as to the required actions in case of emergency, although it is hard to take into account all the nuances of the unpredictable and harsh climate conditions of the Arctic.

Finally, cable operations cause both noise and vibration, which are likely to disturb coastal birds and animals. This will certainly not be welcomed by local inhabitants, who are used to living in harmony with nature.

The role of international law is to help balance these aspects and ensure stable communications so as to facilitate the development of the digitalisation in the Arctic. Thus, special attention should be paid to the precautionary principle. A general understanding of this principle is set out within the Rio Declaration on Environment and Development, which recognises it as a principle 15: "In order to protect the environment, the precautionary approach shall be widely applied by States according to their capabilities. Where there are threats of serious or irreversible damage, lack of full scientific certainty shall not be used as a reason for postponing cost-effective measures to prevent environmental degradation" (The United Nations Conference on Environment and Development, 1992).

The principle applies widely in matters related to the pollution of the marine environment and the conservation of marine living resources, which are topics traditionally covered by the law of the sea. It appears that it also has sufficient grounds to apply to the carrying out of activities associated with laying submarine cables in the Arctic, particularly in light of upcoming projects. With the application of the precautionary principle, activities related to submarine cables may be considered an excellent example of promoting a balance between the preservation of the environment and the facilitation of human well-being in the north. This principle should be ensured by private sector organisations engaged in cable operations (by applying cost-effective means, using the most up-to-date knowledge available, etc.) and supervised by the governmental authorities of the Arctic states controlling the successful implementation of the principle at the legal level.

7.5.2 *Absence of a Leading Authority*

The lack of a central international authority responsible for submarine cables is considered to be one of the most complex disadvantages of the legal regime governing submarine cables under international law (Rapp, 2010). This tendency is followed by domestic legal systems, where a special authority for dealing with submarine cables rarely exists. Several international institutions were created by the UNCLOS to deal with different matters related to the law of the sea (e.g. International Seabed Authority or International Tribunal for the Law of the Sea). However, submarine cables have not been assigned a special status, and no international authority has been created to cover submarine-cable-related issues. In this sense, submarine cables remain a "neglected" area of international law (Beckman, 2010).

To install or repair a submarine cable in another state's maritime zone (most commonly in the territorial sea or on the continental shelf), it is necessary to bring a foreign cable ship, a team of professionals and equipment to such a maritime space. This requires the regulation of immigration as well as technical, environmental and other matters to ensure that such operations are possible. All these activities must be regulated and implemented by states in their domestic legal systems, and all states are concerned that their interests should be respected. From this perspective, states perform controlling functions. Since there is no central authority nor any supranational government in this sector, states must control cable-related activities intended to provide their territory with communications facilities. Projects involving the laying of submarine cables in the Arctic are large-scale projects designed to traverse the maritime zones of several states. This means that the cable installer must deal with the domestic laws of each state. National requirements may differ significantly. Hence, more time is required to deal with each case. This issue may delay a project and render it less attractive to investors. For instance, Ari-Jussi Knaapila expressed his opinion regarding the Arctic Connect Project and underlined how "the biggest challenge is permitting [construction] from the Russian side, because we're talking about mainly using the Russian EEZ [exclusive economic zone] regions" (Bannerman, 2018, p. 38).

In the field of submarine cables, the only institution that is currently responsible for cable issues is the International Cable Protection Committee (ICPC). The ICPC is an extraordinary actor involved in submarine cable matters worldwide. Since there is no specifically established international organisation for dealing with submarine cables in either the system of the UN or among the independent international organisations, the ICPC took the initiative in terms of collecting materials and proposing solutions for certain technical and legal problems as well as serving as a forum for negotiations between the private sector, scientists, scholars, government officials and policymakers. Holding the legal status of a private company, the ICPC operates as a forum in which current submarine cable issues are discussed. Although the ICPC performs functions and competencies similar to those of an intergovernmental organisation, it does not satisfy the definition of an intergovernmental organisation according to international law, which can be found in Article 2.1 (i) of the Vienna Convention on the Law of Treaties 1969. Additionally, it lacks certain of the powerful instruments inherent within a standard international organisation. First, its decisions are non-binding. Second, it does not address local cases but instead focuses on general recommendations. Third, its work is based on membership, and those involved in submarine cable projects in the Arctic are not necessarily members of the ICPC.

Due to the realities of the Arctic, a special authority created to issue recommendations, guidelines or best practices regarding the laying and maintaining of submarine cables in this region would represent a big step forward in the digitalisation process and the development of telecommunications infrastructure. The Task Force on Telecommunications Infrastructure in the Arctic (TFTIA) was established to develop and further advise the Arctic Council on issues concerning the development of telecommunications, although thus far no document explicitly addressing legal matters associated with submarine cables has been issued by it. Addressing legal issues through a specifically authorised and established body would facilitate the development of cable projects in the Arctic and thus render them even more attractive. There is no need to create a separate authority, as the TFTIA could take on this role and contribute to the development of this issue. In cooperation with the Arctic states, which could provide it with up-to-date and translated legislation presented in a comprehensive manner, the TFTIA could offer recommendations concerning the legal regime of submarine cables in the different maritime zones of the Arctic states. The ICPC has attempted to collect national legislation in its Members' Database, although the initiative does not cover all the Arctic states. Moreover, not all the documentation is translated into English and, further, access to the database is limited.

The creation of a regional and more detailed Arctic collection of domestic laws would facilitate the development of submarine cables projects in the north and make the process easier for the private sector. It could become a reality in the framework of a leading international authority responsible for submarine cables issues. The possibility of avoiding delays due to limited available knowledge and faster negotiations at the universal level in the case of damage to cables are only a few of the advantages that would stem from developing this initiative.

7.5.3 Competitive Uses of the Arctic Ocean

Submarine telecommunications cables are not the only type of activity performed in the Arctic waters. There are numerous activities taking place on both the surface and the seabed. They could be performed in different maritime zones, although they mainly take place on the continental shelves of Arctic states and the high seas. Apart from submarine cables, the UNCLOS recognises the following freedoms of the sea: navigation, overflight, construction of artificial islands and other installations, fishing and scientific research. However, since most of the activities related to cable installation are conducted directly on the seabed, the exploration and exploitation of seabed resources are of particular interest in this context.

The freedom to lay submarine cables on the continental shelf is contained within Article 79 (2) of the UNCLOS. It provides an interesting norm regarding the regulation of two adjacent activities on the continental shelf: the laying of submarine cables and the exploration and exploitation of seabed resources. It provides that "subject to its right to take reasonable measures for the exploration of the continental shelf [...] the coastal State may not impede the laying or maintenance of such cables...". Thus, on the continental shelf, these two activities are regulated by the convention, while a coastal state has priority in terms of exercising its rights to explore the continental shelf. If it already conducts operations concerning the exploration and exploitation of seabed resources in a certain area within its continental shelf, a coastal state may block any project involving the laying of a submarine cable in the same area, and doing so would be in accordance with the UNCLOS. The other issue in this regard is that it is not entirely clear what "reasonable measures" means and how such measures should be evaluated.

The reasonable measures that the coastal state has the right to take in order to develop the resources of its continental shelf are based on domestic legislation, declarations and national policies concerning the development of the natural resources of such a state. In addition, the political context must be taken into account. The resources of the continental shelf are extremely valuable, meaning that states are unlikely to agree to suspend the development of natural resources in the location where a submarine cable is planned. There is reason to believe that in this case the emphasis on the protection of natural resources would take precedence over the social aspect.

On the high seas, the approach is different. Article 87 (2) of the UNCLOS is based on mutual respect for the rights and freedoms of the contractor engaging in the exploration of the resources of the Area and the state with the aim of laying a submarine cable in the same area.

In the Arctic Ocean, regardless of the maritime area, those sites where oil and gas have been discovered could be crossed by new submarine cables. A comparative study of two maps, namely an oil and gas location map (The Harvard WorldMap Project, 2019) and a map of submarine cables (TeleGeography, 2019), reveals that these areas overlap in the Arctic. For example, the Quintillion cable will run through areas in the Beaufort Sea (Koakoak and Mackenzie Delta) where potential crossovers could occur. The Cable Svalbard Undersea System runs close to the Snøhvit oil and gas site in the Barents Sea. The Terra cable system in Alaska crosses two oil and gas locations (McArthur River and Kenai). Finally, the ROTACS cable project, if it is ever built, is intended to be laid through certain areas of oil and gas in the Barents and Kara seas.

For a long time, the problem of correlation between the activity of laying submarine cables and the development of seabed resources was not among the matters widely discussed in relation to the law of the sea. However, with the increasing digitalisation of the northern regions, the issue has become more relevant. Even if the cable itself does not cross the exact area where oil and gas activities are taking place, all the preliminary stages, such as the seabed survey, marine scientific research and route planning, could provoke intersections between two competing activities. The survey for a future cable might reveal that the available and most optimal cable route would pass through oil and gas reserves.

This could be a concern with regards to the future development of the submarine cable industry in the Arctic because two cases of conflicts between submarine cables and the exploration of the seabed have already occurred in practice. Cables laid in the Indian and Pacific oceans (SAFE and HONOTUA, respectively) cross sites where exploration activities are currently taking place (Burnett, Lodge, Le Gurun, & Leonard De Juvigny, 2015). In the Arctic, the potential for crossovers is quite high, taking into account the planned projects and the geography of their routes. The ICPC has issued several (non-binding) documents to address the issue of intersections (ICPC Members' Database, 2018). Thus, these two cases and the ICPC's recommendations should at least be taken into account, and the Arctic states should be aware of the potential problem.

7.6 Conclusion

Over time, human activities in the Arctic have developed so that new forms of use of the northern territories, waters and seabed have gradually been introduced. With both time and the development of new technologies permitting the exploration of the Arctic region from different perspectives, these new forms of use received the opportunity to be implemented. One new method of development concerns the construction of submarine cables bringing telecommunications to the north.

Submarine cables projects receive more attention from businesses, governments and the scientific community nowadays. Plans to lay submarine cables through the Arctic are becoming a reality. Some cable projects have already been finished, some are in progress and several others are planned.

The connection of the northern communities to the global submarine telecommunications cables network represents a significant opportunity from the human security perspective. On the one hand, fast and reliable internet access facilitates the economic growth of remote territories, promotes cultural and educational values, and makes connectivity easier. From this point of view, residents of the Arctic become more secure as the digitalisation of the region contributes to the development of local infrastructure, thereby making life in the north easier.

On the other hand, people may become more dependent on telecommunications and, thus, more vulnerable to its failures and limitations, which undermines human security. Moreover, the appearance of new and unusual activities in the Arctic, such as the laying of submarine cables, can have a negative impact on the local population due to interrupting their cultural and socio-economic foundations and changing their traditional way of life.

However, by following the carefully developed strategy of balancing social needs, environmental knowledge, governance and other uses of the seabed, the development of telecommunications in the Arctic could occur without disrupting other activities in the region.

The modern legal regime governing submarine cables requires a new assessment, especially in light of new submarine cables projects in the Arctic, due to their unique characteristics when compared to cables laid in other parts of the world. The disadvantages that exist in the contemporary legal regime must be eliminated at both the international and regional levels. The Arctic region could become an example of the successful development of the regulation of submarine cables at the local level through the cooperation and joint efforts of Arctic states.

References

Anthony, S. (2012). $1.5 billion: The cost of cutting London-Tokyo latency by 60ms. *Extreme tech*. Retrieved from https://www.extremetech.com/extreme/122989-1-5-billion-the-cost-of-cutting-london-toyko-latency-by-60ms.

Arctic Council. (2017). *Telecommunications infrastructure in the Arctic: A circumpolar assessment.* Retrieved from https://oaarchive.arctic-council.org/bitstream/handle/11374/1924/2017-04-28-ACS_Telecoms_REPORT_WEB-2.pdf?sequence=1.

Arctic Economic Council. (2016). *Arctic broadband, recommendations for an interconnected Arctic.* Retrieved from https://arcticeconomiccouncil.com/wp-content/uploads/2017/02/AEC-Report_Final-LR-1.pdf.

Bannerman, N. (2018). Connecting beneath the ice. *Capacity.* Retrieved from http://edition.pagesuite-professional.co.uk/html5/reader/production/default.aspx?pubname=&edid=7fa976b8-a5af-4cbb-9058-a95f627e8119&pnum=40.

Beckman, R. (2010). *Submarine cables—A critically important but neglected area of the law of the sea.* ISIL Conference. Retrieved from https://studylib.net/doc/18317499/submarine-cables-%E2%80%93-a-neglected-area-of-the-law-of-the-sea.

Burnett, D. (2010). Recovery of cable repair ship cost damages from third parties that injure submarine cables. *Tulane Maritime Law Journal, 35*(1), 103–125.

Burnett, D., & Carter, L. (2017). International submarine cables and biodiversity of areas beyond national jurisdiction the cloud beneath the sea. In D. Burnett & L. Carter (Eds.), *International submarine cables and biodiversity of areas beyond national jurisdiction: The cloud beneath the sea* (pp. 1–72). Leiden: Brill.

Burnett, D., Lodge, M., Le Gurun, G., & Leonard De Juvigny, A. (2015). *Submarine cables and deep seabed mining advancing common interests and addressing UNCLOS "Due regard" obligations* (Technical Study: No. 14). International Seabed Authority. Retrieved from https://www.isa.org.jm/sites/default/files/files/documents/techstudy14_web_27july.pdf.

Carter, L., & Burnett, D. (2011). *About submarine telecommunications cables.* International Cable Protection Committee. Retrieved from https://www.iscpc.org/documents/?id=1753.

Carter, L., Burnett, D., Drew, S., Marle, G., Hagadorn, L., Bartlett-McNeil, D., & Irvine, N. (2009). *Submarine cables and the oceans—Connecting the world.* UNEP-WCMC Biodiversity Series 31. Retrieved from https://www.unep-wcmc.org/system/dataset_file_fields/files/000/000/118/original/ICPC_UNEP_Cables.pdf?1398680911.

Cinia. (2019). *Cinia company.* Retrieved from https://www.cinia.fi/en/company.html.

Davenport, T. (2015). Submarine cables, cybersecurity and international law: An intersectional analysis. *Catholic University Journal of Law and Technology, 24*(1), 57–109.

Delaunay, M. (2014). The Arctic: A new Internet highway? *Arctic yearbook*. Retrieved from https://www.arcticyearbook.com/briefing-notes2014/117-the-arctic-a-new-internet-highway.

Delaunay, M. (2017). Briefing note submarine cables: Bringing broadband Internet to the Arctic, a life changer for northerners? *Arctic Yearbook*. Retrieved from https://www.arcticyearbook.com/briefing-notes2017/278-submarine-cables-bringing-broadband-internet-to-the-arctic-a-life-changer-for-northerners.

Deloitte Insights. (2018). *Embracing the BRI ecosystem in 2018 Navigating pitfalls and seizing opportunities*. Retrieved from https://www2.deloitte.com/insights/us/en/economy/asia-pacific/china-belt-and-road-initiative.html.

Ekwere, K. (2016). Submarine cables and the marine environmental: Enhancing sustainable and harmonious interactions. *China Oceans Law Review, 1*(2016), 129–186.

Global Legal Group. (2010a). *The international comparative legal guide to: Telecommunication laws and regulations 2010—Finland*. Retrieved from https://www.roschier.com/sites/default/files/iclg_telecommunications_laws_and_regulations_2010.pdf.

Global Legal Group. (2010b). *The international comparative legal guide to: Telecommunication laws and regulations 2010—Sweden*. Retrieved from https://www.roschier.com/sites/default/files/telecommunication_laws_and_regulations_2009.pdf.

Heezen, B., & Johnson, G. (1969). Alaskan submarine cables: A struggle with a harsh environment. *Arctic, 22*(4), 413–424.

Hsu, J. (2016). An Internet cable will soon cross the Arctic circle. *Scientific American*. Retrieved from https://www.scientificamerican.com/article/an-internet-cable-will-soon-cross-the-arctic-circle/.

ICPC Members' Database. (2018). *Legal database/national legislation/Sweden*. Retrieved from https://www.iscpc.org/.

Koh, T. (1982). *A constitution for the oceans*. Retrieved from http://www.un.org/depts/los/convention_agreements/convention_overview_convention.htm.

Ministry of Enterprise and Innovation of Sweden. (2016). *A completely connected Sweden by 2025—A broadband strategy* (Item No. N2016/08008/D). Retrieved from https://www.government.se/496173/contentassets/afe9f1cfeaac4e39abcdd3b82d9bee5d/sweden-completely-connected-by-2025-eng.pdf.

Ministry of the Interior of Iceland. (2003). *Electronic Communications Act No. 81*. Retrieved from 18 April 2019. https://www.unodc.org/res/cld/document/isl/1940/electronic-communications-act_html/Iceland_Electronic_Communications_Act_No._81_26_March_2003.pdfAccessed.

Ministry of Transport and Communications of Finland. (2016). *Information security strategy for Finland: The world's most trusted digital business environment*. Retrieved from http://julkaisut.valtioneuvosto.fi/bitstream/handle/10024/75353/9-2016_Information_Security_Strategy_for_Finland.pdf.

Norwegian Ministers. (2012). *Cyber security strategy for Norway*. Retrieved from https://www.regjeringen.no/globalassets/upload/fad/vedlegg/ikt-politikk/cyber_security_strategy_norway.pdf.

Norwegian Ministry of Local Government and Modernisation. (2015–2016). *Digital agenda for Norge: IKT for en enklere hverdag* [*Digital agenda for Norway in brief*]. Retrieved from https://www.regjeringen.no/contentassets/07b212c03fee4d0a94234b101c5b8ef0/en-gb/pdfs/digital_agenda_for_norway_in_brief.pdf.

Opperud, H. (2010). Telecommunications regulation Norway. *Lex Mundi*. Retrieved from http://www.lexmundi.com/Document.asp?DocID=1871.

OSPAR Commission. (2009). *Assessment of the environmental impacts of cables*. Biodiversity Series 1. Retrieved from https://qsr2010.ospar.org/media/assessments/p00437_Cables.pdf.

OSPAR Commission. (2019). *Introduction for the executive secretary*. Retrieved from https://www.ospar.org/about/introduction.

Perez-Alvaro, E. (2013). Unconsidered threats to underwater cultural heritage: Laying submarine cables. *Rosetta*. Retrieved from http://www.rosetta.bham.ac.uk/issue14/perezalvaro.pdf.

Pierce, E. (2017). *Quintillion subsea cable system*. Retrieved from https://arcticeconomiccouncil.com/wp-content/uploads/2017/06/TOW-Summit-Pierce.pdf.

Pro-Arctic. (2016). *Helsinki hochet prolozhyt arkticheskyi cabel v Asii* [Helsinki wants to lay Arctic cable in Asia]. Retrieved from http://pro-arctic.ru/05/05/2016/news/21410.

Quintillion. (2016). *System*. Retrieved from http://qexpressnet.com/system/.

Rapp, R. J. (2010). *Submarine cables: Critical infrastructure. United States interests in prompt adherence to the law of the sea convention*. Retrieved from https://docplayer.net/8835367-Submarine-cables-critical-infrastructure-supplier-perspective.html.

Skepys, B. (2012). Is there a human right to the Internet? *Journal of Politics and Law, 5*(4), 15–29.

State Duma of the Russian Federation. (1995). *Federal law on the continental shelf of the Russian federation No. 187-FL*. Retrieved from http://extwprlegs1.fao.org/docs/html/rus21902E.htm.

Subsea World News. (2018). WFN strategies nets EAUFON subsea cable survey deal. *Subsea World News*. Retrieved from https://submarineworldnews.com/2018/02/21/wfn-strategies-nets-eaufon-submarine-cable-survey-deal/.

Task Force on Telecommunications Infrastructure in the Arctic. (2018). *Task force on telecommunications infrastructure in the Arctic*. Retrieved from https://arctic-council.org/index.php/en/task-forces/342-tftia.

TeleGeography. (2019). *Submarine cable map*. Retrieved from https://www.submarinecablemap.com/.

The Harvard WorldMap Project. (2019). *Oil & gas map*. Retrieved from https://worldmap.harvard.edu/maps/6718.

The Task Force on Telecommunications Infrastructure in the Arctic. (2018). https://arctic-council.org/index.php/en/task-forces/342-tftia.

The United Nations Conference on Environment and Development. (1992). *The Rio declaration on environment and development*. Retrieved from http://www.unesco.org/education/pdf/RIO_E.PDF.

Titarenko, E. (2017). Polarnet Proekt zaruchilsya podderzkoi [Polarnet project enlisted support]. *ComNews*. Retrieved from https://www.comnews.ru/content/108946/2017-07-31/polarnet-proekt-zaruchilsya-podderzhkoy.

United Nations. (1963). *Order No. 191 on the delimitation of the territorial sea of Greenland*. Retrieved from http://www.un.org/Depts/los/LEGISLATIONANDTREATIES/PDFFILES/DNK_1963_Order.pdf.

United Nations. (1982). *Convention on the Law of the Sea (UNCLOS)*. 1833 United Nations Treaty Series 3 21 ILM 1261. Retrieved from http://www.un.org/depts/los/convention_agreements/texts/unclos/unclos_e.pdf.

United Nations. (2019). *Chronological lists of ratifications, accessions, and successions to the convention and the related agreements*. Retrieved from http://www.un.org/depts/los/reference_files/chronological_lists_of_ratifications.htm.

United States. (2019). *U.S. code, title 47, chapter 2—Submarine cables*. Retrieved from https://www.law.cornell.edu/uscode/text/47/chapter-2.

Van Logchem, Y. (2014). Submarine telecommunication cables in disputed maritime areas. *Ocean Development and International Law, 45*, 107–122.

Wilson, S. (2013). Risks to submarine cables in the Arctic. *SubOptic forum*. Retrieved from https://suboptic.org/wp-content/uploads/2014/10/Poster_170.pptx.

Worzyk, T. (2009). *Submarine power cables: Design, installation, repair, environmental aspects*. Berlin: Springer-Verlag.

CHAPTER 8

Connecting the Arctic While Installing Submarine Data Cables Between East Asia, North America and Europe

Juha Saunavaara

8.1 Introduction

Questions concerning connectivity in the Arctic have drawn unprecedented international attention in recent years, with the Arctic Ocean, which used to be one of the few sea regions without submarine communications cables (Delaunay, 2014), having been identified as a potential location for new routes that could revolutionise global submarine data cable networks. There are many reasons behind the ongoing changes that are exemplified by the Arctic Connect Project,[1] which is led by the Cinia Group, and by the Quintillion Subsea Holdings project.[2] On the demand side, global internet traffic has increased rapidly. In particular,

[1] A project originating in Finland that aims to create a submarine fibre-optic cable connection between East Asia, Europe and North America through the Northeast Passage.

[2] A project originating in Alaska that aims to create a submarine fibre-optic cable connection between East Asia, Europe and North America through the Northwest Passage.

J. Saunavaara (✉)
Hokkaido University, Sapporo, Japan
e-mail: juha.saunavaara@arc.hokudai.ac.jp

M. Salminen et al. (eds.),
Digitalisation and Human Security, New Security Challenges,
https://doi.org/10.1007/978-3-030-48070-7_8

the growth rate of data transfer between Asia and Europe is high. On the supply side, the decline in ice cover in the Arctic Sea as well as the technological solutions used by Quintillion to install a fibre-optic cable near Alaska serve to reduce uncertainty in terms of feasibility and supply. Clearly, geography plays an important role as well. If one wants to connect East Asia, North America and Europe with a submarine fibre-optic cable, the Arctic Ocean offers a shortcut, meaning that the physical cable connection is shorter and the network latency (i.e. the time required for a data packet to be transferred between two nodes in a network) is decreased. In addition, both projects recognise the need to provide internet access and connections to remote Arctic areas, including the European High North.

The Arctic submarine data cable projects can be understood as ways of providing fast and affordable internet to people living in the Arctic or as ways of connecting global economic centres outside the Arctic and supporting business activities in fields such as financial technology, artificial intelligence (AI), the Internet of Things (IoT), software as a service (SaaS) and cloud computing. Approaching the different ways of interpreting the objectives and motives of these projects through the lens of human security draws attention to the role offered to local people and communities. When questioning whether local inhabitants are active agents or merely passive objects of development, one may also question whether local stakeholders and the leaders of the Arctic communications cable projects share a common understanding of the positive and negative potentials connected with the initiatives.

To clarify the role of the European High North and other Arctic regions in projects to install submarine data cables in the Arctic Ocean, this chapter seeks answers to the following questions. How do the companies in charge of such projects describe the relationship between their projects and connectivity within the European High North and other Arctic regions? How do the representatives of different regions and communities along the planned routes of the cable projects envision the relationship between those projects, improved connectivity in their regions and the overall improvement of human wellbeing? What kind of cooperation takes place between the companies leading the projects and regional actors? This chapter considers both argumentation directed towards the public and more private negotiation processes using sources such as published statements, reports and policy papers, newspapers and other types of media, interviews and participatory observations. While the European High North is a core region in relation to the Arctic Connect Project, this chapter also references other geographical locations so as to better understand the recent developments in connectivity in the Arctic, the features of the international submarine fibre-optic cable business and the processes that may occur in the European High North during upcoming projects.

8.2 The Characteristics of Submarine Fibre-Optic Cable Projects

Large-scale submarine fibre-optic cable projects intended to connect Europe, North America and East Asia often feature in professional discussions concerning the digitalisation of the European High North and the Arctic more generally. Although the Arctic Connect Project seems to be relatively well known in northern Finland and Norway (Saunavaara, 2018), it is necessary to clarify a number of features of submarine communications cable projects in general. When analysing local people's interests, needs, hopes and fears regarding submarine fibre-optic cables as well as their potential to influence decisions concerning large-scale digital infrastructure projects, one needs to understand the role of these cables as pillars of the global flow of information. Despite the general belief concerning the prevalence and importance of wirelessness, submarine fibre-optic cables have transported more than 99% of all transoceanic digital communications in recent years, with only 0.5% of the internet being transmitted through satellites (Rauscher, 2010; Starosielski, 2015; Uljas, 2018a).

Submarine communications cables have the capability to connect areas separated by great distances, although their local physical presence and visibility are limited. The cable itself is only 20–50 millimetres in diameter. Different wire armouring and burial techniques can be used to protect the cable from both human activity and environmental threats, such as the movement of ice in Arctic waters, and the submarine part of the connection is, in theory, invisible (Carter, Burnett, & Davenport, 2014; Woolston, 2018). The landing stations form a critical part of the communications network, connecting intercontinental submarine cables to national terrestrial networks. Nevertheless, they are usually small and modest-looking constructions, and they are typically located in relatively remote areas (Starosielski, 2015). In other words, while fibre-optic cable connections may appear comparable to roads, railway lines or gas pipelines when viewed on a map, they actually represent an infrastructure that is easy to forget after the installation process has been completed.

Submarine communications cables and landings can currently be found in surprisingly few places around the world. This situation has not changed, although the problems linked to a lack of diversity and robustness as well as the potentially catastrophic impacts of natural and human disasters at a choke point[3] (i.e. where a great number of international submarine fibre-optic cable paths meet) have been quite widely discussed, for example, at the 2009 Reliability of Global Undersea Communications Cable Infrastructure Summit (Rauscher, 2010). In Starosielski's (2015) detailed explanation of the reasons for the limited geographic dispersion of signal paths, she refers to the enormous capacity and expense of each system, the limited number of sites where existing systems can be interconnected that are insulated from harmful human activity, and the desire to avoid risks that are bound to the development of new, untested routes in relation to which maintenance costs may be more expensive and market demand may be uncertain.

The physical characteristics that render submarine fibre-optic cables difficult to see with the naked eye and people's imagination contribute to their apparent invisibility. The improvement of connectivity—which could be achieved through the installation of landings for a submarine fibre-optic cable—is an abstract phenomenon that enables but does not ensure positive developments. Therefore, it is often difficult for people who lack awareness of specific projects and knowledge concerning telecommunication infrastructure more generally to envision or evaluate the overall impact of Arctic submarine cable projects on their everyday lives or their community (Joensuu, 2018; Yamamoto, 2017a). The apparent invisibility of fibre-optic cables partially explains why they do not foster great hopes or trigger fear or concern among the people who live in close proximity to them.

Fibre-optic cables differ from satellites in that they are a point-to-point technology. As it is expensive and time-consuming to connect people who live in the most remote areas, where the market demand and the possibility of conducting profitable business are extremely low, such people may be left behind if there is no regulation or incentive

[3] These locations include the Luzon Strait, the Suez Canal–Red Sea–Mandab Strait passage and the Strait of Malacca.

in place to force or motivate the construction of physical connections. Intercontinental submarine fibre-optic cables can be described as international information highways. If one is driving and needs to use a highway, one usually drives on smaller roads first before entering the highway via a ramp. In the digital world, the infrastructure leading to the highway is usually built and operated by local telecommunications companies. As new submarine fibre-optic cable systems allow local providers to offer better services, the actual effects of the construction of a new submarine system depend on the actors in charge of the development of terrestrial networks.

Submarine communications cables are expensive long-term investments that are expected to function for around 25 years. State and regional administrations make significant contributions to policies facilitating digitalisation within national borders, and they may even take part in cross-border cooperation (Bauer, 2010; Eskelinen, Frank, & Hirvonen, 2008; Salminen & Hossain, 2018). However, large-scale submarine communications cable projects have evolved into multinational projects planned and implemented by private enterprises (Burnett, Davenport, & Beckman, 2014; Malecki & Wei, 2009). The Arctic fibre-optic cable projects are no exception, although the Finnish government has played an active role in the initiation of the Arctic Connect Project, while Cinia is a state-owned company. Cinia's role is to guide the project to a point at which a new international project company that completes the development and implementation phase can be established (Joensuu, 2017; Saunavaara, 2018).

While these kinds of characteristics do not nullify the importance of public–private partnerships and cooperation with national, regional and local authorities, they do help to explain the significance of non-Arctic partners. Japan and China, for example, have shown interest and been involved in the Arctic Connect Project. More specifically, Japan's hosting of the Arctic Economic Council's 3rd Top of the World: Arctic Broadband Summit as well as its organisation of other international events concerning Arctic communications cables in Hokkaido reflect the country's interest in the topic. In addition, China's Arctic Policy, which was published in January 2018, specifically mentions submarine fibre-optic cables next to topics such as shipping routes, resource development and environmental protection, describing it as a field in which China promotes practical cooperation at the bilateral and multilateral levels (The State Council Information Office of the People's Republic of China, 2018).

It is worth noting that the actors involved with Arctic submarine communications cables do not claim to be able to solve all the connectivity problems in the Arctic by themselves. On the contrary, a consensus regarding both the need for a system of systems and the complementary role played by satellites and microwaves can be observed in forums in which issues concerning Arctic connectivity are debated. Different types of technologies can support each other, and the presence of a submarine cable landing station can encourage, for example, satellite companies to construct their own landing stations in the same area (Joensuu, 2018; Klouda, 2017).

8.3 Human Security and Submarine Communications Cables

Questions concerning national security and cybersecurity have always been relevant to the development of submarine communications networks. Security concerns have not only affected decisions concerning routes and landings, but also hindered plans to utilise the submarine fibre-optic infrastructure for scientific purposes (Starosielski, 2015). The projects in the Arctic are no exception. While Team Telecom[4] has designated the Quintillion network a critical infrastructure for US national security (Woolston, 2018), in April 2018 the Russian Ministry of Defence announced its own plan to lay a transarctic fibre-optic cable between Severomorsk and Vladivostok. The latter initiative is reportedly intended to serve the navy and coastal troops as well as to increase Russia's national defence capabilities. While little is currently known about the progress of the project, which highlights the Russian interest in Arctic cables also demonstrated by the unrealised Russian Optical Trans-Arctic Submarine Cable System (ROTACS) project, the idea of utilising submarine communications to survey the underwater environment, as has been discussed by scholars, seems to have been co-opted for

[4]Team Telecom is comprised of staff members from the US Department of Homeland Security, the Department of Justice and the Federal Bureau of Investigation as well as representatives from the Department of Commerce, the Department of Defense, the Department of State, the Department of the Treasury and the Office of the United States Trade Representative.

military use (Navy Recognition, 2018; Nilsen, 2018). However, studies exploring submarine communications cables from the perspective of human security are rare or even non-existent.

The aim of human security can be defined in various ways. The definitions offered by Alkire (2003) range from the "protection of the vital core of all human lives from critical pervasive threats" to "the creation of political, economic, social, cultural, and environmental conditions in which people live knowing that their vital rights and freedoms are secure" (p. 2). These definitions refer to both actions that provide human security through the prevention of the materialisation of identified threats and the need to ensure that one's actions do not threaten human security (Alkire, 2003). According to Salminen and Hossain (2018), a human security approach to digitalisation and cybersecurity mandates that one to pay attention to "the people and communities experiencing digitalization in their everyday life" (p. 1). Thus, it significantly differs from traditional approaches emphasising the importance of "economic growth, the advancement of information society, national security, and enhancement of service provisioning" (p. 1). Rather than understanding people who live in the Arctic as objects or passive receivers of development and security who automatically benefit from policies that facilitate digitalisation, the human security approach focuses on digital opportunities and threats identified and experienced by local people and communities (Salminen & Hossain, 2018). This approach can also be utilised as an analytical framework for understanding the meaning(s) of a single large-scale digital infrastructure project, such as an international submarine fibre-optic cable, from the perspective of the people and communities in the European High North and other parts of the Arctic. Although it is an extremely flexible term, human security offers lenses that can help one to focus on the most significant phenomena and vocabulary that can be used to conceptualise observed reality, even if the term is not used by the agents themselves.

In addition to understanding the typical features of submarine fibre-optic cable projects, one should be aware of the local conditions in the European High North—which are also described elsewhere in this volume—before attempting to explain what the construction of Arctic Connect might mean for human security in the region or why local people perceive the project in a certain way. From the perspective of global communications networks, the European High North is a remote peripheral region. Although the Luleå region has developed into a globally significant hub for data centres (Dawn-Hiscox, 2018), the existing networks were not planned and built with an international information highway in mind. Moreover, as noted by Salminen and Hossain (2018), national information infrastructure and services have not been designed from the perspective of, or even taken into consideration, the needs of people and communities in the European High North.

While the following sub-sections elaborate on the various positive effects of the Arctic Connect Project on human security in the European High North, it must be acknowledged that the project also has limitations, which can be identified in Quintillion's project. Quintillion is a privately owned business that has spearheaded the process of connecting Japan and London via Alaska and the Northwest Passage since it acquired the assets of a Canadian company, Arctic Fibre, in 2016. Phase 1 of Quintillion's project—a regional network containing a submarine fibre-optic cable around Alaska with landings in six hub communities and a terrestrial connection between Prudhoe Bay and Fairbanks—was completed after two and a half years of construction. The submarine fibre-optic network went live on December 1, 2017, and the company has since moved on to planning Phase 2 (a submarine cable connecting Alaska and Japan) and phase 3 (a submarine cable connecting Alaska and London) (Saunavaara, 2018; Woolston, 2018).

There is much evidence to suggest that Quintillion has revolutionised internet access in Alaska, although there are still limitations that should not be overlooked. According to one estimate, the 250-million-dollar (approximately 220-million-euro) project, including 1200 miles of submarine fibre-optic cable, would only reach 12,500 Alaskan residents (Wood, 2018). Regardless of whether or not this number is accurate,[5] it demonstrates the peculiarities of the Arctic region as well as the need for cooperation between companies that provide different kinds of connectivity solutions. If the Arctic Connect plan is implemented, it will directly affect the people who live in the European High North by improving the terrestrial connection between the new landing planned in Kirkenes, Norway, and the existing C-Lion1 cable landings connecting southern Finland and Rostock, Germany.[6] In addition, as seen in the case of

[5] The source does not reveal how the number of residents was calculated. Quintillion has not replied to the author's request for a comment on the estimate published by StateScoop.

[6] It is worth noting that there are plans to build a new submarine fibre-optic cable connecting the current landings of C-Lion1 with cities on the shores of the Bothnian Bay.

Alaska, it may inspire other companies to improve their own terrestrial systems.

Quintillion and Cinia—and the forthcoming Arctic Connect Project company—are middle-mile companies that do not offer end-user services. Rather than selling services to private consumers, they deal with telecommunications companies, which in turn service local people, businesses and the public sector. According to Quintillion, their system has reduced the cost of wholesale service in Alaska by up to 90%—and by an average of 50%—when compared to the cost of existing microwave and satellite wholesale services. However, the company does not control consumer prices (Falsey, 2017; Joensuu, 2018; Pierce, 2017; Woolston, 2018).

8.4 The Role of the Arctic as Described by the Companies

The Arctic Connect Project will connect Europe, East Asia and North America with an 18,000-kilometre-long submarine fibre-optic cable system running through the Northeast Passage. The system itself is designed to consist of multiple routes and six fibre pairs. While some of the fibre pairs would not have landings in the Arctic Ocean, others could land in three to five locations in Arctic Russia. This design is intended to enable different cybersecurity solutions for different partners. The most important selling point of the project is its capacity to offer a new, shorter route for data traffic between the three continents, which are home to 85% of the world's population (Joensuu, 2017, 2018; Saunavaara, 2018; Uljas, 2018a).

Both the Arctic Connect Project and the Quintillion project exist because they offer shorter network latency and new routes that would bring long-awaited diversity to the relatively centralised global submarine cable network. The main audiences for this kind of message are potential collaborators, customers and investors worldwide. In this context, the projects are strongly connected to issues such as the development of the industrial internet, big data, the utilisation of 5G technologies, the advancement of cloud computing and SaaS solutions, the potential of autonomous vehicles, and existing or future need for low-latency trading and financial technology (Joensuu, 2017; Uljas, 2018a, 2018b; Woolston, 2018).

However, both Cinia and Quintillion have identified various benefits that their projects would also bring for people who live in the Arctic, including the European High North. When describing the expected outcomes of the projects, representatives of the companies often talk about the importance of connectivity in general and also mention the opportunities afforded by better broadband access, including those related to e-learning, e-healthcare services and other public services. The sparsely populated Arctic regions, which often lack physical service networks such as those available in big cities, are described as very high potential users of digital services (Joensuu, 2017; Uljas, 2018a, 2018b; Woolston, 2018). When describing their projects' connections with community security, the companies tend to concentrate on possibilities rather than threats. The decision to focus on the possibility of e-services rather than on the potential disappearance of the physical service network is understandable, as they do not want to be associated with processes that are generally considered undesirable and in which they actually play no role.

The economic security aspect of the human security discourse is made explicit in arguments concerning the new business opportunities afforded by improved connectivity. This kind of argumentation often refers to the changes caused by digitalisation to industries, the economic activity that is already present in the Arctic, the possibility of attracting new businesses as well as improved telecommunication possibilities. While such discussions tend to remain rather abstract, the increased possibility of attracting new data centre investments is usually mentioned as the most concrete example of a positive outcome of improved international connectivity via submarine communications cables (Joensuu, 2017; Uljas, 2018a, 2018b; Woolston, 2018). The link between data centres and new submarine fibre-optic cables has been acknowledged around the globe (Linden, 2017; Starosielski, 2015), but the vision concerning the new data centres is especially easy to market in the European High North. Indeed, some of the world's largest data centres are already located in Nordic countries because they offer a cold climate, cheap and often green energy, an educated workforce and a stable society. The field, which is already huge (the carbon footprint of the data centre industry overtook that of the aviation industry in 2013 and data centres account for over 7% of the world's energy demand), is expected to grow enormously over the coming years (Gudjonsson, 2018). In addition, well-known studies have highlighted the significant impact that large-scale data centres can have on a regional economy in the form of investments, temporary and permanent employment opportunities and the consumption of local supplies and services, especially during the construction phase (Copenhagen Economics, 2017; The Boston Consulting Group, 2014).

The marketing message of Cinia differs from that of Quintillion in terms of the availability, speed and price of broadband services in the Arctic. The actors involved in the Arctic Connect Project rarely refer to these factors, while the representatives of Quintillion often emphasise their proven capability to bring fibre-optic connections to communities that previously lacked such infrastructure and hence were dependent on slower and more expensive microwave and satellite services (Falsey, 2017; Pierce, 2017; Quintillion, 2018; Woolston, 2018). There are several interconnected reasons bound to the natures of these projects that explain these differences. First, Arctic Connect is an intercontinental project that is not divided into phases, meaning that there is no regional or solely Arctic phase equivalent to Quintillion's phase 1. Second, the discussion concerning the Arctic part of the Arctic Connect cable system has mainly concentrated on northern Finland and Norway, where existing broadband services may not be as highly developed as in the southern parts of those countries but are still much better than those in most other places in the Arctic. Attempts to overcome the existing caps on broadband accessibility in the European High North seem to be bound to national broadband promotion policies rather than to any international submarine communications cable project. In addition, the planned landings in Arctic Russia would be constructed in areas characterised by significant industrial activity. While the new connections would offer opportunities to local people, the focus is on benefits to industries (Joensuu, 2018).

8.5 Local Wishes and the Lack of Human Security Concerns

Various regional development plans for northern Finland (Saunavaara, 2018) as well as national and regional digitalisation and cybersecurity frameworks from Finland, Norway and Sweden (Salminen & Hossain, 2018; Zojer, 2019), show that the importance of improved connectivity, which is often mentioned in discussions concerning the Arctic data cable projects, is widely accepted. Within most documents concerning these plans and frameworks, the importance of improved connectivity is emphasised and linked to improved quality of life, the possibility of finding new business opportunities and attracting new investments, and the capability to offer new types of public services. Furthermore, the belief that the implementation of the Arctic Connect Project would improve northern Finland's and northern Norway's ability to attract new data centre investments is shared by many local actors representing both the public and private sectors (Hansen, 2018; Rämet, 2018; Rautajoki, Sørheim Nilsen, Wallin, & Tuovinen, 2017; Riipi, 2017). It is, however, worth noting that these conceptions are not unique to the European High North.

One of the key objectives of the Comprehensive Economic Development Strategy for Alaska 2017–2022 is improved access to state-wide broadband. The Strategy describes access to reliable high-speed internet service as the foundation for participation in the contemporary economy as well as a way to increase both efficiency and quality of life. In addition, the document refers to a World Bank analysis showing that there is a 1.2% increase in per capita gross domestic product (GDP) growth for every 10% increase in the penetration of broadband services (Alaska Department of Commerce, Community, and Economic Development, 2017). The state-wide Broadband Task Force published a report in October 2014 that considered job creation, education, libraries, health care and public safety, among other topics, and stated that there was no longer a need to debate the benefits of broadband and the role it plays in the economy (Alaska Broadband Task Force, 2014).

Although the seemingly uncontested trust in the positive correlation between improved connectivity and economic and social development seen in different parts of the circumpolar north may be surprising, it is actually part of a wider global discourse. The idea that the competitiveness of businesses, their productivity and the comparative advantages of regions depend on telecommunications technologies is not new (Capello & Nijkamp, 1996). At the beginning of the 2000s, the general development discourses consistently identified information and communication technologies (ICT) as a requirement for both economic growth and improvement of social conditions. The lack of ICT, especially limited internet connectivity, was understood to be an important factor contributing to the widening gap between developed and developing regions. While some researchers (Avgerou, 2003) criticised the oversimplified tool-and-effect link between ICT and economic development (i.e. the general association of ICT with positive socioeconomic effects), which is mentioned in many influential reports published by international development agencies, the internet and broadband are still described as gateway and general-purpose technologies. This means that they have fundamentally transformed the way in which both people and businesses operate as well as how and where economic activity is organised (Czernich, Falck, Kretschmer, & Woessman, 2011; Grubesic & Mack, 2016).

The lack of opposing arguments regarding Arctic submarine communications cables can be identified as a feature accompanying the positive expectations. Different preferences with regards to technical solutions and critical discussions concerning the costs and benefits are always present in debates regarding all large-scale infrastructure projects. However, neither the Arctic Connect Project nor the Quintillion project seem to have triggered local objections, which often occur in relation to other types of industries and infrastructure projects in the Arctic (see, e.g., Lassila, 2018; Taksami, 2018). These projects have not faced, for example, questions concerning traditional livelihoods, traditional land use patterns, the rights of indigenous peoples or environmental issues. In other words, although the different aspects of human security, which range from economic and political security to health and environmental security, may sometimes conflict (Zojer, 2019), major conflicts of interest have been avoided in the Arctic submarine communications cable projects to date.

Aside from the lack of criticism in newspapers and other types of traditional media, the conclusions drawn from the analysis of social media, which offers new ways for people to express their opinions and attempt to influence ongoing or planned projects, can be used to support the argument concerning the lack of opposition. The lack of proper connectivity still prevents the participation of many people who live in the Arctic, but Twitter discussions can be used to understand different views on Arctic data cable projects. For example, discussions featuring the hashtag "#Koilliskaapeli" (a Finnish term often used when referring to the Arctic Connect Project) can be compared with discussions featuring the hashtag "#Jäämerenrata" (a Finnish term often used to refer to the plan to build a new railway from northern Finland to a port in the Arctic Ocean). While both projects, at least currently, aim to connect northern Finland with Kirkenes, the former discussions include only a dozen or so tweets, while the latter include almost ten times more. The differences are even more significant when retweets and likes, which indicate the attention each discussion has received, are considered. The content of these discussions are also different, as most tweets about the railway project feature heavy criticism and polarised discussion. Meanwhile, the tweets concerning the Arctic Connect Project are positive and describe the advancement of the project.

When analysing the lack of criticism and human security concerns, attention needs to be paid to at least three key issues. First, the Arctic Connect Project—as well as Phases 2 and 3 of Quintillion's project—is still in the planning stage. More critical voices may emerge when the planned route is made available to the public. Yet, the comparison with the railway project reveals that other projects in the European High North have faced much greater resistance even though they are further away from the decision-making and construction phases. Second, the lack of opposition is most likely due to the invisibility of the fibre-optic networks, as discussed earlier. The environmental effects of submarine fibre-optic cables, which are also discussed in Chapter 7 of this volume, are very small and, therefore, the linkage between one's everyday life and the planned international submarine communications cable project is difficult to envision. Third, local actors' limited ability to contribute to the funding of the planned project, which is discussed below, likely explains the lack of critical public discussion concerning its price. Furthermore, it can be speculated that Arctic Connect is seen as a public project among Finns due to Cinia's ownership structure and the open support of the Finnish Government, which may affect its public image. It remains to be seen whether the assessments of human security threats will be different after the international organisation implementing the project is established.

8.6 Cooperation and Local People's Ability to Influence Digital Development

Representatives of the companies in charge of submarine fibre-optic cable projects often refer to the importance of public–private partnerships as well as to the need for cooperative relations with regional and local decision makers (Linden, 2017; Woolston, 2018). As many local stakeholders have also been eager to announce their support for such projects, it is worth determining what kind of cooperation is possible and what is actually taking place.

While the need to secure sufficient funding is one of the fundamental starting points for all Arctic submarine communications cable initiatives, the Arctic communities, which are relatively small, cannot play a major role in this process. The total budget of the Arctic Connect Project, including the backhaul routes and landing stations, is approximately 700 million euros. As most investments in the project are expected to originate outside the Arctic (Joensuu, 2017; Uljas, 2018a), it depends on decisions made by non-Arctic actors. The situations in other parts of the Arctic are similar. For example, although Alaska-based actors are involved, most of Quintillion's funding originates from the 48 states of the continental United States (Downing, 2017; Kang, 2017).

However, the budget for the development phase of Arctic Connect, during which questions concerning the legal structure and permits, financial model, environmental studies, technical solutions and business plan for the implementation phase must be answered, is expected to be less than 1% of the total budget (Joensuu, 2017). While that is still a lot of money from the perspective of local municipalities and other regional actors, it is low enough that they can play a role. Currently, the cities of Rovaniemi and Oulu, the Kainuu region and local companies have committed to helping plan the route of Arctic Connect. While the investment of C-Fiber Hanko indicates the commitment of hub communities in southern Finland, Cinia's cooperation with Bredbåndsfylket in Norway, IT Norbotten in Sweden and Polarnet in Russia shows that there is also cross-border interest in the project within the European High North (Bredbåndsfylket, 2017; Cinia, 2018a, 2018b; Joensuu, 2017; Uljas, 2018a).

There also exist other views on the role of Arctic actors in funding Arctic submarine communications cable projects. When the Northwest Passage cable project was still led by Arctic Fibre, the then leader of the project envisioned that the investments of the Canadian High Arctic Research Center Service, Canada's Department of Defence and oil companies that wish to drill in the area would help to fund the network. The funding generated along the cable's route, rather than simply at its endpoints, was emphasised, while the need for governmental support was considered crucial. Representatives of Arctic Fibre expected that Canada could finance approximately one-third of the project, which would amount to a cost of 640 million Canadian dollars (about 430 million euros) (Bennett, 2016; Starosielski, 2015). These plans never materialised, and the project only advanced after Quintillion secured private funding and took the lead in the project.

A cooperative attitude at the beginning of a project is important because it may forecast a relatively smooth process when planning becomes more concrete and issues such as permits and zoning come to the fore (Linden, 2017; Sedrak, 2018). Submarine communications cable projects outside the European High North or Arctic have demonstrated landing stations to be one of the most likely sources of problems. There exist various examples indicating that the attitudes of local governments, communities, businesses and NGOs can both encourage and prevent, or at least impede, the landing of submarine cables in the region. Therefore, cable companies' tendency to utilise existing routes and landings, which leads to the concentration of critical infrastructure, can at least partially be explained by their desire to choose paths along which negotiations with local stakeholders have already been completed (Starosielski, 2015).

Environmentalists interested in protecting the local flora and fauna and/or the scenic and visual qualities of coastal areas, as well as people who consider beaches to be an integral part of local cultural practices and leisure, may resist the construction of landings. However, past cases have shown that organisations representing the interests of the fishing industry may play the most influential role in defining the routes of submarine cables (Sato, 2018; Starosielski, 2015; Yamamoto, 2017b).

Human activity leads to the vast majority of cable faults, and fishing is the cause of more than half of all human-activity-related faults (Wargo & Davenport, 2014). Yet, fishermen's hesitancy regarding the no-anchor zones around the cable route and other regulations limiting fishing activities is understandable. The fact that Quintillion has managed to land its cable in six new Alaskan communities shows that negotiations between cable companies and local actors can be successful. However, again in Alaska, fishing industry officials have expressed concern about trawl nets and other fishing gear damaging the submarine cable in the Bering Sea, where the current network is to be supplemented with a cable to East Asia (Paulin, 2017). There is some existing landing station infrastructure in Kirkeness and Teriberka, where the Arctic Connect Project is planning to land, and no major fishing activity takes place in these areas. However, a cost–benefit analysis will determine whether the existing landing stations are to be used. Currently, the project is in such an early phase that concrete discussions with local municipalities concerning, for example, zoning have not yet begun (Joensuu, 2018). Therefore, the current stage of the project explains the lack of arguments intended to convince local stakeholders that Arctic Connect does not threaten either the environment or human security in the communities that will host the landing stations.

The bargaining power of local stakeholders, their ability to become engaged and their capability to have an impact vary in different parts of the Arctic and around the world. For example, it is well known that it is difficult to land a submarine fibre-optic cable in California due to the strict regulations and the power of environmental activists. In addition, in many parts of the world, cable companies have paid compensation to the fishing industry through formal and informal channels or organised goodwill campaigns, including those funding local schools and issuing scholarships, in order to secure local support for their projects. Further, landings may be easier to secure and fewer permits may be needed in regions that are poorer or where the formal or informal political structures restrict local actors' ability to resist or negotiate (Starosielski, 2015). In Norway and Russia, where Arctic Connect is intended to land, the local stakeholders' statuses and capabilities will most likely differ, not only because of differences in national legislation,[7] but also because of differences in the implementation of regulative documents. A previous study (Nenasheva, Bickford, Lesser, Koivurova, & Kankaanpää, 2015) on the European High North and Russia showed that even if the legal tools for public participation in different countries are the same conceptually, they manifest very differently. According to Nenasheva et al. (2015), these differences arise from the legal framework, stakeholders' expectations and the cultural backgrounds of those involved.

Finally, it is worth repeating that local stakeholders may not only be agents of resistance, as they may also recognise the positive potential of (international) submarine fibre-optic cables and landings. These local actors can be proactive, thereby ensuring flexibility in terms of zoning or investing in digital infrastructure so as to achieve the benefits of improved connectivity. The establishment of C-Fiber Hanko, which brings together local enterprises and the municipalities of Hanko and Raseborg, is an example of local stakeholders' willingness and capability to act in such a way that led to the establishment of a new landing station connected to the C-Lion1 submarine cable system (Saunavaara, 2018).

[7]See Chapter 7 in this volume for more detail on the legal regimes.

8.7 Conclusion

The need to improve connectivity in the Arctic has been emphasised by a number of actors. The issues concerning the digital infrastructure and services in the Arctic have now moved from the margins to the mainstream, inspiring the Arctic Council to establish a dedicated task force, the Arctic Economic Council to establish a working group, the University of the Arctic to establish a thematic network, Finland to prioritise connectivity in its chairmanship programme for the Arctic Council 2017–2019, different actors to incorporate questions concerning connectivity into their Arctic strategies and declarations, and the organisers of certain major Arctic events to select connectivity as one of their principal themes. Discussions concerning or referring to ongoing Arctic submarine communications cable projects are one of the most visible and concrete parts of the new Arctic connectivity discourse. Although the vocabulary of human security has not been utilised in most of these discussions, the issues and arguments are clearly connected to that concept.

Most statements made in debates about connectivity in the Arctic have approached digital infrastructure and services from a technical, economic or strategic perspective, as noted by Salminen and Hossain (2018). They argue that this kind of approach may ignore the interests, needs and fears of local stakeholders in the Arctic. Such neglect can cause uncertainty or even anger among people who feel that they cannot influence the digital development that affects their daily lives. Nevertheless, the general attitude towards the Arctic submarine fibre-optic cable projects has been positive in both the European High North and Alaska. Resistance and arguments reflecting the negative emotions caused by such projects have remained minor, while many desirable impacts of the projects have been recognised. For example, the projects have been viewed as potential providers of economic, health and community security rather than as a threat or a source of negative security. While the actual effects of the still uncertain implementation of the Arctic Connect Project in the European High North depend on other actors' interests and capabilities to take advantage of the opportunities afforded by the project, it is widely believed that these effects will contribute to the digital development of the region.

If attention is only paid to the raison d'être of the ongoing international submarine fibre-optic cable projects intended to connect Europe, East Asia and North America through the Arctic, the role of local people and their needs will barely be visible (the regional Phase 1 of Quintillion's project being an exception to this). Local people and communities are, however, some of the many crucial stakeholders who have the means to promote and advance, or to oppose, these projects. Many of the issues that have caused friction between locals and cable companies in other geographical contexts will materialise in the European High North once the implementation of the Arctic Connect Project is ensured, the planning of the route progresses and negotiations between different stakeholders become more concrete. Yet, the developments observed thus far indicate that Arctic submarine communications cable projects could demonstrate the possibility to reconcile different interests and facilitate dialogue between local stakeholders and companies from different parts of the world with different financial interests. Should this occur, these digital infrastructure projects might be able to serve as benchmark cases for ongoing and future transport infrastructure projects in the Arctic, for which reconciliation will likely prove more difficult.

References

Alaska Broadband Task Force. (2014, October). *A blueprint for Alaska's broadband future*. A report from the Statewide Broadband Task Force. Retrieved from http://www.alaska.edu/oit/bbtaskforce/docs/Statewide-Broadband-Task-Force-Report-FINAL.pdf.

Alaska Department of Commerce, Community, and Economic Development. (2017). *Northern opportunity: Alaska's economic development strategy*. Retrieved from https://northernopportunity.com/wp-content/uploads/2017/10/Final-Statewide-CEDS.pdf.

Alkire, S. (2003). *A conceptual framework for human security*. Working Paper 2. University of Oxford, Centre for Research on Inequality, Human Security and Ethnicity. Retrieved from http://citeseerx.ist.psu.edu/viewdoc/download?doi=10.1.1.580.2805&rep=rep1&type=pdf.

Avgerou, C. (2003). The link between ICT and economic growth in the discourse of development. In M. Korpela, R. Montealegre, & A. Poulymenakou (Eds.), *Organizational information systems in the context of globalization* (pp. 373–386). New York, NY: Springer.

Bauer, M. (2010). Regulation, public policy, and investment in communications infrastructure. *Telecommunications Policy, 34*, 65–79. https://doi.org/10.1016/j.telpol.2009.11.011.

Bennett, M. (2016, May 31). Arctic fiber cable proposed to link Asia and Europe. *Anchorage Daily News*. Retrieved from https://www.adn.com/arctic/article/arctic-fiber-cable-proposed-link-asia-and-europe/2012/01/30/.

Bredbåndsfylket. (2017). *Arctic Connect*. Retrieved from https://www.bbft.no/nyheter/arctic-connect.

Burnett, D., Davenport, T., & Beckman, R. (2014). Why submarine cables? In D. R. Burnett, R. C. Beckman, & T. M. Davenport (Eds.), *Submarine cables: The handbook of law and policy* (pp. 1–15). Leiden and Boston: Martinus Nijhoff Publishers.

Capello, R., & Nijkamp, P. (1996). Telecommunications technologies and regional development: Theoretical considerations and empirical evidence. *The Annals of Regional Science, 30*(1), 7–30. https://doi.org/10.1007/BF01580535.

Carter, L., Burnett, D., & Davenport, T. (2014). The relationship between submarine cables and the marine environment. In D. R. Burnett, R. C. Beckman, & T. M. Davenport (Eds.), *Submarine cables: The handbook of law and policy* (pp. 179–212). Leiden and Boston: Martinus Nijhoff Publishers.

Cinia. (2018a). *The Arctic connect project to power business in northern Finland*. Retrieved from https://www.cinia.fi/en/archive/the-arctic-connect-project-to-power-business-in-northern-finland.html.

Cinia. (2018b). *Investment in Arctic connect boosts the economy in western Uusimaa*. Retrieved from https://www.cinia.fi/en/archive/investment-in-arctic-connect-boosts-the-economy-in-western-uusimaa.html.

Copenhagen Economics. (2017). *Finland's economic opportunities from data centre investments*. A study prepared for Google. Retrieved from https://www.copenhageneconomics.com/dyn/resources/Publication/publicationPDF/5/395/1492669001/copenhagen-economics-2017-finlands-economic-opportunities-from-data-centre-investments.pdf.

Czernich, N., Falck, O., Kretschmer, T., & Woessman, L. (2011). Broadband infrastructure and economic growth. *The Economic Journal, 121*, 505–532. https://doi.org/10.1111/j.1468-0297.2011.02420.x.

Dawn-Hiscox, T. (2018). *Facebook plans third data center in Luleå, Sweden*. DCD. Retrieved from https://www.datacenterdynamics.com/news/facebook-plans-third-data-center-in-luleå-sweden/.

Delaunay, M. (2014). *The Arctic: A new internet highway?* Arctic Yearbook 2014. Retrieved from https://arcticyearbook.com/arctic-yearbook/2014/2014-briefing-notes/112-the-arctic-a-new-internet-highway.

Downing, S. (2017, August 4). Quintillion CEO out, new one in, as project nears completion. *Must Read Alaska*. Retrieved from http://mustreadalaska.com/quintillion-ceo-new-one-project-nears-completion/.

Eskelinen, H., Frank, L., & Hirvonen, T. (2008). Does strategy matter? A comparison of broadband rollout policies in Finland and Sweden. *Telecommunications Policy, 32*, 412–421. https://doi.org/10.1016/j.telpol.2008.04.001.

Falsey, J. L. (2017, May 9). Fiber-optic broadband set to go live in northern Alaska by year's end. *Anchorage Daily News*. Retrieved from https://www.adn.com/business-economy/2017/05/08/quintillion-ceo-fiber-based-broadband-set-to-go-live-in-alaska-by-end-of-the-year/.

Grubesic, T. H., & Mack, E. A. (2016). *Broadband telecommunications and regional development*. London and New York: Routledge.

Gudjonsson, H. (2018, June 27). *Speech at the 3rd top of the world: Arctic broadband summit*. June 27–28, Sapporo, Japan.

Hansen, D. K. (2018, January 23). *Speech at the Arctic Frontiers 2018*. January 21–26, Tromso, Norway.

Joensuu, J. P. (2017, October 17). *Speech at Arctic data cables, digitalization and regional development seminar*. Sapporo, Japan.

Joensuu, J. P. (2018, August 28). Phone interview with J. Saunavaara.

Kang, C. (2017, December 2). Melting Arctic ice makes high-speed Internet a reality in a remote town. *The New York Times*. Retrieved from https://www.nytimes.com/2017/12/02/technology/from-the-arctics-melting-ice-an-unexpected-digital-hub.html.

Klouda, N. (2017, June 14). More rural broadband options coming through fiber, satellites. *Alaska Journal of Commerce*. Retrieved from http://www.alaskajournal.com/2017–06-14/more-rural-broadband-options-coming-through-fiber-satellites#.W7MkhPZuJdg.

Lassila, M. (2018). Mapping mineral resources in a living land: Sami mining resistance in Ohcejohka, northern Finland. *Geoforum, 96*, 1–9. https://doi.org/10.1016/j.geoforum.2018.07.004.

Linden, S. (2017, May 23). Personal interview with J. Saunavaara.

Malecki, E. J., & Wei, H. (2009). A wired world: The evolving geography of submarine cables and the shift to Asia. *Annals of the Association of American Geographers, 99*, 360–382. https://doi.org/10.1080/00045600802686216.

Navy Recognition. (2018). Russian navy to lay fiber optic cables to connect Arctic and far East. *Navy Recognition*. Retrieved from http://www.navyrecognition.com/index.php/news/defence-news/2018/april-2018-navy-naval-defense-news/6161-russian-navy-to-lay-fiber-optic-cables-to-connect-arctic-and-far-east.html.

Nenasheva, M., Bickford, S. H., Lesser, P., Koivurova, T., & Kankaanpää, P. (2015). Legal tools of public participation in the environmental impact assessment process and their application in the countries of the Barents Euro-Arctic region. *Barents Studies, 1*(3), 13–35.

Nilsen, T. (2018, April 24). Russia plans to lay trans-Arctic fiber cable linking military installations. *The Independent Barents Observer*. Retrieved from https://thebarentsobserver.com/en/security/2018/04/russia-slated-lay-military-trans-arctic-fibre-cable.

Paulin, J. (2017, December 7). Fiber-laying ship calls on internet-starved Aleutians. *Alaska Journal of Commerce*. Retrieved from http://www.alaska-journal.com/2017-07-12/fiber-laying-ship-calls-internet-starved-aleutians#.W7boy_ZuJdg.

Pierce, E. (2017, June 14). *Speech at the 2nd top of the world: Arctic broadband summit*. Oulu, Finland.

Quintillion. (2018). *Quintillion fiber system in service*. Retrieved from http://qexpressnet.com/.

Rauscher, K. F. (2010). *ROGUCCI, the report*. Issue 1, rev. 1. IEEE Communications Society. Retrieved from http://www.ieee-rogucci.org/files/The%20ROGUCCI%20Report.pdf.

Rautajoki, T., Sørheim Nilsen, N. K., Wallin, S., & Tuovinen, J. (2017). *Arctic connect—Arctic cooperation for a new digital highway to Asia*. Written statement. Lapland Chamber of Commerce, the Regional Business Association of Tromsø/Troms Chamber of Commerce, Swedish Lapland Chamber of Commerce and Oulu Chamber of Commerce. Retrieved from https://arcticeconomiccouncil.com/wp-content/uploads/2017/06/Arctic-Connect.pdf.

Riipi, M. (2017, June 6). Personal interview with J. Saunavaara.

Rämet, J. (2018, March 5). Personal interview with J. Saunavaara.

Salminen, M., & Hossain, K. (2018). Digitalization and human security dimensions in cybersecurity: An appraisal for the European High North. *Polar Record, 53*(2), 1–11. https://doi.org/10.1017/S0032247418000268.

Sato, Y. (2018, June 27). *Speech at the 3rd top of the world: Arctic broadband summit*. June 27–28, Sapporo, Japan.

Saunavaara, J. (2018). Arctic subsea communications cables and the regional development of northern peripheries. *Arctic and North, 32*. https://doi.org/10.17238/issn2221-2698.2018.32.63.

Sedrak, K. (2018, June 27). *Speech (video) at the 3rd top of the world: Arctic broadband summit*. June 27–28, Sapporo, Japan.

Starosielski, N. (2015). *The undersea network*. Durnham and London: Duke University Press.

Taksami, N. (2018). A survey of Finnish media debates on the Arctic corridor railway planned to connect the silk road and the polar silk road. Briefing Note. In *Arctic Yearbook 2018. Special Section: China & the Arctic* (pp. 118–122). Retrieved from https://arcticyearbook.com/images/yearbook/2018/China-and-the-Arctic/AY2018-Special-Section-Complete.pdf.

The Boston Consulting Group. (2014). *Digital infrastructure and economic development: An impact assessment of Facebook's data center in northern Sweden*. Facebook. Retrieved from http://www.investindalarna.se/app/uploads/2016/11/boston-consulting-group-facebook.pdf.

The State Council Information Office of the People's Republic of China. (2018). *China's Arctic policy*. Retrieved from http://english.gov.cn/archive/white_paper/2018/01/26/content_281476026660336.htm.

Uljas, A. (2018a, June 27). *Speech at the 3rd top of the world: Arctic broadband summit*. June 27–28, Sapporo, Japan.

Uljas, A. (2018b, January 23). *Speech at the Arctic Frontiers 2018*. Tromso, Norway.

Wargo, R., & Davenport, T. (2014). Protecting submarine cables from competing uses. In D. R. Burnett, R. C. Beckman, & T. M. Davenport (Eds.), *Submarine cables: The handbook of law and policy* (pp. 255–280). Leiden and Boston: Martinus Nijhoff Publishers.

Wood, C. (2018, May 15). Alaska partners with nonprofit to mend its K-12 broadband gap. *StateScoop*. Retrieved from https://statescoop.com/alaska-partners-with-nonprofit-to-mend-its-k-12-broadband-gap.

Woolston, K. (2018, June 27). *Speech at the 3rd top of the world: Arctic broadband summit*. June 27–28, Sapporo, Japan.

Yamamoto, T. (2017a, January 27). Personal interview with J. Saunavaara.

Yamamoto, T. (2017b, October 17). *Speech at Arctic data cables, digitalization and regional development seminar*. Sapporo, Japan.

Zojer, G. (2019). The interconnectedness of digitalization and human security in the European high north: Cybersecurity conceptualised through the human security lens. *The Yearbook of Polar Law, 10*, 297–320.

PART IV

Society and Environment

CHAPTER 9

Cybersecurity of Digital Citizens in the Remote Areas of the European High North

Ville Kivivirta, Leena Viinamäki and Arto Selkälä

9.1 Introduction

This chapter focuses on how digital citizens in the remote areas of the European High North experience the "terrain" of cybersecurity by analysing the views of both citizens and the regional public administration. The "terrain" of cybersecurity has been conceptualised in the literature as "part of the milieu through which life is lived and the field of intervention at which the apparatuses of security are directed" (Simon & de Goede, 2015, p. 84). The context of cybersecurity in the remote areas of Finland in which the present study was conducted is such that

V. Kivivirta (✉) · A. Selkälä
University of Lapland, Rovaniemi, Finland
e-mail: ville.kivivirta@ulapland.fi

A. Selkälä
e-mail: arto.selkala@ulapland.fi

L. Viinamäki
Lapland University of Applied Sciences, Rovaniemi, Finland
e-mail: Leena.Viinamaki@lapinamk.fi

M. Salminen et al. (eds.),
Digitalisation and Human Security, New Security Challenges,
https://doi.org/10.1007/978-3-030-48070-7_9

without functional cyberinfrastructure, access to many public services is almost impossible, although at the same time citizens in those areas are accustomed to dysfunctional infrastructure, such as slower internet connections and even occasional power cuts. In Europe, the emphasis on distributed cyberinfrastructure and the dangers of disconnectedness are now referred to by the European Commission (2009) as "peculiar governance challenges" (p. 1). In terms of contemporary security practices, disconnectedness is now said to define danger (Reid, 2006, p. 30).

The present research was conducted as part of the *Digital Citizenship and Accessibility of Services in Rural Areas* research project, funded by the Finnish Ministry of Agriculture and Forestry, which investigated digitalisation in the remote regions of Finland (see Viinamäki et al., 2017). Digitalisation refers to the process of integrating various areas of life into digital information and communication technologies (ICT) infrastructures. Our empirical study focuses on the views of individual citizens and officials acting in their authoritative role. We hope that our research will counterbalance the present discourse, which treats actors in sparsely populated areas as merely objects of development and, sometimes, as a burden to state-centred cybersecurity.

We did not define cybersecurity in advance. Instead, we tested the common definitions of cybersecurity in our empirical analysis (Table 9.1). Rather than defining cybersecurity for the respondents, we approached the issue from the following viewpoints in our surveys: (1) accessibility, including the acceptability and the technical reliability of e-services; (2) equality, including the accessibility of e-services in different regions and among different population groups; (3) infrastructure, including the network infrastructure and personal devices; and (4) qualification, including the ability to use applications, portals and databases. These are all preconditions for the use of e-services, especially in the sparsely populated areas of the European High North. Cybersecurity is thus considered in this article in conjunction with what it means to be a citizen in the currently digitalising European High North.

As shown in Table 9.1, the definitions of cybersecurity vary depending on the context and focus. Some publications refer to it as *cybersecurity*, while others prefer *cyber security* (European Union Agency for Cybersecurity, 2015, pp. 10–12). National cybersecurity strategies do not often address human security issues (Dunn Cavelty, 2014), and many commonly used definitions of the term are either very broad or highlight the technological dimensions. In the Nordic context, for example, in Finland, cybersecurity is often conceptualised as "the desired end state in

Table 9.1 Ten definitions of cybersecurity

Source	*Definition*
University of Maryland University College (2018). ***Introduction to cyber security***. http://umuc.edu/academic-programs/cyber-security/about.cfm	"Cyber security, also referred to as information technology security, focuses on protecting computers, networks, programs, and data from unintended or unauthorized access, change, or destruction. Government agencies, the military, corporations, financial institutions, hospitals, and other groups collect, process, and store a great deal of confidential information on computers and transmit that data across networks to other computers. With the growing volume and sophistication of cyberattacks, ongoing attention is required to protect sensitive business and personal information, as well as safeguard national security"
Techopedia. Home/Dictionary/Tags/Security/Cybersecurity. https://www.techopedia.com/definition/24747/cybersecurity	"Cybersecurity refers to preventative methods used to protect information from being stolen, compromised or attacked. It requires an understanding of potential information threats, such as viruses and other malicious code. Cybersecurity strategies include identity management, risk management and incident management. … Cybersecurity is a very broad category which encompasses numerous hardware and software technologies, and can be applied on any level, including personal, corporate or governmental devices or networks. Passwords are a cybersecurity tool that people encounter nearly every day. Other common cybersecurity tools include: • Anti-virus/anti-malware software; • Software patches; • Firewalls; • Two-factor authentication; • Encryption. A cybersecurity plan is critical for any company with highly sensitive information"

(continued)

Table 9.1 (continued)

Source	*Definition*
The Economic Times. Business. News›Definitions›Security›Cyber Security https://economictimes.indiatimes.com/definition/cyber-security	"**Definition**: Cyber security or information technology security are the techniques of protecting computers, networks, programs and data from unauthorized access or attacks that are aimed for exploitation **Description**: Major areas covered in cyber security are: 1. Application Security; 2. Information Security; 3. Disaster recovery; 4. Network Security. Application security encompasses measures or counter-measures that are taken during the development life-cycle to protect applications from threats that can come through flaws in the application design, development, deployment, upgrade or maintenance. Some basic techniques used for application security are: a. Input parameter validation, b. User/Role Authentication & Authorization, c. Session management, parameter manipulation & exception management, and d. Auditing and logging. Information security protects information from unauthorized access to avoid identity theft and to protect privacy. Major techniques used to cover this are: a. Identification, authentication & authorization of user, b. Cryptography"
Definition of Cybersecurity. Gaps and overlaps in standardization. V1.0. (2015). https://www.enisa.europa.eu/publications/definition-of-cybersecurity	Components constituting the definition of cybersecurity: 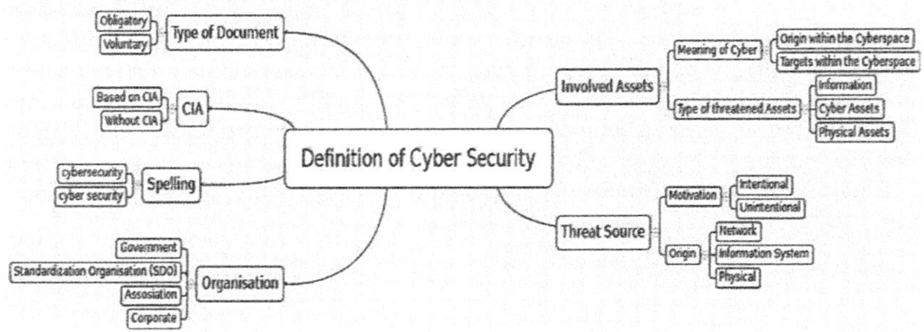

(continued)

Table 9.1 (continued)

Source	*Definition*
International Telecommunication Union—Standardization Sector (ITU-T X 1205). (2018). https://www.itu.int/en/ITU-T/studygroups/2013-2016/17/Pages/cybersecurity.aspx	"Cybersecurity is the collection of tools, policies, security concepts, security safeguards, guidelines, risk management approaches, actions, training, best practices, assurance and technologies that can be used to protect the cyber environment and organization and user's assets. Organization and user's assets include connected computing devices, personnel, infrastructure, applications, services, telecommunications systems, and the totality of transmitted and/or stored information in the cyber environment. Cybersecurity strives to ensure the attainment and maintenance of the security properties of the organization and user's assets against relevant security risks in the cyber environment. The general security objectives comprise the following: •Availability; •Integrity, which may include authenticity and non-repudiation; •Confidentiality"
National Institute of Standards and Technology, U.S. Department of Commerce. Computer Security Resource Center. https://csrc.nist.gov/glossary/term/cybersecurity	"Prevention of damage to, protection of, and restoration of computers, electronic communications systems, electronic communications services, wire communication, and electronic communication, including information contained therein, to ensure its availability, integrity, authentication, confidentiality, and nonrepudiation"
Fischer, E. A., Congressional Research Service. (2016). *Cybersecurity Issues and Challenges: In Brief*. https://fas.org/sgp/crs/misc/R43831.pdf	"A broad and arguably somewhat fuzzy concept, cybersecurity can be a useful term but tends to defy precise definition. It usually refers to one or more of three things: •A set of activities and other measures intended to protect—from attack, disruption, or other threats—computers, computer networks, related hardware and devices software and the information they contain and communicate, including software and data, as well as other elements of cyberspace •The state or quality of being protected from such threats •The broad field of endeavor aimed at implementing and improving those activities and quality"

(continued)

Table 9.1 (continued)

Source	*Definition*
Craigen, D., Diakun-Thibault, N., & Purse, R. (2014). *Defining Cybersecurity*. https://timreview.ca/sites/default/files/article_PDF/Craigen_et_al_TIMReview_October2014.pdf	"Cybersecurity is the organization and collection of resources, processes, and structures used to protect cyberspace and cyberspace-enabled systems from occurrences that misalign de jure from de facto property rights"
Salminen, M., & Hossain, K. (2018). Digitalisation and human security dimensions in cybersecurity: An appraisal for the European High North. *Polar Record, 54*(2), 108–118. https://doi.org/10.1017/s0032247418000268	"Within the rearticulated research framework, cybersecurity may be redefined to integrate human security dimensions, and digitalisation refocused to support development defined by the people and communities themselves"
Hatleback, E. N. (2018). The protoscience of cybersecurity. *Journal of Defense Modelling and Simulation: Applications, Methodology, Technology, 15*(1), 5–12	"The activity of defending the technological assets of some entity from an adversary's attempts to meddle with those assets in some way"

Source Author's own creation

which the cyber domain is reliable and in which its functioning is ensured" (Finland's Cyber Security Strategy [FCSS], 2013, p. 1). Further, some definitions of cybersecurity do underline issues related to human security and local communities (see Salminen & Hossain, 2018; Zojer, 2019).

The present study focuses on digital citizens because, as Schou and Hjelholt (2018) conclude, being able to use digital technologies is a normative component of citizenship these days. From the perspective of social theory, citizenship is a problematic and loaded concept, which is influenced by various and often implicit background assumptions (Dwyer, 2004; Roche, 1987; Selkälä, 2013). There has been an ongoing ideological debate as to who exactly counts as a citizen with full rights and what that their role in society actually is. Citizens, who are also customers, can be conceptualised by dividing citizenship into various levels, with full citizenship requiring specialist knowledge of active participation as a customer (cf. Arnstein, 1969). This means that digital citizens are not merely passive users of services who consume cybersecurity. As such, citizens also fulfil various roles that are being reshaped due to digitalisation.

The remainder of this chapter is structured as follows. First, we discuss empirical and methodological issues. Second, the context of cybersecurity in the European High North is explained in light of official statistics and the research literature. Third, the results of the empirical analysis are presented. Finally, the implications of the results are discussed, especially from the perspective of cybersecurity.

9.2 Research Methods and Empirical Data

We analysed empirical data so as to study various views relating to micro-level cybersecurity. The empirical material was gathered by conducting a large-scale paper-and-pencil survey ($n = 1063$) among citizens in Lapland, Ostrobothnia and North Karelia, excluding the regional capitals. A web-based Webropol questionnaire was also made available to public sector professionals in charge of e-services in order to map their views regarding e-services and to collect their development ideas. These views and ideas were then examined by means of a three-phase gap analysis, which allowed us to compare the views of public service professionals and citizens as customers and e-service users. Additionally, we organised workshops based on the initial results and conducted interviews with the Sami people and third-sector actors.

The research team used data triangulation to acquire a deeper understanding of the digitalisation process currently taking place in Finland

Table 9.2 Triangulation research design

Research data	*N*	*n*	*Designated function in triangulation research design*
Postal survey of citizens from three selected regions of Finland (Ostrobothnia, North Karelia, Lapland) conducted in October 2016	3000	1063	Citizens' comments on the ongoing digitalisation process in everyday life from their own perspective
Web-based questionnaire conducted via Webropol for public sector professionals from January–February 2017	229	48	Public sector professionals' comments on the ongoing digitalisation process from an administrative perspective
Group interview data from North Karelia and Lapland workshop meetings and interview data obtained via e-mail from Ostrobothnia	3	–	Comments of key informants in the regional workshops on the ongoing digitalisation process from their own perspective (e.g. third sector, local and regional authorities)
Statistical data (Eurostat, Statistics Finland)	23	–	Contextual data regarding how citizens utilise digitalisation

Source Author's own creation

(Table 9.2). Our triangulation research design was based on (1) statistical data (macro-level data), (2) group interview data obtained from the North Karelia and Lapland workshop meetings and interview data obtained via e-mail from Ostrobothnia (meso-level data), and (3) questionnaire data collected from citizens and public sector professionals (micro-level data) (Mertens & Hesse-Biber, 2012; Viinamäki & Kilpiäinen, 2015, pp. 15–16) (see Table 9.2).

We used MA multiple constituency evaluation to analyse the survey data (Bamberger, 2012; Kuorilehto, 2014; Vartiainen, 2007). We initially analysed the answers given by the public sector professionals, who were regarded as *service providers*, and the answers of the citizens, who were regarded as *service users*. Based on our results, however, it appears that these roles are not necessarily clear-cut in reality. Nevertheless, it is safe to assume that public sector professionals have gained insight into organising digitalisation at a regional level, while citizens possess empirical knowledge of digitalisation in everyday life. The multiple constituency evaluation design of our research also allowed for a gap analysis to be conducted among service providers and service users.

Table 9.3 Number of participants and the response rates

	Lapland	*Ostrobothnia*	*North Karelia*	*Combined*
Invited Respondents (n)	1000	1000	1000	3000
Participants (n)	371	319	364	1054
Response Rates (%)	37	32	36	35
Participants (Weighted n)	358	309	356	1023

Note Response rates based on RR2-definition of American Association for Public Opinion Research (2008, p. 34)
Source Author's own creation

9.2.1 Data

We utilised the available statistical data and also gathered empirical material by conducting a paper-and-pencil survey ($n=1063$) among citizens in Lapland, Ostrobothnia and North Karelia, excluding the regional capitals. A web-based questionnaire was also made available to public sector professionals in charge of e-services so as to map their views regarding e-services and to collect their development ideas.

9.2.2 Survey Respondents/Sample Units

Citizens from three selected regions of Finland, namely Ostrobothnia, North Karelia and Lapland, participated in the research.

9.2.3 Sampling Procedure

During the first stage, 3000 sample units were evenly allocated to the three selected regions (Table 9.3). During the second stage, a two-phase stratified random sampling with proportional allocation was conducted based on municipality (first phase) and age (second phase) within each regional stratum. The selected stratified sampling procedure and the selected weighting procedure were applied to control the distribution of the final data, especially in terms of the age variable, as younger people are overrepresented in relation to the use of digital services. The sampling frame from which the respondents were selected is maintained by the Population Register Centre of Finland.

9.2.4 *The Representativeness of Non-Weighted Data and Weighting*

We compared the municipality and age distributions of the non-weighted survey data with the corresponding population distributions within each regional sample in order to determine the unit non-response error. We found the Lapland and North Karelia samples to be representative in terms of municipality. However, in the Ostrobothnia sample, the citizens of two municipalities, namely Isokyrö and Laihia, were overrepresented (10–14%), while the citizens of Närpiö, Pedersöre, Uusikaarlepyy and Vöyri were underrepresented (4–7%). Additionally, citizens aged between 18 and 34 years were underrepresented (6–9%) in all the regional samples. The observed unit nonresponse error was reduced by weighting the data within each regional sample by municipality and age (see Heeringa, West, & Berglund, 2010, pp. 104–105; Thompson 2012). Before analysing the results, we will first describe the context of the European High North, which will be particularly useful for those unfamiliar with the region.

9.3 Cybersecurity in the European High North

Although we did not define the term "cybersecurity", the respondents did not answer our surveys in a vacuum. Cybersecurity can be viewed as a macro-level issue, focusing on the key national functions that rely on ICT; as a meso-level issue, focusing on how organisations and institutions deal with cybersecurity; or as a micro-level issue, focusing on how individuals utilise and rely on ICT. The context of the European High North is important because, although the region is remote, sparsely populated and far from the capital cities, ICT and digitalisation are nevertheless playing an important role in the transformation of the region's economy, working life and public sector operations. However, the pace of digitalisation is not the same everywhere, and the process itself is shaped by various geographical and social inequalities (Townsend, Sathiaseelan, Fairhurst, & Wallace, 2013). Still, digitalisation is mostly situated in an urban context, as illustrated by the literature concerning smart and creative cities (Roberts, Beer, Philip, & Townsend, 2017). This raises an important question: how can we use the existing research, which was principally conducted in Western cities and densely populated areas, to analyse the digital citizens of the European High North?

In the prior research, the so-called "digital divide" has been studied among various populations, and the differences between urban and rural areas have been considered (Van Dijk, 2008). Previous studies also addressed the intensity of ICT usage, people's polarised attitudes towards digitalisation (Hodge, Carson, Carson, Newman, & Garrett, 2016) and people's experiences of using e-services in the countryside (Selkälä, Viinamäki, Suikkanen, & Vasari, 2016). Our research focuses on three Finnish regions. Lapland is characterised by sparsely populated areas, the development of joint service points, networked security initiatives and the Arctic dimension. Ostrobothnia is a bilingual rural region that has well-resourced regional authorities and relatively short distances between populated areas when compared to Lapland and North Karelia. North Karelia is also a sparsely populated region, which is characterised by various digitalisation projects that have been conducted with considerable public funding. All three regions have benefitted from national cyberinfrastructure development initiatives. Lapland and North Karelia have also served as test beds for digitalisation, especially in the fields of digital education, e-health and intelligent transport systems. In remote areas, long distances render e-services an attractive option for all citizens, and the Finnish public sector has reduced its physical presence in many parts of Lapland and North Karelia while investing heavily in the development of e-government.

In some parts of the countryside, especially in Lapland and North Karelia, efforts have been made to develop the digital infrastructure, for example, through 5G, so as to guarantee people access to the fixed broadband in towns and along the main roads, which is viewed as important for the local economy (see Map 9.1). The problem is that there are people living outside these locations, which forces one to stray beyond an infrastructure-based narrative when analysing cybersecurity in such areas. The cyberspace where cybersecurity is studied cannot be conceptualised without considering the material infrastructure that supports it (cf. Mitchell, 1995).

Map 9.1 presents the population density,[1] the demographic dependency ratio[2] and the fixed broadband coverage in the Finnish municipalities[3] in 2016. According to the Finnish Communications Regulatory Authority (2017), around 1.74 million households in Finland have access to fixed broadband with a download speed of 30 Mbps or more. This represents 66% of all Finnish households. Among all the Finnish regions, people in Uusimaa and Ostrobothnia enjoy the best internet conditions, with more than 75% of households having access to fixed broadband with a speed of at least 30 Mbps (Map 9.1). In Ostrobothnia, the provision of 100-Mbps broadband is equally extensive across the region. In Uusimaa, however, less than half of all households have access to 100-Mbps broadband, which falls far below the coverage of 30-Mbps broadband (Finnish Communications Regulatory Authority, 2017).

The Nordic Region is already one of the most highly digitised parts of the world, although there is some variation in the levels of broadband accessibility in the different regions of Finland (Nordic Council of Ministers, 2018). Compared to other European Union (EU) countries, Finns also possess better digital skills, while their use of internet services is widespread to such an extent that online banking has been almost universally adopted in the country. The online provision of Finnish public services is very highly rated (Digital Economy and Society Index, 2017). Table 9.4 shows the realisation of digital citizenship and digital activity in the Nordic countries (excluding Iceland) and in the 28 EU countries.

Aside from analysing the digitalisation process at the national level, it is also important to consider the regional and local levels. Map 9.1 shows the preconditions for the use of e-services, which vary from one municipality to another, partly depending on population density and structure, with both contributing to internet accessibility, equality and

[1] The population density gives the population number per square kilometre. Hence, the population density equals the population divided by the total land area (km^2). Data obtained from Sotkanet.fi Statistics and Indicator Bank© National Institute for Health and Welfare (2005–2018).

[2] The demographic (or population) dependency ratio is the number of people aged under 15 and over 64 per hundred working-age people aged 15–64. The greater the number of children and/or retirement-age people, the higher the dependency ratio. Data obtained from Sotkanet.fi Statistics and Indicator Bank© National Institute for Health and Welfare (2005–2018).

[3] Data obtained from Finnish Communications Regulatory Authority (2017).

Table 9.4 Digital citizenship and digital activity in the nordic countries and the 28 EU[a]

	28 EU Countries	*Denmark*	*Finland*	*Sweden*	*Norway*
Connection to the internet and computer use in 2016					
Level of internet access (percentage of households)	85	94	92	94*	97
Type of connection to the internet: broadband (percentage of households)	83	92	91	89*	96
Type of connection to the internet: mobile broadband (percentage of households)	41	33	85	59*	64
Internet use in 2016					
In the last 12 months (percentage of individuals)	84	97	94	95*	98
Daily (percentage of individuals)	71	89	85	85*	92
Internet activities in 2016					
Online banking (percentage of individuals)	49	88	86	83*	91
Finding information about goods and services (percentage of individuals)	66	86	85	83*	88
Sending/receiving e-mails (percentage of individuals)	71	93	85	88*	92
Activities via the internet not performed due to security concerns in 2015 (latest available statistical year)					
Prevented individual from ordering or buying goods or services for private use (percentage of individuals who used the internet within the last year)	19	22	26	34	27
Prevented individual from carrying out banking activities such as account management (percentage of individuals who used the internet within the last year)	18	12	7	15	7
Prevented individual from providing personal information to online communities for social and professional networking (percentage of individuals who used the internet within the last year)	29	32	42	54	46
Prevented individual from communicating with public services and administration (percentage of individuals who used the internet within the last year)	10	7	5	16	11

(continued)

Table 9.4 (continued)

	28 EU Countries	*Denmark*	*Finland*	*Sweden*	*Norway*
E-government activities of individuals via websites in 2016					
Interaction with public authorities (percentage of individuals who used the internet within the last year)	58	91	87	82*	87
Obtaining information from public authorities' websites (percentage of individuals who used the internet within the last year)	50	88	82	78*	79
Downloading official forms (percentage of individuals who used the internet within the last year)	35	55	67	48*	63
Submitting completed forms (percentage of individuals who used the internet within the last year)	34	73	64	50*	63

Note * = break in time series. There were no statistical data available from the Republic of Iceland

[a]*Source* Eurostat (n.d.) statistics from years 2015 or 2016

Source Author's own creation

infrastructure through market mechanisms (Official Statistics of Finland, 2018). Closer scrutiny of the municipalities shows that significant differences can be found between even neighbouring municipalities. Some areas, such as Utsjoki (98% access to 30-Mbps broadband) in the north and Valtimo and Rautavaara in the eastern part of Finland, which have deliberately championed fast broadband access, have achieved their goals through participation in the national broadband strategy and use of the associated state aid. Moreover, given that the Finnish municipal system is constructed around strong, autonomous municipalities, another explanation for the high percentage of fast broadband access concerns municipal attractiveness.

According to official statistics, digital activity in the Nordic countries was more intensive than the average digital activity seen in the 28 EU countries in 2016 (Table 9.4). In particular, the intensity of internet use, internet activities and individuals' e-government activities via websites was greater in the Nordic countries than in the other 28 EU countries. For example, the use of online banking was most common in Norway, where 91% of the population reported using online bank services. By comparison, the corresponding average figure for the 28 EU countries was only 49%. The shunning of internet activities due to security concerns varies according to the type of activity. In both the Nordic countries and the 28 EU countries, citizens were found to be most cautious about submitting personal information to online communities during social and professional networking and when purchasing goods or services.

Overall, the European digital infrastructure can be considered sufficient for digital citizens, although the increasing digitalisation also challenges us to advance cybersecurity. In fact, the Nordic countries can be seen as forerunners when it comes to e-services and awareness of cybersecurity issues. Yet, interacting with public authorities and obtaining information via their websites were the most popular digital activities among all EU citizens.

In 2001, the Organisation for Economic Co-operation and Development (OECD) defined the term "digital divide" as "the gap between individuals, households, businesses and geographic areas at different socio-economic levels with regard both to their opportunities to access ICT and to their use of the Internet for a wide variety of activities" (p. 5). On the one hand, the gap refers to the division between those who have access to computers and the internet and those who do not, while on the other hand, it refers to the diverse types and levels of internet use, motivation and skills. Clearly, a digital divide exists between urban and remote areas (European Parliament, 2015; Eurostat, 2010, 2014; Findahl, 2013; IKT-Einsatz in Haushalten, 2012; (N)ONLINER Atlas, 2012; Philip, Cottrill, Farrington, Williams, & Ashmore, 2017; Statistics Sweden, 2012; Tilastokeskus, 2015).

Public administration research has highlighted the increasingly fragmented and inter-organisational context of the provision of public services (Haveri, 2006). Citizens play a role in this provision as consumers as well as customers, but they are also co-producers and co-developers of services related to digital governance. The role of citizens goes hand in hand with the service-dominant approach that privileges co-production (Osborne, Radnor, & Nasi, 2012). The implementation of the service-dominant approach has faced many difficulties and, from the perspective of co-production, public service delivery is not even possible without demanding digital citizens (Osborne, Radnor, & Strokosch, 2016). While the public management literature has highlighted the importance of networked governance, the governance literature in the social sciences field has also acknowledged the importance of resilience, as adaptation and flexibility represent a way to respond to shocks, requiring a mode of living in which one is prepared for unknown risks (Duffield, 2012).

Cybersecurity is not always experienced in the cities of the European High North. This shifts the focus away from the polis and the nation state and towards a network or association of various actors, thereby changing the traditional conceptualisations used when studying communities. If remote regions do not possess the required information infrastructure and resources and/or the necessary digital skills, their cybersecurity may be compromised, which may in turn lead to problems in relation to digital governance (Viinamäki et al., 2017). Furthermore, cybersecurity issues can hamper the achievement of policy goals. Today, technologies such as e-health require citizens to access public services via web portals. Furthermore, since both customers and practitioners use many databases, the complexity of cybersecurity has been multiplied (Filkins, 2014; KPMG, 2015).

In Finland, regional authorities play a significant role in managing and allocating funding for the development of information infrastructure and digital services. They co-ordinate cybersecurity by applying national strategies locally and ensuring that legislative standards are met.

Traditionally, the development of e-government has mostly been guided by the internal interests of the public sector as well as the available technological opportunities (Verdegem & Verleye, 2009), while cyberspace has been viewed as merely an environment for global digital telecommunications (Deibert 2017; International Telecommunication Union, 2014). However, in the contemporary digital world, governance takes place in the multi-agent settings of the infosphere, where citizens are increasingly seen as inforgs or natural agents that exist alongside artificial agents (Floridi, 2014). Therefore, to study the actors involved in contemporary digital capitalism, one has to acknowledge that the agent arrives with markets as networks, thereby making agency a collective and distributed sociotechnical arrangement (Callon, 2008). This networked context is also characteristic of the European High North, in part because a physical presence has been replaced by a digital presence in many public services.

9.4 Analysing Micro-Level Cybersecurity

We chose to focus on the active users of digital services because their importance has increased in the networked, service-based economy of remote regions. Although we too live in the European High North, our analysis is based on empirical research. According to the Malcovich Bias (Glusman, 2010), there is a tendency to assume that everyone uses technology in the same way that we do. We aimed to avoid this tendency by examining statistics and by analysing the micro-level phenomena that reveal the terrain of cybersecurity. In addition, we sought to investigate the views of public sector professionals in their authoritative roles, as they define the field of intervention towards which the apparatuses of security might be directed at a regional level. Although contemporary public administration itself rarely builds the services that it provides, our survey did not include the views of the private corporations that are typically consulted during the construction of public digital services.

Some citizens warmly welcome e-services, as they can be used regardless of time and space, while others use them only because they have to. Many services with public funding require that independent conditions be taken into consideration, and this is particularly true of e-services (e.g. Henman, 2010; Homburg, 2008). Internet user intensity is dependent on the reliability and speed of the network connection. When choosing a service channel, citizens evaluate both time and cost savings. Some public e-services also require a fast internet connection. Our questionnaire contained the following background question (column variable): "*How do you characterise the internet connection that you primarily use (i.e. reliability and speed)?*", while the leading question concerning patterns of use was: "*Have you used the internet in the last 12 months to perform the following?*" (Table 9.5).

People who actively use multiple e-services can be viewed as active digital citizens. Among our respondents, 97.2% had used internet banking, 89.9% had done internet shopping and 89.8% had read online magazines. These active digital citizens stated that the reliability and speed of their internet connection was at least good (90.0–97.9%). More than half of the respondents who had not used many e-services considered the reliability and speed of their internet connection to be poor (58.3–76.0%). Hence, it appears that there is a connection between the reliability and speed of the available internet connection and the internet qualification ratings. Additionally, the extent of e-service use (short- or long-term use) affected the requirements set for the internet connection, making e-learning and remote work practically impossible for those without a reliable internet connection. In terms of remote work and e-learning, it is essential that the internet connection is fast, reliable and secure. Such activities, as well as internet banking, involve the transmission of confidential data concerning the private activities and interests of individuals. From the perspective of the labour market and the e-learning market, the credibility of the information society depends on private cybersecurity (Zojer, 2019). Operational reliability and internet use patterns are also linked to citizens' internet qualification ratings as well as to the available internet infrastructure.

In terms of performance, the respondents were asked, "*How do you characterise the internet connection that you primarily use (i.e. reliability and speed)?*" and the officials were asked, "*How do you characterise the internet connections (i.e. reliability and speed) in the region where your organisation operates in terms of how it enables e-services?*" (Table 9.6).

The citizen respondents to the survey considered that their internet connections were at least fair. More than half of them (51.3%) considered their internet connection to be good, while less than one-third (27.0%) considered it to be very good and less than one-fifth (17.8%) considered it to be excellent. None of the officials considered the internet connections to be only poor or fair. The respondents' views of the benefits and problems of e-services were based on the functioning of

Table 9.5 Relationship of network reliability and speed with internet use patterns

	Network Speed and Reliability							
	Poor		*Fair*		*Good*		*Total*	
	n	%	*n*	%	*n*	%	*n*	%
Remote work, $\chi^2=0.665$, p=n.s.								
No	18	75.0	121	67.6	294	69.8	433	69.4
Yes	6	25.0	58	32.4	127	30.2	191	30.6
Total	24	100	179	100	421	100	624	100
Voice communication, $\chi^2=14.933$, $p<0.01$								
No	18	78.3	105	56.1	208	44.8	331	49.1
Yes	5	21.7	82	43.9	256	55.2	343	50.9
Total	23	100	187	100	464	100	674	100
Writing on forums, $\chi^2=3.857$, p=n.s.								
No	19	82.6	114	62.3	278	62.6	411	63.2
Yes	4	17.4	69	37.7	166	37.4	239	36.8
Total	23	100	183	100	444	100	650	100
Listening or downloading music, $\chi^2=30.614$, $p<0.001$								
No	14	56.0	53	26.2	83	15.9	150	20.0
Yes	11	44.0	149	73.8	439	84.1	599	80.0
Total	25	100	202	100	522	100	749	100
Downloading software, $\chi^2=34.009$, $p<0.001$								
No	18	78.3	67	35.4	116	24.9	201	29.7
Yes	5	21.7	122	64.6	349	75.1	476	70.3
Total	23	100	189	100	465	100	677	100
Searching for a place to study, $\chi^2=1.597$, p=n.s.								
No	19	76.0	126	70.4	325	75.2	470	73.9
Yes	6	24.0	53	29.6	107	24.8	166	26.1
Total	25	100	179	100	432	100	636	100
Using cloud services, $\chi^2=6.078$, $p<0.05$								
No	17	73.9	87	46.8	224	50.7	328	50.4
Yes	6	26.1	99	53.2	218	49.3	323	49.6
Total	23	100	186	100	442	100	651	100
E-mail, $\chi^2=9.080$, $p<0.05$								
No	4	12.1	7	3.0	16	2.7	27	3.2
Yes	29	87.9	223	97.0	576	97.3	828	96.8
Total	33	100	230	100	592	100	855	100

(continued)

Table 9.5 (continued)

	Network Speed and Reliability							
	Poor		Fair		Good		Total	
	n	%	*n*	%	*n*	%	*n*	%
Buying and/or selling goods, $\chi^2 = 9.052$, $p < 0.05$								
No	4	12.9	32	15.2	43	7.9	79	10.1
Yes	27	87.1	179	84.8	499	92.1	705	89.9
Total	31	100	211	100	542	100	784	100
Finding information (note: expected frequencies below 1)								
No	3	9.4	4	1.7	16	2.7	23	2.7
Yes	29	90.6	228	98.3	582	97.3	839	97.3
Total	32	100	232	100	598	100	862	100
Looking for work, $\chi^2 = 2.997$, p = n.s.								
No	14	58.3	105	56.1	218	49.2	337	51.5
Yes	10	41.7	82	43.9	225	50.8	317	48.5
Total	24	100	187	100	443	100	654	100
Reading online magazines, $\chi^2 = 2.990$, p = n.s.								
No	6	19.4	21	9.5	57	10.0	84	10.2
Yes	25	80.6	200	90.5	515	90.0	740	89.8
Total	31	100	221	100	572	100	824	100
Online banking, $\chi^2 = 27.405$, $p < 0.001$								
No	6	17.1	6	2.5	13	2.1	25	2.8
Yes	29	82.9	231	97.5	602	97.9	862	97.2
Total	35	100	237	100	615	100	887	100
E-learning, $\chi^2 = 4.909$, p = n.s.								
No	18	69.2	96	52.5	265	60.8	379	58.8
Yes	8	30.8	87	47.5	171	39.2	266	41.2
Total	26	100	183	100	436	100	645	100
Using social networking services (e.g. Facebook, Twitter), $\chi^2 = 6.271$, $p < 0.05$								
No	13	48.1	67	33.5	141	28.0	221	30.3
Yes	14	51.9	133	66.5	362	72.0	509	69.7
Total	27	100	200	100	503	100	730	100

Source Author's own creation

the internet connections as well as on the network infrastructure in the remote areas.

In terms of the internet infrastructure and accessibility, the speed and reliability of internet connections are key factors when citizens choose whether or not to use e-services (Selkälä et al., 2016). If the use of a given

Table 9.6 Assessing the regional network speed and reliability

Reliability of Network	*Citizens*		*Officials*	
	n	%	*n*	%
Very poor	3	0.4	–	–
Poor	32	3.6	–	–
Fair	246	27.0	14	29.2
Good	466	51.3	15	31.3
Excellent	161	17.8	19	39.6
Total	909	100	48	100

Source Author's own creation

service is so complicated that it requires thematic expertise, a great deal of time and electronic data transfer (e.g. sending an e-mail attachment), citizens may still prefer face-to-face customer service. We studied the views of both citizens and officials regarding the benefits and preconditions of ICT use in the European High North. In terms of the preconditions, the citizens were asked, "*What do you think about the following claims considering the preconditions for the use of e-services?*". The officials were asked, "*Evaluate the significance of the following factors for the users of e-services*". This allowed us to utilise a gap analysis, as shown in Table 9.7.

Both the citizens and the officials who responded to our surveys stressed the importance of reliable internet connections. The citizens (95.1%) and the officials (100%) strongly agreed that it is important to have a reliable internet connection that functions without technical difficulties when using e-services. Further, the citizens (83.3%) and the officials (87.5%) strongly agreed that e-services are user-friendly. Most of the citizens (80.1%) and officials (89.6%) also strongly agreed that e-services are useful because they are available regardless of the time and place.

User-friendliness and the possibility of using e-services regardless of the time and place were also considered important. This type of feedback may serve as a valuable indicator when addressing issues concerning internet equality and qualification (see European Commission, 2012, 2015). The citizens were generally more critical than the officials with regards to the benefits of e-services and, unlike the officials, some of them disagreed or strongly disagreed with the statements in our survey. The difference between the public sector professionals and the citizens is likely due to their different societal positions. However, digitalisation challenges not only the content of public services, but also the actor roles

Table 9.7 Benefits and preconditions of ICT use

Citizens'												
Statement	*Strongly Agree*		*Agree*		*Neither Agree nor Disagree*		*Disagree*		*Strongly Disagree*		*Total*	
	n	%	*n*	%	*n*	%	*n*	%	*n*	%	*n*	%
Reliability of the internet connection without technical difficulties is important when using e-services	844	95.1	35	3.9	6	0.7	1	0.1	1	0.2	888	100
User-friendliness is good when using e-services	753	83.3	137	15.2	9	1.0	3	0.4	2	0.3	904	100
E-services are good because they work regardless of the time and place	721	80.1	137	15.2	30	3.3	9	1.0	4	0.4	901	100
Officials'												
Statement	*Strongly Agree*		*Agree*		*Neither Agree nor Disagree*		*Disagree*		*Strongly Disagree*		*Total*	
	n	%	*n*	%	*n*	%	*n*	%	*n*	%	*n*	%
Reliability of the internet connection without technical difficulties is important when using e-services	48	100	–	–	–	–	–	–	–	–	48	100
User-friendliness is good when using e-services	42	87.5	6	12.5	–	–	–	–	–	–	48	100
E-services are good because they work regardless of the time and place	43	89.6	4	8.3	1	2.1	–	–	–	–	48	100

Source Author's own creation

of both citizens and public officials. In addition, the citizens' behaviour can be viewed as contrary to the dominant cyberdiscourse, which is based on officials' or businesses' interests and which presupposes that citizens ensure their own cybersecurity when using e-services (Finnish Communications Regulatory Authority, 2018).

We also studied the views of citizens and officials regarding the features and impacts of e-services. The citizens were asked, "*What do you think about the following statements concerning the use of e-services?*" and the officials were asked, "*What do you think about the following statements?*". Again, a gap analysis regarding the views of the citizens and the officials was performed (see Table 9.8).

The citizen respondents were generally more critical than the officials with regards to the features and impacts of e-services. For instance, one-third of the citizens (35.1%) strongly agreed that the use of e-services had made the use of public services easier in their home municipality, while half of the officials (50%) strongly agreed that the use of e-services had made the use of public services easier for citizens in their municipality. More than half of the citizens (57.1%) and most of the officials (70.8%) strongly agreed that cybersecurity is an important factor when a citizen chooses whether to use an e-service or visit a physical service point. Three-quarters of the citizens (77%) and four-fifths of the officials (81.3%) strongly agreed that user-friendliness is an important feature of e-services for the disabled (e.g. people with a visual impairment or reduced mobility). Finally, one-third of the citizens (35.9%) strongly agreed that e-services have made their life easier in their home municipality, while half of the officials (50%) strongly agreed that e-services have made the lives of citizens easier.

The citizens' and officials' views regarding security were in accordance with the views of both groups on the possibilities and preconditions associated with e-services, although the officials were more positive about the possibilities. The gaps between the views of the citizens and the officials were the smallest in the replies to the statement "*It is important to pay attention to user-friendliness in relation to the disabled*" and the largest in the replies to the statements involving factors that affect living in the home municipality. One major precondition for the use of e-services is identification. The citizens and the officials had differing views on issues related to identification, perhaps because citizens use their own computer and, therefore, have a personal interest in them, while officials are only

Table 9.8 Views of citizens and officials regarding the features and impacts of E-Services

Citizens'												
Statement	*Strongly Agree*		*Agree*		*Neither Agree nor Disagree*		*Disagree*		*Strongly Disagree*		*Total*	
	n	*%*	*n*	*%*	*n*	*%*	*n*	*%*	*n*	*%*	*n*	*%*
Use of e-services have made the use of public s ervices easier in my home municipality	294	35.1	366	43.7	114	13.6	30	3.6	34	4.1	838	100
Cybersecurity is an important factor when choosing between an e-service and a visit to a physical service point	493	57.1	227	26.3	102	11.8	19	2.2	23	2.6	864	100
User-friendliness for those with a disability (e.g. people with a visual impairment or reduced mobility) is important when offering e-services	671	77.0	136	15.6	39	4.5	12	1.4	13	1.5	871	100
E-services have made my life easier in my home municipality	306	35.9	338	39.7	148	17.3	25	2.9	35	4.2	852	100
Officials'												
Statement	*Strongly Agree*		*Agree*		*Neither Agree nor Disagree*		*Disagree*		*Strongly Disagree*		*Total*	
	n	*%*	*n*	*%*	*n*	*%*	*n*	*%*	*n*	*%*	*n*	*%*
E-services have made the use of public services easier for citizens	24	50.0	24	50.0	–	–	–	–	–	–	48	100
Cybersecurity is an important factor when citizens choose between an e-service and a visit to a physical service point	34	70.8	10	20.8	2	4.2	2	4.2	–	–	48	100
It is important to pay attention to user-friendliness in relation to users with a disability (e.g. people with a visual impairment or reduced mobility)	39	81.3	9	18.8	–	–	–	–	–	–	48	100
E-services have made the lives of citizens easier	24	50.0	23	47.9	1	2.1	–	–	–	–	48	100

Source Author's own creation

Table 9.9 The greatest difficulties associated with E-Services

Citizens	*Officials*
• "Remembering passwords, complicated logging into the services, slow websites" • "Information security issues; how to make sure that information is secure and confidential, hacking and cyberattacks seem to happen all the time" • "Connections do not work in the countryside (work slowly or not at all)"	• "Actors have services that work with different principles. Different identification types, for instance" • "Producing the service in a customer-oriented way, understandably, securely/safely. Digital services should be technically easy to use" • "Functional network connections and a secure electrical grid. Ensuring information security and the related issues together with the producer"

Source Author's own creation

liable for their actions in the workplace (see also European Commission, 2010, pp. 37–38; Lindgren & Jansson, 2013).

Regarding the greatest difficulties encountered when using e-services, the citizens were asked, "*What is the most difficult aspect of using e-services?*", while the officials were asked, "*What is the most difficult aspect of e-services from the perspective of the producer of the service?*". These difficulties were related to cybersecurity and the replies were examined by means of gap analysis (see Table 9.9).

Both the citizens and the officials stressed the importance of security issues when using digital gateways, internet portals, online portals, web portals, etc. In addition, they mentioned internet accessibility, infrastructure and qualification issues as the biggest difficulties associated with the use of e-services. The citizens also brought up the urban–rural dimension of e-service use, stating that in the countryside there are fewer internet access providers than there are in the urban areas because of the lower population density and the associated market mechanisms. The officials pointed out the lack of connectivity between the different governmental information systems. The officials also (at least officially) stated that they were confident that e-service reforms would bring about cost savings. However, the citizens in the Finnish parts of the European High North seemed to be slightly more doubtful about security issues when using e-services, especially in terms of information security (Buess, Iselin, & Bieri, 2017; *Deutsche Institut für Vertrauen und Sicherheit im Internet* [DIVSI], 2013; eGovernment MONITOR, 2017).

9.5 Conclusions

We have now mapped the "terrain" of cybersecurity in the European High North using statistics and empirical analyses. Based on our results, people in the European High North believe that they can use e-services and behave in ways that promote cybersecurity, while the role of public e-service providers is considered to involve guaranteeing both the infrastructure and the services that support the formation of digital citizenship. The frontline operators in cyberspace are businesses (Deibert & Rohozinski, 2010), but the societies of the European High North cannot be analysed without considering the changing landscape of governance and the role of citizens in that landscape. Businesses regard citizens as customers and labour, although overemphasising the role of citizens as customers narrows the perspective: citizens play a wider societal role that the customer concept is capable of capturing (Stout, 2010).

We argue that cybersecurity should be redefined to include the human security dimension, recognising that threats to human beings may arise from a number of sources (Buzan & Hansen, 2009). The definitions of cybersecurity that underline the human security dimension as well as the local community context are in line with our results (see Salminen & Hossain, 2018; Zojer, 2019). Our analysis also shows that the personal circumstances of the internet user, including their socio-economic background, gender, level of education, employment status, age and place of residence, represent basic components of individual cybersecurity behaviour. It also appears that the cybersecurity of citizens in the European High North is connected to the preconditions for e-service usage. Accessibility, equality, infrastructure and qualification are all necessary parts of e-service usage and cybersecurity (Fig. 9.1). However, it is important to note that they vary by municipality and population group (including people's socio-economic background, gender, level of education and age). These preconditions all affect the ways in which citizens experience the importance of cybersecurity, its requirements and the level of know-how needed to achieve it.

The role of citizens and the context shaping citizenship are also changing. The citizenship literature often highlights the role of the space of assemblage (Ong, 2006). Therefore, it is evident that cyberspace plays an important role in the formation of citizenship in the European High North. Fustel de Coulanges (2001 / 1864) made a basic distinction between the *civitas* and the *urbs*, defining *civitas* as the religious and

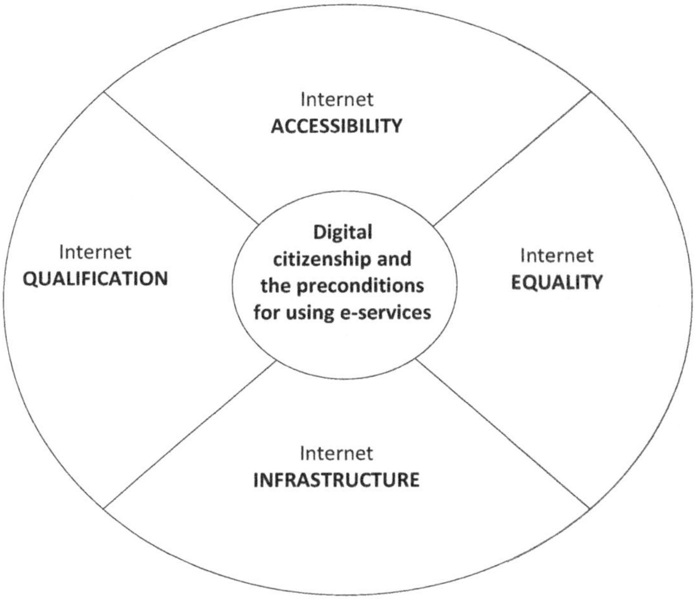

Fig. 9.1 The preconditions for using e-services (*Source* Author's own creation)

political association of families and tribes, and *urbs* as the place of assembly, the dwelling place and the sanctuary of this association. As such, the *civitas* is an association with shared social organisation and government, but it is not necessarily related to any particular place. The *urbs*, however, requires a specific site, typically an urban settlement of the *civitas*, with its own foundation rituals and organisational layout. What is noteworthy here is that taking care of the security of the *urbs* differs from providing security for the *civitas*. In the European High North, the *urbs* can be very different when compared to the typical Western city where digitalisation research is commonly conducted, although standardised digital infrastructure together with the *civitas* nevertheless exist. Due to the importance of assemblage in relation to citizenship formation, we argue that the digitalisation of everyday life produces types of governmentality whereby the role of the digital citizen is emphasised, both as a co-producer and as a co-developer of public services.

Our empirical analysis shows that the gap between citizens and officials in regional public administration is smaller than originally

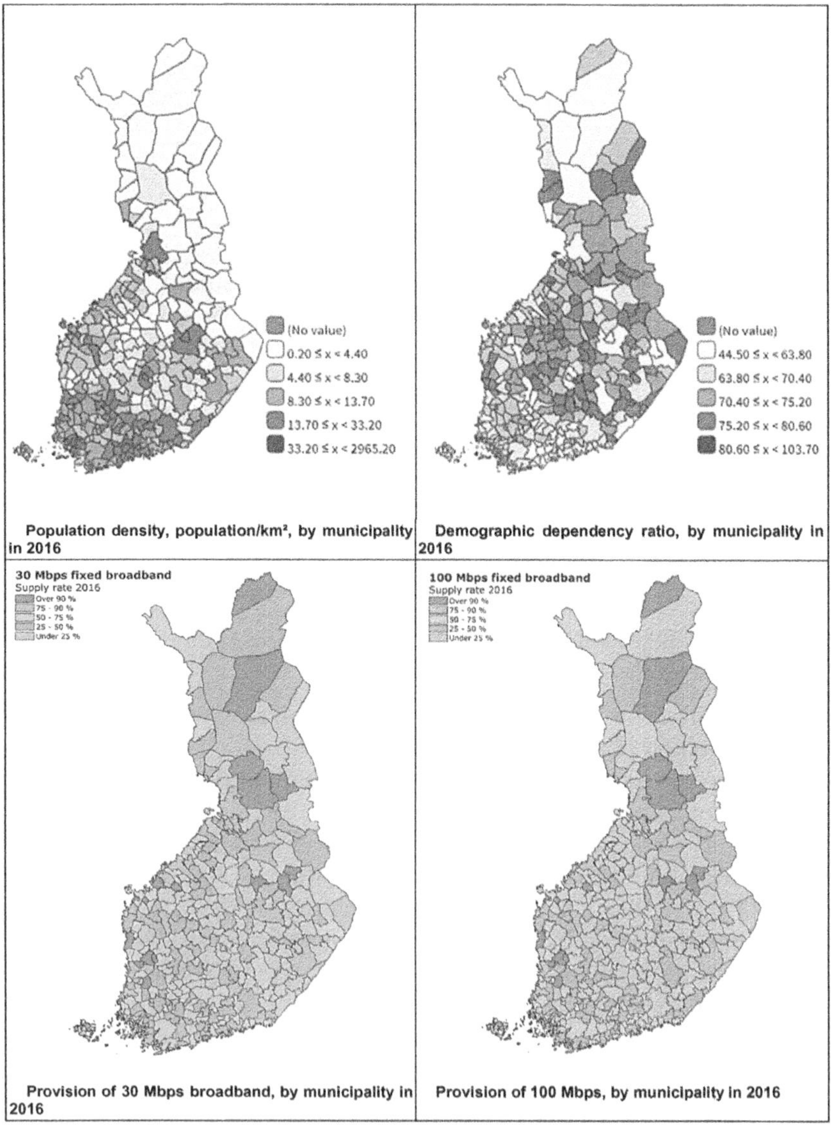

Map 9.1 Population density, population/km^2, demographic dependency ratio and fixed broadband coverage in the Finnish municipalities in 2016

anticipated. This may be because, in the European High North, regional public administration is still relatively close to the citizens, with the officials themselves living and working in the region. Still, based on our gap analysis, the citizens were slightly more critical towards both the reliability of the ICT infrastructure and the benefits of e-services than the officials were. From the perspective of the e-service providers, technical reliability and firewall security form the foundation necessary for citizens to behave in ways that promote cybersecurity and to increase trust in e-service usage. When digital citizens in remote areas face problems with cybersecurity, the social acceptability of e-services may decrease. Today, digital citizens are expected to understand what cybersecurity means and how they should behave in order to avoid problems related to it. To mitigate the risks of marginalisation, the cybersecurity challenges that citizens encounter in remote areas should be further addressed. There is also room for more empirical research concerning the human security dimensions of cybersecurity.

References

American Association for Public Opinion Research (AAPOR). (2008). *Standard definitions: Final dispositions of case codes and outcome rates for surveys* (5th ed.). Lenexa, KS: AAPOR.

Arnstein, S. (1969). A ladder of citizen participation. *Journal of the American Institute of Planners, 35*(4), 216–224.

Bamberger, M. (2012). *Introduction to mixed methods in impact evaluation*. Impact Evaluation Notes 3. Retrieved from https://www.interaction.org/resources/training/guidance-note-3-introduction-mixed-methods-impact-evaluation.

Buess, M., Iselin, M., & Bieri, O. (2017). *Nationale E-Government-Studie*. E-Government in der Schweiz aus Sicht der Bevölkerung, der Unternehmen und der Verwaltung. Demo SCOPE AG/Interface Politikstudien Forschung Beratung GmbH, Adligenswil/Luzern. Retrieved from https://www.egovernment.ch/de/dokumentation/controll#nationale-e-government-studie-2017.

Buzan, B., & Hansen, L. (2009). *The evolution of international security studies*. Cambridge: Cambridge University Press.

Callon, M. (2008). Economic markets and the rise of interactive agencements: From prosthetic agencies to habilitated agencies. In T. Pinch & R. Koray (Eds.), *Living in a material world: Economic sociology meets science and technology studies* (pp. 29–56). Cambridge, MA: MIT Press.

Craigen, D., Diakun-Thibault, N., & Purse, R. (2014). *Defining cybersecurity*. Retrieved from https://timreview.ca/sites/default/files/article_PDF/Craigen_et_al_TIMReview_October2014.pdf.

Deibert, R. (2017). Cyber security. In M. D. Cavelty & T. Balzacq (Eds.), *Routledge handbook of security studies* (2nd ed., pp. 1725–1782). Abingdon: Routledge.

Deibert, R., & Rohozinski, R. (2010). Liberation vs. control in cyberspace. *Journal of Democracy, 21*(4), 42–57.

Deutsche Institut für Vertrauen und Sicherheit im Internet (VIVSI). (2013). *DIVSI Entscheider-Studie zu Vertrauen und Sicherheit im Internet.* (German Institute for trust and security on the Internet (VIVSI). DIVSI decision-study on trust and security on the Internet.) Retrieved from https://www.divsi.de/wp-content/uploads/2013/07/DIVSI_Entscheiderstudie-130205-Druckfassung.pdf.

Digital Economy and Society Index. (2017). *Finland*. Retrieved from https://ec.europa.eu/digital-single-market/en/news/digital-economy-and-society-index-desi-2017.

Duffield, M. R. (2012). Challenging environments: Danger, resilience, and the aid industry. *Security Dialogue, 43*(5), 475–492.

Dunn Cavelty, M. (2014). Breaking the cyber-security dilemma: Aligning security needs and removing vulnerabilities. *Science and Engineering Ethics, 20*(3), 701–715.

Dwyer, P. (2004). *Understanding social citizenship: Themes and perspectives for policy and practice*. Bristol: Policy Press.

eGovernment MONITOR. (2017). *Nutzung und Akzeptanz digitaler Verwaltungsangebote—Deutschland, Österreich und Schweiz im Vergleich* (Use and acceptance of digital administration offers—Germany, Austria and Switzerland in comparison). Retrieved from https://www.egovernment.ch/de/dokumentation/controll.

European Commission. (2009, June 17). *Workshop on the establishment of a European public-private partnership for resilience*. Brussels.

European Commission. (2010). *User expectations of a life events approach for designing e-Government services. Final Report*, EU. Retrieved from http://ec.europa.eu/digital-agenda/sites/digital-agenda/files/smart2009-0075.pdf.

European Commission. (2012). *Proposal for a Regulation of the European Parliament and of the Council on electronic identification and trust services for electronic transactions in the internal market*. COM(2012) 238 final. 2012/0146 (COD). Retrieved from http://eurlex.europa.eu/LexUriServ/LexUriServ.do?uri=COM:2012:0238:FIN:en:PDF.

European Commission. (2015). *A digital single market strategy for Europe*. COM/2015/0192 final. Retrieved from https://eur-lex.europa.eu/legal-content/EN/TXT/?uri=COM%3A2015%3A192%3AFIN.

European Parliament. (2015). *Bridging the digital divide in the EU.* Retrieved from http://www.europarl.europa.eu/RegData/etudes/BRIE/2015/573884/EPRS_BRI(2015)573884_EN.pdf.

European Union Agency for Cybersecurity. (2015). *Definition of cybersecurity—Gaps and overlaps in standardization.* Retrieved from https://www.enisa.europa.eu/publications/definition-of-cybersecurity.

Eurostat. (2010). *Eurostat Jahrbuch der Regionen 2010* (Eurostat regional yearbook 2010). Retrieved from http://ec.europa.eu/eurostat/documents/3217494/5727349/KS-HA-10-001-DE.PDF/8e20d449-c6f7-49b1-945b-8cf735573d24.

Eurostat. (2014). *Informationsgesellschaftsstatistiken auf regionaler Ebene. Eurostat Jahrbuch der Regionen EU* (Eurostat Information Society Statistics at regional level. Eurostat regional yearbook). Retrieved from http://ec.europa.eu/eurostat/statistics-explained/index.php/Information_society_statistics_at_regional_level/de.

Filkins, B. (2014, February). *Health care cyberthreat report: Widespread compromises detected, compliance nightmare on horizon—A Sans analyst white paper.* Retrieved from https://www.sans.org/reading-room/whitepapers/analyst/health-care-cyberthreat-report-widespread-compromises-detected-compliance-nightmare-horizon-34735.

Findahl, O. (2013). *Svenskarna och intenet 2013* (Swedes and the Internet 2013. Internet Infrastructure Foundation). Stiftelsen för internetinfrastruktur. Retrieved from https://www.iis.se/docs/SOI2013.pdf.

Finland's Cyber Security Strategy (FCSS). (2013). *Government Resolution* (24 January 2013). Secretariat of the Security Committee. Retrieved from http://www.defmin.fi/files/2378/Finland_s_Cyber_Security_Strategy.pdf.

Finnish Communications Regulatory Authority. (2017). *Fibre guarantees access to high-speed broadband.* Retrieved from https://www.viestintavirasto.fi/en/statisticsandreports/reviewsandarticles/2017/fibreguaranteesaccess-tohigh-speedbroadband.html.

Finnish Communications Regulatory Authority. (2018). *Finnish users' passwords in clear text revealed.* Retrieved from https://www.viestintavirasto.fi/en/cybersecurity/alerts/2018/varoitus-2018-02.html.

Fischer, E. A. (2016). *Cybersecurity issues and challenges: In brief.* Retrieved from https://fas.org/sgp/crs/misc/R43831.pdf.

Floridi, L. (2014). *The fourth revolution: How the infosphere is reshaping human reality.* Oxford: Oxford University Press.

Fustel de Coulanges, N. (2001 / 1864). *The ancient city: A study on the religion, laws, and institutions of Greece and Rome.* Kitchener, ON: Batoche Books.

Glusman, A. (2010). *The Malcovich Bias.* Retrieved from http://glusman.blogspot.fi/2010/02/malcovich-bias-over-years-ive-noticed.html.

Hatleback, E. N. (2018). The protoscience of cybersecurity. *Journal of Defense Modeling and Simulation: Applications, Methodology, Technology, 15*(1), 5–12.

Haveri, A. (2006). Complexity in local government change. *Public Management Review, 8*(1), 31–46.

Heeringa, S., West, B., & Berglund, P. A. (2010). *Applied survey data analysis*. Boca Raton, FL: Taylor & Francis.

Henman, P. (2010). *Governing electronically—e-government and the reconfiguration of public administration policy and power*. Basingstoke: Palgrave Macmillan.

Hodge, H., Carson, D., Carson, D., Newman, L., & Garrett, J. (2016). Using Internet technologies in rural communities to access services: The views of older people and service providers. *Journal of Rural Studies, 54*, 469–478.

Homburg, V. (2008). *Understanding e-government: Information systems in public administration*. Abingdon: Routledge.

IKT-Einsatz in Haushalten. (2012). *Einsatz von Informations- und Kommunikationstechnologien in Haushalten 2012*. Herausgegeben von STATISTIK AUSTRIA (ICT usage in households. (2012). Use of information and communication technologies in households 2012. Edited by Statistics Austria). Retrieved from http://www.e-government.steiermark.at/cms/dokumente/10103295_34808287/8f9aabca/IKT-Einsatz_in_Haushalten_2012.pdf.

Informationsgesellschaftsstatistiken auf regionaler Ebene (2014). International Telecommunication Union (ITU). (2012). *Understanding cybercrime: Phenomena, challenges and legal response*. Retrieved from www.itu.int/ITU-D/cyb/cybersecurity/legislation.html.

International Telecommunication Union (ITU). (2018). *Definition of cybersecurity*. Retrieved from https://www.itu.int/en/ITU-T/studygroups/2013-2016/17/Pages/cybersecurity.aspx.

KPMG. (2015). *Health care and cybersecurity: Increasing threats require increased capabilities*. Retrieved from http://advisory.kpmg.us/content/dam/kpmg-advisoviso-ry/PDFs/ManagementConsulting/2015/KPMG-2015-Cyber-Healthcare-Survey.pdf.

Kuorilehto, R. (2014). *Moniasiantuntijuus sosiaali- ja terveydenhuollon perhetyössä. Monitahoarviointi Q-metodologialla* (Transprofessionality in social and health care family work. Multiple constituency evaluation with Q methodology). Acta Universitatis Ouluensis D Medica 1239. Retrieved from http://jultika.oulu.fi/files/isbn9789526203966.pdf.

Lindgren, I., & Jansson, G. (2013). Electronic services in the public sector: A conceptual framework. *Government Information Quarterly, 30*(2), 163–172.

Mertens, D. M., & Hesse-Biber, S. (2012). Triangulation and mixed methods research: Provocative positions. *Journal of Mixed Methods Research, 6*(2), 75–79.

Mitchell, W. (1995). *City of bits: Space, place, and the infobahn*. London: MIT Press.

National Institute of Standards and Technology, U.S. Department of Commerce. Computer Security Resource Center. (n.d.). *Cybersecurity*. Retrieved from https://csrc.nist.gov/glossary/term/cybersecurity.

(N)ONLINER Atlas. (2012). *Eine Topographie des digitalen Grabens durch Deutschland. Nutzung und Nichtnutzung des Internets, Strukturen und regionale Verteilung*. Eine Studie der Initiative D21, TNS Infratest. Viewed 16 March 2016. Retrieved from http://www.initiatived21.de/wp-content/uploads/2012/06/NONLINER-Atlas-2012-Basiszahlen-f%C3%BCr-Deutschland.pdf.

Nordic Council of Ministers. (2018). *State of the Nordic region*. Retrieved from http://norden.diva-portal.org/smash/get/diva2:1180241/FULLTEXT01.pdf.

OECD. (2001). *Understanding the digital divide*. Retrieved from http://www.oecd.org/sti/1888451.pdf.

Official Statistics of Finland. (2018, December 31). *Population projection: Population structure*. Helsinki: Statistics Finland.

Ong, A. (2006). Mutations in citizenship. *Theory, Culture & Society, 23*(2–3), 499–531.

Osborne, S., Radnor, Z., & Nasi, G. (2012). A new theory for public service management? Toward a (public) service-dominant approach. *American Review of Public Administration, 43*(2), 135–158.

Osborne, S., Radnor, Z., & Strokosch, K. (2016). Co-production and the co-creation of value in public services: A suitable case for treatment? *Public Management Review, 18*(5), 639–653.

Philip, L., Cottrill, C., Farrington, J., Williams, F., & Ashmore, F. (2017). The digital divide: Patterns, policy and scenarios for connecting the 'final few' in rural communities across Great Britain. *Journal of Rural Studies, 54*(2017), 386–398.

Reid, J. (2006). *The biopolitics of the war on terror*. Manchester: Manchester University Press.

Roberts, E., Beer, D., Philip, L., & Townsend, L. (2017). Rural resilience in a digital society. *Journal of Rural Studies, 54*, 355–359.

Roche, M. (1987). Citizenship, social theory, and social change. *Theory and Society, 16*(3), 363–399.

Salminen, M., & Hossain, K. (2018). Digitalisation and human security dimensions in cybersecurity: An appraisal for the European High North. *Polar Record, 54*(2), 108–118.

Schou, J., & Hjelholt, M. (2018). *Digitalization and public sector transformations*. London: Palgrave Macmillan.

Selkälä, A. (2013). *Kansalaisuuden hallinta suomalaisessa sosiaali- ja terveyspolitiikassa* (The governance of citizenship on Finnish municipal social and health policy). Acta Universitatis Lapponiensis 248. Rovaniemi: Lapland University Press.

Selkälä, A., Viinamäki, L., Suikkanen, A., & Vasari, P. (2016). e-Kansalaisuus syrjäseudulla: Internetin Käyttökokemuksia Lapista. *Yhteiskuntapolitiikka, 81*(3), 332–342.

Simon, S., & de Goede, M. (2015). Cybersecurity, bureaucratic vitalism and European emergency. *Theory, Culture & Society, 32*(2), 79–106.

Statistics Sweden. (2012). *Privatpersoners användning av datorer och internet* (Private use of computers and the internet). Retrieved from http://www.scb.se/statistik/_publikationer/LE0108_2012A01_BR_IT01BR1301.pdf.

Stout, M. (2010). Climbing the ladder of participation: Establishing local policies for participatory practice. *Public Administration and Management, 15*(1), 46–97.

Techopedia. (n.d.). *Cybersecurity*. Retrieved from https://www.techopedia.com/definition/24747/cybersecurity.

Telecommunication Standardization Sector (ITU-T). (2018).The Economic Times. (n.d.). *Definition of 'cyber security'*. Retrieved from https://economictimes.indiatimes.com/definition/cyber-security.

Thompson, S. K. (2012). *Sampling*. Hoboken, NJ: John Wiley.

Tilastokeskus. (2015). *Väestön tieto- ja viestintätekniikan käyttö* (Statistics Finland: The use of ICT and population data). Retrieved from http://tilastokeskus.fi/til/sutivi/2015/sutivi_2015_2015-11-26_fi.pdf.

Townsend, L., Sathiaseelan, A., Fairhurst, G., & Wallace, C. (2013). Enhanced broadband access as a solution to the social and economic problems of the rural digital divide. *Local Economy, 28*(6), 580–595.

University of Maryland University College. (2018). *Introduction to Cyber Security*. Retrieved from http://umuc.edu/academic-programs/cyber-security/about.cfm.

Van Dijk, J. (2008). One Europe, digitally divined. In A. Chadwick & P. N. Howard (Eds.), *The handbook of internet politics* (pp. 288–304). London: Routledge.

Vartiainen, P. (2007). Monitahoarvioinnin periaatteet ja prosessit (Principles and processes of multi-faceted evaluation). In L. Viinamäki & E. Saari (Eds.), *Polkuja soveltavaan yhteiskuntatieteelliseen tutkimukseen (Paths for applied social research)* (pp. 152–172). Helsinki: Kustannusosakeyhtiö Tammi.

Verdegem, P., & Verleye, G. (2009). User-centered E-Government in practice: A comprehensive model for measuring user satisfaction. *Government Information Quarterly, 26*, 487–497.

Viinamäki, L., & Kilpiäinen, S. (2015). Methodological background for the socio-economic part of the mine research. In L. Viinamäki (Ed.), *Socio-economic challenges in the mining industry—Four cases from the Barents region* (pp. 15–16). Publications of Lapland UAS. Publication series B. Reports 16/2015. Retrieved from http://urn.fi/URN:ISBN:978-952-316-096-5.

Viinamäki, L., Kivivirta, V., Selkälä, A., Voutilainen, O., Syväjärvi, A., & Suikkanen, A. (2017). *Ajasta ja paikasta riippumatta. Digikansalaisuus ja palveluiden saavutettavuus maaseudulla—hankkeen loppuraportti* [*Regardless of time and place. Final report of the Digital citizenship and acceptability of the services in the rural areas project*]. LAPIN AMKIN JULKAISUJA. Sarja A. Referee-tutkimukset 1/2017. Retrieved from http://www.theseus.fi/handle/10024/137218.

Zojer, G. (2019). The interconnectedness of digitalisation and human security in the European High North: Cybersecurity conceptualised through the human security lens. *The Year Book of Polar Law, 10*, 297–320.

CHAPTER 10

Analysis of Online Social Networking When Studying the Identities of Local Communities

Maxim Shishaev, Andrey Fedorov and Igor Datyev

10.1 Introduction

The sustainability of local communities is seen as a key component of security from almost all perspectives and, therefore, it is closely bound up with human security issues. According to the Copenhagen School (Buzan, Wæver, Waever, & de Wilde, 1998), societal security is considered to be one of the five most important sectors of security, and it

M. Shishaev (✉)
Murmansk Arctic State University, Murmansk, Russia
e-mail: shishaev@iimm.ru

M. Shishaev · A. Fedorov · I. Datyev
Institute for Informatics and Mathematical Modeling, Kola Science Centre of the Russian Academy of Sciences, Apatity, Russia

A. Fedorov
e-mail: fedorov@iimm.ru

I. Datyev
e-mail: datyev@iimm.ru

M. Salminen et al. (eds.),
Digitalisation and Human Security, New Security Challenges,
https://doi.org/10.1007/978-3-030-48070-7_10

is defined as "the ability of a society to persist in its essential character under changing conditions and possible or actual threats" (Wæver, Buzan, Kelstrup, Lemaitre, & Carlton, 1993, p. 23). Identity is positioned as the ultimate societal security criterion. According to Wæver (as cited in Hama, 2017, p. 5), societal security is "about identity, about the self-conception of communities and of individuals identifying themselves as members of a community". Thus, belonging to a society is an important component of a human's personal wellbeing, which tightly connects societal and human security issues and justifies the identities of communities being the focus of topical studies. The problems associated with a community identity are also relevant within the framework of the traditional view of security (realism), as the nation can be viewed as a specific kind of community, thereby ensuring societal security as one of the necessary conditions for national sovereignty. Finally, according to the human security perspective, the maintenance of traditional relationships and values is considered to be a major issue, especially for traditional communities, particularly ethnic minority groups. According to the *Human Development Report* issued by the United Nations Development Programme (UNDP, 1994), community security is one of the seven key components of human security. This renders the investigation of the identities of local communities a very important task.

The identity of a community is a multifaceted construction, as it has many dimensions and varies over time. To study its dynamics, it is necessary to have access to large amounts of constantly updated and relevant data, which limits the possibilities of using traditional methods for researching social phenomena. In the era of digitalisation, such data, whether directly or indirectly characterising the identities of communities, circulate and accumulate in ever-increasing volumes within online social networking services. This means that an online social networking analysis represents a promising way to monitor and analyse the identity of a local community. However, there are still a number of technical, legal and ethical problems and restrictions that need to be considered in this regard.

Digitalisation has caused rapid changes in both societies and human activities in the European High North and it has also created new opportunities and new threats for local communities. Despite being sparsely populated, the territories of the European High North and the western part of the Russian Arctic are well covered by telecommunications services. The availability of virtual communications increases individuals' opportunities for unhindered communication, thereby promoting a higher level of personal freedom and establishing the prerequisites for the formation of new identities in a virtual environment. However, such opportunities also have a downside: the emergence of new group identities inevitably poses a threat to existing ones. The level of this threat depends on the differences in the contents of the identities, including the values shared within a group. The identities of the northern communities seem particularly vulnerable in this regard because they are formed largely on the basis of territorial specificity (traditional households, physical compactness of social groups), while identities generated in a virtual environment are usually trans-territorial. For rural communities in the European High North, territory-specific activities and values are represented by reindeer husbandry, fishing, hunting, picking mushrooms and berries, etc. In addition to this territorial dependency of local communities' identities, the region is characterised by the limited development of transport infrastructure and the associated low availability for local people. This remains a burning issue in the Russian High North, where the possibilities for the free movement of people are much more limited than in the "central" regions. Moreover, the digitalisation of social communications exacerbates the problem of privacy in general. The data circulating in online social networks naturally accumulate and could be used without its owner's knowledge and for undesired purposes.

The first question addressed in this chapter concerns the essence of a conceptual framework or model that is both sufficiently representative and adequate in terms of complexity to be used with the experimental tools under consideration. The second question asks, "What are the opportunities, challenges and barriers associated with analysing the identities of local communities based on online social networks?" This concerns not only the technological and methodological aspects of acquiring and interpreting data concerning real social phenomena from online social networks, but also the difficulty of complying with the legal and ethical restrictions on the use of personal data in a virtual environment.

In this chapter, we consider a local community to be a group of interacting people living in a common location, sharing traditional values and unified by social cohesion, and we assume that communities in the European High North are mostly rural. We focus on the problems and restrictions associated with studying community identity on the basis of an online social networking analysis. By networking, we not only mean a messaging process but also the publication of any information in the virtual world. The online social networking in question is carried out via special platforms known as social networking services (SNSs). Social networking services (also known as social networking sites or social media services) are commonly defined as "Web-based services that allow individuals to create and share an individual profile and lists of established connections with others on the site and to view, navigate, and interact with one's online connections as well as with the connections made by others in the user's network" (Amichai-Hamburger & Hayat, 2017, p. 1). There are a variety of SNSs currently available on the web, which are united by two common features (Obar & Wildman, 2015): SNSs are internet-based applications (Kaplan & Haenlein, 2010) and their backbone is user-generated content (Boyd & Ellison, 2007). People use SNSs to build social networks with other people who share similar interests, activities, backgrounds and/or real-life relations. At the same time, online social networking determines the identity of social groups, which tightly connects them to societal security issues. SNSs both mirror real societies and affect them, and this renders SNSs a very attractive arena in which to study social phenomena, including those concerning human security.

10.2 Conceptual Model of Identity

In this chapter, we consider identity mostly with regard to its social aspect, which defines how people identify themselves in relation to the others according to what they have in common. Melucci (1989) defined collective identity as "an interactive and shared definition produced by several individuals (or groups at a more complex level) and concerned with the orientation of action and the field of opportunities and constraints in which the action takes place" (p. 44). We do not accentuate the conceptual difference between the terms "social identity" and "collective identity" because, within the proposed conceptual model discussed below, the dimensions of the identity space could be represented by both values and social reputation.

Plenty of identity studies have been conducted in the fields of sociology, psychology, political science and philosophy. It is widely believed that Freud (1949) started the so-called "psychological line" in relation to studying identity, which focuses mainly on investigating self-identity or self-concept formation. A significant contribution to the development of the identity concept from the perspective of psychological science was made by Erikson (1968), who considered identity to be a complex entity with a multilevel structure. A lot of modern psychological studies on identity are inspired by the work of Tajfel (1986), who proposed the social identity theory to explain the formation of self-identity through the correspondence of the individual with social groups.

A common feature of the "sociological line" identity studies concerns the consideration of the social environment as a source of norms, symbols and models of behaviour that serve as the basis for the self-determination of a person as a member of a given social group (Nedoseka & Sharova, 2016). In the sociological theories applied to the study of identity, several basic approaches stand out, including functionalism, structural functionalism, symbolic interactionism, phenomenological sociology and integrative approach (Balich, 2013).

There are many research directions representing different views on identity, such as the primordial and existential views in the fields of sociology and psychology, respectively. However, it can be said that the political sciences are still mostly dominated by the constructivist and instrumentalist approaches. In the present chapter, the main research question is intended to address how identities form, evolve and collapse. As one of the key areas of interest for political researchers is specific kinds of communities, as shaped on the basis of common political values, it is the identity of social groups that is taken into account in most studies. For example, the authors of the "radical democracy" concept (Laclau & Mouffe, 1985) promote the constructivist principle of political identities that are not given but are instead formed and reconstructed through debates in the political sphere.

Despite the clear variability in views concerning the identity concept, all the approaches share certain commonalities that are significant for practical modelling purposes. One valuable idea in the context of understanding and modelling identity phenomena is the significance of contacts between persons defining social distance and, therefore, the strength of a social group. According to the symbolic interactionism view, identity only evolves due to the inclusion of an individual in a social group and the resultant communication between group members (Abels, 2000). Huntington (2004) presents an integrative approach to identity studies and states that identities are defined by the selfhood, being simultaneously the result of interaction between a person or a group with other people or groups.

Another view, shared to some extent by most researchers, is that identity is defined by certain components of culture, values, actions, etc., which are either shared or rejected by an individual or a social group. In addition, these components are not only to be stated but also evaluated. Hence, identity is assumed to be defined in some kind of space with a measure. Modern researchers state that both individuals and groups possess identity, which provides grounds for considering personal and group identities in a common feature space (Huntington, 2004). However, many authors conceptualise the social world precisely as space, whether it be the Blauspace (McPherson, 1983) or the social space (Bourdieu, 1989). By introducing some metric criteria to that space, we may be able to identify cluster effects, to assess the distances between points and to recognise the fact and the trajectory of "tendencies". In general, such formalism is adequate for the problem being considered, even if only making adjustments to the character of the space dimensions and space metric.

Another valuable aspect in the pool of different views on group identity is the "antagonistic" definition of the term. Freud (1949) dealt with the issue of the process of constructing the collective identities of "we" through the definition of "they" while noting the immense role played by antagonistic relations in establishing identity among community members, linking the latter to one collective identity (Mouffe, 2005). The works of political scientists such as Zizek, Mouffe and Laclau consider a discursive mechanism for constructing an identity based on social antagonism (Laclau & Mouffe, 1985; Slobodyanik, 2007; Torfing, 1999). In sociology and social psychology, the importance of antagonism to the formation of social groups is also touched upon in the works by Tajfel and colleagues, who formulated the social identity theory. They note that in-group favouritism and out-group derogation play a key role in structuring the social space (Cosmides, Tooby, & Kurzban, 2003; Tajfel & Turner, 1986).

In sum, we can conclude that identity has three key attributes, which provide the necessary conditions for the formation and preservation of group identity: (1) some common values shared by the community, (2) the existence of the "anti-identity" of "they" and (3) sustainable communications between community members. The most commonly used formal apparatus for describing communications within social groups is the network theory developed by a number of modern researchers (Barabási, 2018; Watts & Strogatz, 1998). This theory forms the basis for the social networks analysis methods, certain issues with which are considered in the following section in relation to the task of local community identity analysis. With reference to the first two attributes of identity, it seems that the most adequate general approach to formalising them for analysis and monitoring purposes involves their representation as points and areas of a multidimensional space. Identity has various diverse dimensions, including territorial, ethnic and cultural dimensions. While some components of identity can be quantified (i.e. age, welfare, etc.), others can only be measured by implementing qualitative and not always ordinal scales. Thus, the space of identities is multidimensional and heterogeneous. Beyond that, estimates of the components of identity predominantly appear to be interval values with uncertainty. In accordance with this view, the identity space for a two-dimensional case will look like the one shown in Fig. 10.1. A point of the space denotes personal identity and, in the general case, it would be an area with a non-spherical configuration reflecting uncertainty in terms of determining identity components. The high similarity between the points of a determined metric indicates a possible group identity.

In constructing such a space, we face a number of questions:

- How do we represent points of the identity space with respect to uncertainty?
- How do we identify antagonistic clusters in the space?
- How do we estimate the sustainability of the clusters?

Undoubtedly, it is possible to implement some current ideas and methods to address the issues listed above. In particular, it seems promising to consider the sustainability of a cluster by assessing factors such as its "diameter", the number of points included, and the presence and strength of the links between cluster points (as caused by personified

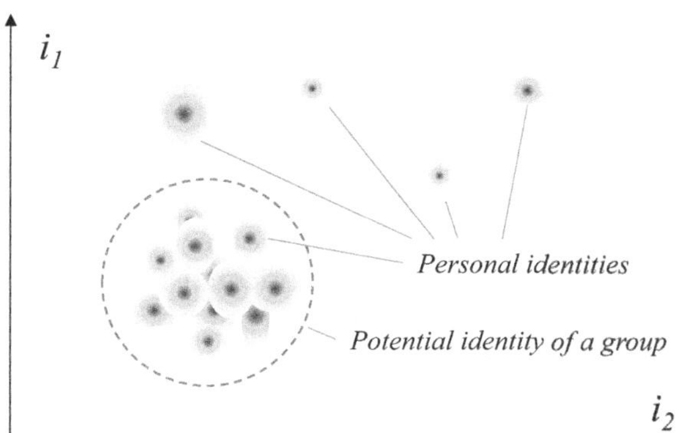

Fig. 10.1 Two-dimensional space of the "shared values" component of identity (*Source* Author's own creation)

identity). It is also assumed that the cluster sustainability is heavily dependent on the appropriate characteristics of the antagonistic cluster, as a "threat" facilitating the formation of a group identity. In this context, we can look at the fuzzy projective spaces, rough, soft, multi- and other atypical sets, and soft calculations as possible formal tools (Pawlak, 1982; Petrovsky, 2003; Zadeh, Abbasov, & Shahbazova, 2015). The efficiency of formalisation tools depends to a large degree on the nature of the initial data and the way it were obtained.

10.3 General Framework for Studying Identity on the Basis of SNS Data

According to Wellman et al. (1996), "computer networks are inherently social networks, linking people, organizations, and knowledge" (p. 213). Such computer-supported social networks have become important bases for new kinds of human activities: virtual communities, computer-supported cooperative work and telework. This has led to the emergence of a huge volume of work devoted to the study of the laws governing the formation and functioning of virtual social networks formed on the basis of an SNS. Alongside consideration as an independent object of research, SNSs create conditions for the use of the data accumulated in them for the analysis of real social objects and systems. Clearly, if members of the local community are represented in the online world of an SNS, the key attributes of the group identity are reflected in the online network communications, user-generated content, metadata and spontaneous data accumulated within the service. By looking at and analysing SNS data related to the community, we can reach some conclusions concerning its state and dynamics.

The success of SNSs is obvious from their current dominance worldwide, with Facebook, for example, having a massive 2.13 billion active users per month and an average of 1.4 billion per day in 2017 (Facebook Newsroom, 2018). LinkedIn, a career-oriented SNS, generally requires that a member personally knows another member in real life prior to contacting them online. Some services require members to have had a pre-existing connection before getting in touch with each other. Any large SNS generates big data (profiles, links and content). By analysing such data, we can derive a lot of useful information about different communities and topics as well as about a certain user's personality (Khotilin

& Blagov, 2016; Tan, Blake, Saleh, & Dustdar, 2013). Commercial organisations have demonstrated significant interest in SNSs as a tool for interacting with an audience, especially in terms of implementing users' feedback. Using specialised services and tools, companies analyse information about users and their activities and then personalise offers for individual segments of the target audience, thereby increasing the conversion rate and reducing the cost of advertising campaigns. For example, Bakaev and Blessing (2017) suggest the use of psychology and patterns of human behaviour to increase the effectiveness of such tools and services. Their method is based on analysing the relationships and communities of a given profile, and it allows us to identify different profiles belonging to the same user.

The interest of SNSs in users stems from the fact that they react promptly to real-life events and share their impressions as quickly as possible. Since posted messages are not subjected to deep censorship, they represent an extremely valuable material for content analysis when seeking to draw conclusions about ongoing social processes, emerging trends, people's attitudes towards certain subjects and phenomena, and their participation in political life. In addition to microblogging by individuals, the main social media, companies, political parties and associations are also present on platforms such as Twitter. The analysis of information posted by them is also of great interest from the perspective of comprehension the development of social processes. An example of an application that allows Twitter messages to be accumulated in a database on a local computer for subsequent classical content analysis is the Scai4Twi system (Shalak, 2018).

The first and most obvious task when studying identities of local communities on the base of information from SNSs is to obtain data characterising identity. In this regard, there are no significant technical problems and there are a number of quite effective information technologies and tools already available. As the analysis of SNSs deals with the space of virtual personalities, we need specific approaches that allow us to analyse real social entities based on observations of their virtual activity. This can be achieved by combining the analysis of different data related to the studied object within SNSs: user-generated content, metadata concerning the network and spontaneous (i.e. collected in a natural way during the operation of the SNS) data consisting of user profile attributes, timestamps, the structure of the links of the considered object, etc. One specific problem associated with using such data for our task is linking the "virtual observations" performed as part of online social networking activities with the real-world social objects and processes. Alongside the technological problems concerning the use of SNS data, it is necessary to take into account the legal and ethical constraints, including those connected to the hidden commercial use of user-generated data. The general framework for studying identity on the basis of SNS data is presented in Fig. 10.2.

In accordance with the proposed model, for the effective monitoring of identity, a tool is required that, using the minimum time and material resources, is able to gather (on the basis of SNS data) information on the following parameters:

- The substantial attributes of the shared values of the studied community, as expressed in the form of an emotional relationship between community members and objects and activities.
- The sets of subjects constituting antagonistic communities of anti-identity carriers ("they").
- The intensity and other parameters of communications within and outside the community, including the characteristics of the contact networks of community members.

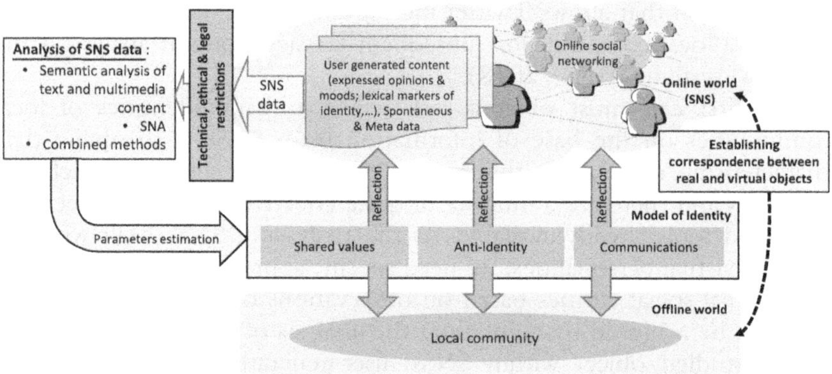

Fig. 10.2 The general framework for studying identity on the basis of SNS data (*Source* Author's own creation)

The first priority is to obtain raw data from the SNS, taking into account the technical capabilities as well as any legal and ethical restrictions. Technical constraints are not a problem, as most modern SNSs are supplied with advanced application programming interfaces (APIs) that provide access to the platform data. Ethical and legal restrictions regarding the use of personal data are more complex issues, which will be discussed in the following sections. In the next step, the SNS data should be analysed so as to interpret the results into the parameters of the identity model (an estimation of the model parameters should be performed).

10.4 Overview of Methods for SNS Data Analysis When Studying Identity

There are a wide variety of methods and technologies available for analysing SNS data. In general, these methods use the entire spectrum of data available for analysis, including spontaneous data and metadata, and they are based on various models, including specific, task-oriented ones. Among the available methods, it is possible to distinguish two groups of methods that can be effectively used for the analysis of SNS data: (1) methods for the semantic analysis of user content, and (2) methods of social network analysis, which are based on a general theory of networks and use information on various kinds of interconnections between SNS subjects as the input data.

The first group of methods includes two subcategories, namely semantic text analysis and multimedia content analysis, which are radically different in terms of the nature of the data being analysed and, accordingly, the core models and methods used. The task of analysing multimedia content is, technically speaking, the task of image recognition, which consists of identifying objects of interest in images or videos. The main methods for this are pattern recognition, digital geometry and signal processing. Image recognition is intensively used in SNS data analysis applications (e.g. Shyak, 2018). In the context of analysing the identities of local communities, pattern recognition can be used to identify elements of the cultural identities of subjects (e.g. national clothes), their social connections (appearance on joint photos and videos) and their common interests (the same activity). One peculiar kind of multimedia data analysis involves the technology of artificial emotional intelligence designed to recognise emotions in speeches, images and videos. This technique can be used to determine the values forming an identity.

The semantic analysis of texts currently comprises a huge number of methods and technologies used in various problem areas. In general, the task of semantic analysis is to interpret the text into a system of concepts and relations, which limits the subsequent use of its components (elements of the text or corresponding abstractions) in accordance with the rules adopted in this subject area. For the semantic analysis of texts, various models are used—special grammar, lexical-syntactic patterns, statistical text models, etc.—which provides different levels of semantic depth. Of particular interest in the context of the interpretation of SNS data in the identity model is the specific category of methods of semantic analysis associated with natural language texts: sentiment analysis (analysis of the emotional colouring of the text) and the methods of opinion mining based on it. The task of opinion mining involves searching the text of triples ("entity-orientation-holder"), to which time stamps can be added to analyse the dynamics of opinions. The range of methods used for sentiment analysis is very wide, ranging from latent semantic analysis to language models. The sentiment analysis of SNS data finds application in relation to a variety of issues, including community analysis (Deitrick & Hu, 2013; Jain, Metta, & Roy, 2018). When estimating the parameters of an identity model, a sentiment analysis is applicable for identifying the meaningful attributes of personal values and their carrier subjects.

In addition to user-generated content, SNSs accumulate large amounts of additional information, that is, metadata necessary to implement the functionality of the SNS platforms, as well as "spontaneous" data, such as timestamps, geotagging and other information related to user activity on the network. This information can be used to improve the efficiency with which problems can be solved based on SNS data (in comparison to "pure" semantic analysis or network analysis). For this purpose, algorithms and technologies are used that combine the capabilities of various approaches to analysing SNS data.

The second group of methods applicable to the parametrisation of an identity model on the basis of SNS data comprises the methods of social network analysis (SNA). An SNA is an interdisciplinary academic approach that emerged from the fields of social psychology, sociology and statistics. Various theoretical frameworks, such as graph theory, balance theory, social comparison theory and the social identity approach, have been implemented for the purposes of SNA (Kilduff & Tsai, 2003).

The SNA approach is currently used in a variety of different scientific disciplines and it forms one of the main paradigms in modern sociology, as it constitutes a part of the evolving field of network science (Borgatti, Mehra, Brass, & Labianca, 2009; Easley & Kleinberg, 2012).

There are a number of works in the network science field, and in the SNA field in particular, devoted to sociological research (e.g. Barabási, 2018; Granovetter, 1973; Moreno, 1956; Carrington, Scott, & Wasserman, 2005). Jacob Moreno (2014) was one of the founders of the SNA discipline, the branch of sociology that deals with the quantitative evaluation of an individual's role in a group or community via the analysis of the network of connections between that individual and others. Moreno (2014) also developed sociometry, group psychotherapy, psychodrama and sociodrama. Granovetter (1973) is best known for his work in both social network theory and economic sociology, particularly his theory on the spread of information in social networks, which is known as "The Strength of Weak Ties". In the field of economic sociology, Granovetter (1985) introduced the concept of "embeddedness" and the idea that economic relations between individuals or firms are embedded in actual social networks and, therefore, do not exist in an abstract idealised market. The Barabási–Albert model is a random scale-free network model used to demonstrate a preferential attachment or a "rich-get-richer" effect. In this model, an edge is most likely to attach to nodes with higher degrees (Albert & Barabási, 2002). Scott's (1979) research in stratification and political economy has explored the changing patterns of ownership in contemporary capitalist economies as well as the ways in which these changes are reflected in elite recruitment. In his research on the networks of shareholding and interlocking directorships, Scott (1986) developed the models of structures and mechanisms through which this constraining power operates.

The above-mentioned works have formed a strong scientific basis for applying the network approach to sociological research. There have been many studies devoted to SNA that have focused on various aspects of online social networking analysis and network community mining in general. At the same time, the SNA and general network approaches are used in a number of methods and applications, both specialised and more general: recommendation systems, link analysis, expert identification, influence propagation, trust and distrust relationship prediction, opinion mining, mood analysis and community detection (Anjaria, Mohana, & Guddeti, 2014; Yuan, Cheng, Zhang, Liu, & Lu, 2015;

Zhang, Fang, Chen, & Tang, 2015). The kinds of network analysis methods that are most relevant to our research question are discussed below.

Today, the research direction referred to as semantic SNA, which combines the technologies of the semantic web and the classical SNA, is being developed to improve the analysis of SNSs. A widely known example from Russia is a major research project concerning the semantic analysis of online social networking data, namely the "Krybrum" (or "Cribrum") monitoring system (Ashmanov and Partners, 2018). The system tracks over 800,000 internet sources, approximately 150,000,000 individual authors and more than 20,000 online media. Further, it collects 100% of messages in Russian that are posted on the Twitter microblogging platform, as well as on VK and Instagram. Cribrum uses its own unique technologies to collect, store and analyse users' big data, searching for the morphology of the Russian language as well as automatically highlighting the emotional colouring of statements.

Aside from the study of SNSs as a special type of virtual social environment, researchers have also focused on the relationship between online (existing in the virtual world of SNSs) and offline (existing in the real world) social processes, entities and systems. Such tasks are crucial in the context of our research questions. An example of a research study addressing this problem can be seen in the work by Smirnov et al. (2016). The authors focused on the particular task of finding the profiles of given (real) people in the considered SNSs. The research was conducted on VKontakte SNS (VK, the Russian SNS equivalent to Facebook) data, taking as an example the communities of a secondary school and a university (over 16,000 people in total). The authors proposed the procedure of matching VK profiles to real students on the basis of personal identification data (direct matching) as well as by using the aggregate social context of each person. Direct matching permitted them to identify the profiles of around 18% of students. A special technique described in the article increased this number to 88% for school pupils and 93% for university students. The authors compared the age, gender and grade point average of the identified students with those whom the authors did not find on VK. They also compared the structures of the social relationships and retrieved data from VK concerning the expected structures of the students' social ties. The authors found that the structure of "virtual" social relationships reproduced both the socio-demographic division of students into grades, years of study or majors as well as the spatial division into different school buildings or university campuses. The findings of their study contribute to our understanding of how reliable data derived from this particular SNS is, how its accuracy can be improved as well as how it can be used to study real social objects on the basis of virtual world observations.

Another issue when linking online and offline "worlds" is determining the geographical locations of virtual entities. Zamyatina and Yashunsky (2018) focused on the practical usage of Russian Federation Virtual Population project data. The project represents a new database relative to the VK social network. The project website is a customisable resource designed to study the VK structure. The dataset relative to the network structure can be found on the website, as can an interactive map builder that summarises the data as a map and provides different geographical, temporal and demographical filters. The website possesses a large dataset that allows for the investigation of the VK audience profile, taking into account parameters such as gender, age, education and migratory behaviour. The website boasts of its data concerning users' virtual friendships, which seems to be the first large-scale research into whether virtual friendship relationships are territorially dependent. The database appears to be useful for studying not only the virtual space but also offline social processes such as youth migration at the micro level, meaning that in this case it serves as an implicit data source.

Some authors propose the use of geo-visualisation technologies for tasks relating to the study of SNSs. Ma (2012) considered a visualisation environment to incorporate social network graphics with geo-visualisation methods. A number of works concerning different approaches and applications of the geo-visualisation of social media data, including some combining geo-visualisation and sentiment analysis, are reviewed in Croitoru, Crooks, Radzikowski, and Stefanidis (2017).

A number of similar tasks related to the identification and analysis of the structures of real and virtual communities are considered within community mining methods and applications. A network community refers to a group of abstract network nodes within which the connecting links are dense but between which they are sparse. A network community mining (NCM) problem can be defined as a question of finding all such communities within a given network. A large variety of problems can be translated into NCM problems, including SNA. Network communities in different application contexts may imply different meanings and serve different purposes. For example, they may be sets of electronic units closely placed on a very large-scale integration circuit board, collections of processes that frequently communicate with each other, segments of an image, circles of a society within which people share common interests and maintain frequent contacts, classes of web pages related to common topics or groups of proteins exhibiting similar functions. Discovering such hidden patterns in networks will enable us to better understand the structural and/or functional characteristics of networks and to utilise them more effectively. Yang, Liu, and Liu (2010) introduced NCM problems and discussed various approaches to tackling them.

Two widely considered problems in relation to NCM concern the dynamics or evolution of the community and the overlap of communities. The evolution of the community appears, for example, when a recommendation system analyses the available data and suggests something the actor might be interested in and, as a result, a new link is introduced into the network. Overlapping communities are made possible if a node is a member of more than one community. Chitra Devi and Poovammal (2016) discussed various modularity-based approaches for detecting overlapping communities in social networks. The identification and monitoring of community evolution represent important tasks within the field of NCM. Most networks, such as social media, blogs and co-authorship networks, are dynamic, as they tend to evolve gradually due to frequent changes in the activity and interaction of their individuals. Furthermore, the communities inside a dynamic network could grow or shrink, and the community membership of individuals shifts regularly. In these dynamic networks, researchers may be interested in the evolution of communities and the membership of individuals, such as author communities in the blogosphere, the analysis of mobile subscriber networks and the evolution of research communities. To detect the evolutionary behaviour of the network and the communities, it is necessary to use certain special tools, taking into account temporal information. For example, Takaffoli, Sangi, Fagnan, and Zäiane (2011) presented a framework that encompasses a community matching algorithm and an event detection model intended to capture all the possible events that occur for communities. This includes tracing the formation, survival and dissolution of communities as well as the identification of meta-communities (a series of similar communities seen at different time points) for any dynamic social network.

The SNA methods are based on the formal apparatus of the network theory, and the most common way to reflect a network community is a node-link diagram. With this, the interpretation of basic network components can be very different: the nodes may represent the individual users/actors/items/resources, whereas the links could show messages sent between node objects as well as other types of connections, for example, those that indicate “friendship”. By selecting different interpretations of nodes and links, we can fill the network with various semantic loads. This renders SNA a very promising means for extracting estimations of parameters for the identity model from SNS data. Semantic SNA and the information technologies that implement such analyses could be adapted to extract the parameters of “shared values” and “anti-identities” for the model, while NCM methods could be used to extract the model parameters characterising communications within the considered community.

In sum, it can be said that there is currently a wealth of experience in relation to the study of online social networks, which can also be applied to the study of the identities of local communities. Many authors attribute the analysis of SNSs to big data analysis, which suggests the solutions to specific problems to be largely based on the heuristics associated with a combination of different data sources and approaches to their analysis. Ideally, it is necessary to process all the information available within an SNS. Good prospects in this regard can be seen in sentiment analysis, semantic-web technologies and other methods based on the processing of formalised knowledge. However, the use of these broad technical capabilities is only possible if the legal and ethical restrictions on the use of personal data are respected.

10.5 Ethical and Legal Concerns Regarding the Use of SNS Data

The use of SNS data inevitably brings up the question of safety when gathering, storing and processing personal data as well as the subsequent moral and ethical aspects. The main advantage of using SNSs concerns the interactions that occur between people through messaging and hosting data that describes a person, her interests, her personal preferences, etc. In terms of the use of SNSs, this raises questions concerning the scope of the legal and ethical rules regarding the use of personal data and issues concerning privacy.

From one perspective, privacy issues regarding the use of SNSs are barely important or even non-existent. Such a perspective might be based on the assumption that SNSs are designed as a means of open communication and that their use is well regulated by agreements and rules. When initially signing up to use an SNS, the user accepts agreements concerning the purpose of using the network as well as the basic rules for working with information. Overall, the formal description of the SNS's purpose and the tools provided for its use provide the user with a clear idea of the specifics of working with the network itself and the data stored in it. The terms of use for the Vkontakte SNS, for instance, explicitly prohibit users from illegally collecting and processing other users' personal data. In addition, SNSs (e.g. Facebook, VKontakte) also provide a large spectrum of settings that help to divide the published data into several privacy categories (hidden, open to all, open to friends, open to group members, etc.). This way, by arranging his or her communication inside the SNS, a user personally decides what information to provide, to whom and who is responsible for it. The data that users have allowed access to can be viewed by authorised network users directly through the UI (user interface) provided by the SNS or it can be accessed through the API. Hence, in theory, all potential moral and ethical problems associated with accessing private SNS data can be overcome by regulatory documents and officially provided access control tools. However, the "theory" is at odds with practice: despite accepting the agreements, some users may be unfamiliar with the text of the agreements or not fully understand them, while recent scandals involving Facebook have shown that the network uses more data than that provided by the user agreement (e.g. Busemann & Schimroszik, 2018). In addition, even consciously open user data can be used, with deep processing, for purposes that the user does not realise. This is especially relevant in the context of the problem considered in this chapter, as the users of social networks are unlikely to assume that the data they publish can be used to study their identity, including their personal preferences, non-public membership of various social groups, etc.

As with any software product, an SNS has a range of legal subjects that require the basis of their activity to be defined. These subjects are the owners, network administrators, users, etc. Within an SNS, the rights and obligations of these subjects are set by the corresponding agreements and rules (legal information). The rules imply that the main goal of using an SNS lies in its extension by adding and editing an account and a personal page and, thereby, communicating with other users. The same regulations explicitly specify the responsibility of users for all materials posted on the network and establish the requirement to comply with all rules and other legislative documents. The use of a social network is only considered legal if it is done in accordance with the official regulatory means provided by the network's owner. These tools include the web interface, mobile applications and third-party applications developed using the officially provided API.

Regardless of whether the legal information explicitly stipulates this point or not, the current legislation must also be observed when using an SNS. The user agreement establishes that the user of the network is an individual who has passed the corresponding site registration. It also states that during the registration, a user must provide actual and up-to-date information that will serve as the basis for user's personal account and the user's webpage. Such information includes a surname, a name, a unique login and a password (Vkontakte, 2017). Thus, the SNS administration assumes that any information regarding the account directly relates to personal data. Such information can be collected and processed, albeit only when following the relevant domestic laws.

In accordance with the legislation of the Russian Federation, personal data are referred to as any information directly or indirectly related to a specified or determined individual (i.e. the subject of the personal data) (Russian Federation, 2016). The law characterises the collection of personal data as a kind of personal data processing and any action (operation) or set of actions (operations) performed with personal data using (or not using) automated tools. The law also sets out a number of cases that allow for the collection and processing of personal data. In general, it is necessary to obtain the consent of the entity whose data are to be collected and processed. However, such consent is not required if the data are going to be used for professional research and statistical purposes, provided that the rights and legitimate interests of the personal data subject will not be violated. Personal data processing is considered legal if the data are made publicly available by the subject (Russian Federation, 2011).

In May 2018, a new statutory document (General Data Protection Regulation [GDPR]) was introduced into the European legislative system so as to regulate the collection and processing of personal data belonging to citizens of Europe, wherever they are territorially located. This document introduces a strict set of rules and establishes liability for violations related to the processing of personal data. The GDPR requires the prior consent of the owner to use the data. It is necessary to specify the purpose of the data's use and to ensure the upholding of the principle of "data minimisation": personal data are not allowed to be collected in amounts exceeding those required for the declared purpose. The collected data concerning the behaviour of the monitored subjects have also become personal. To ensure that the data are processed in a legal way, a data controller must be allocated, who determines the purpose and tasks of the research as well as the forms and volumes of data necessary to achieve them. The subsequent data manipulation is performed by the data processor according to the instructions of the controller (European Parliament and the Council of the European Union, 2016).

In accordance with the new regulations, the SNSs targeted towards European users have made adjustments to their license user agreements. For example, Facebook has made a number of changes, primarily concerning the developers of software products (applications and services) that use the SNS. More specifically, it has introduced a new requirement to obligatorily warn users that the installed application can collect, process and transfer personal data to third parties (Tiku, 2018; Facebook for developers: Facebook Platform Policy, 2018). Usually, this information is used by Facebook itself and other applications to collect statistics and target advertising. However, all applications must now also provide the user with a clear opportunity to prohibit such actions.

The issue of the use of personal data in online social networks is particularly relevant today. It should be noted that the principles regulating this issue are the subject of continuous development. With the increasing use of social networks and the emergence of new advanced data processing technologies, new aspects of the problem are opening up. A characteristic example of this concerns the face recognition service launched by Facebook, which is now having to make significant corrections to its rules of use following legal issues in several countries.

The face recognition service is very popular among the users of social media. With it, users share photos with friends and search for new friends. This function directly affects the formation and maintenance of social relations through social media. In 2011, Facebook developers switched on the automatic face recognition feature for photos uploaded by users to their pages. In addition to the previously listed tasks, this function allows them to solve the problem of the inconsistent use of someone else's photos. Technically speaking, to implement this function, the developers used neural networks that automatically analysed the photos uploaded by users and then formed a database of special descriptive tags. In general, the collection and storage of such information involves the processing of biometric data. Such activities are differently regulated in different countries. Litigation against Facebook (Shyak, 2018) led to a change so that the "face recognition" option can now be manually turned off by the user, while previously recognised photos can be permanently cleared of identification tags (Facebook Help Centre, 2018). However, the full potential of the algorithms used by Facebook remains unknown, and they can be used implicitly, for example, as part of the selection algorithm and recommendation of friends. The decision to adjust the use of photo-processing algorithms was only made by Facebook developers under the pressure of lawsuits. Yet, each legal case (precedent) of this kind contributes to the formation of ethical issues related to the application of modern information technologies.

While the requirements of the Russian legislation allow for the impersonal use of personal data for research (non-commercial) purposes, the European GDPR regulation establishes tight controls over software products that work with personal user data. In relation to the task of analysing identity on the basis of the proposed framework, in order to satisfy the requirements of the GDPR and the established ethical standards, it is necessary to obtain explicit user permissions to use (in full or in part) the following data: user-generated content, user profile data and spontaneous data (user activity data). The purpose of using the data, that is, identity analysis, should also be explicitly stated and a guarantee must be provided that the user has consciously given the appropriate permissions. Finally, it is necessary to ensure control over compliance with the law when using personal data (i.e. to employ a data protection officer).

10.6 Conclusion

In this chapter, we investigated the opportunities and problems associated with studying local communities' identities on the basis of online SNA. The task of studying identity was considered in the context of societal security issues, which are significant components of the overall human security domain.

An identity is a complex object with various dimensions represented by shared values and actions or by social reputation. The complexity of the modelling and practical analysis of identity is aggravated by the common impossibility of obtaining accurate quantitative parameter estimates. For this reason, when constructing a practice-oriented formal model of identity, it is necessary to use methods for processing information with uncertainty. These could be probabilistic methods, methods of fuzzy or interval logic, soft calculations and something else. A relevant way of representing identity is a multidimensional fuzzy space, the dimensions of which correspond to the attributes of identity and points (areas) of which serve as the attitudes of an individual towards the corresponding values and actions. The proximity of points in the space is one of the necessary conditions for the emergence and existence of a group identity. In addition, some other necessary conditions are the existence of sufficiently reliable and intensive communication between members of the community as well as the existence of a sufficiently significant and compact (in the model space) "anti-identity". Thus, the formal model of identity, as oriented for use in the societal security context, should determine the triple: *group_identity*, *intra-group_communications* and *anti-identity*.

The territorially determined nature of local communities' identities is an important peculiarity in the context of the problem under consideration. In the European High North, the importance of this factor is even more critical due to the limited development of transport communication. It determines the importance of such specific practical tasks of SNA as the establishment of a correspondence between virtual and real social entities and the identification of the regional identity (home region) of the subjects of the virtual space. However, the establishment of such a correspondence significantly affects the problem of privacy, which is aggravated by the fact that local communities in the North are usually few in number.

Social networking services, as an independent sphere of existence of social objects and processes, are currently being actively studied. There are a large number of methods, technologies and applications available for analysing various kinds of objects and processes within an SNS. Modern SNSs provide plenty of efficient tools (API) for extracting the heterogeneous data collected and circulating within an SNS. At the same time, there are plenty of methods suitable for analysing SNS data, both the semantics of the content and the networking aspects. Most commonly, techniques and algorithms are based on certain heuristics that depend on the specific task of analysis. Thus, from a technological point of view, the analysis of SNSs appears to be a promising and effective means of studying real social phenomena, including the identities of local communities.

When implementing this approach, one has to deal with not only technological but also ethical and legal restrictions. The most serious issues are related to the use of personal data within an SNS. In the territories of the European High North, this is governed by Russian national legislation as well as the legal norms of the European Union. Modern intellectual data processing technologies make it possible to extract more and more personal information from SNS data. This significantly exacerbates the problem of ensuring privacy as well as compliance with legal and ethical standards when working with online social networks, which is reflected in the gradual tightening of the regulations governing the use of personal information within a network.

UNDP (1994) introduced a new 'people-centred' concept of human security. It covers a number of human security dimensions, including community security. According to the concept, most people derive security from their membership in a social group. Such groups offer different kinds of practical support for their members making the life of an individual more secure. Nevertheless, the flip side of the coin is the existence of oppressive practices and social restrictions that directly impact personal security in many communities. This shows the complexity and multidimensionality of community security. At the same time, digitalisation reinforces the emergence and rapid development of new kinds of communities and human activities—digital ones that form and evolve within online social networking services. The mentioned complexity and multiplicity in effects appear in this online world as well. The task of studying identity on the basis of online social networks considered in this chapter exposes the dualism of the effects of digitalisation: on the one hand, they open up new opportunities for the study and development of social processes; on the other hand, modern digital technologies dramatically affect privacy.

References

Abels, H. (2000). *Interaction, identity, presentation: Introduction to interpretative sociology*. St Petersburg: Aleteya Publishing House.

Albert, R., & Barabási, A.-L. (2002). Statistical mechanics of complex networks. *Reviews of Modern Physics, 74*(1), 47–97. https://doi.org/10.1103/revmodphys.74.47.

Amichai-Hamburger, Y., & Hayat, T. (2017). Social networking. In P. Rössler (Ed.), *The international encyclopedia of media effects* (2nd ed.) (pp. 1–12). London, UK: Wiley. https://doi.org/10.1002/9781118783764.wbieme0170.

Anjaria, M., Mohana, R., & Guddeti, R. (2014). A novel sentiment analysis of social networks using supervised learning. *Social Network Analysis and Mining, 4*, 181. https://doi.org/10.1007/s13278-014-0181-9.

Ashmanov & Partners. (2018). *Cribrum monitoring system*. Retrieved from https://www.ashmanov.com/tech-i-services/kribrum/.

Bakaev, V. A., & Blessing, A. V. (2017). Analysis of profiles in social networks. In *Proceedings of the 3rd International Conference on Information Technology and Nanotechnology* (pp. 88–91), 21–24 May 2018. Samara, Russia. https://doi.org/10.18287/1613-0073-2017-1903-88-91.

Balich, N. (2013). Social identity: Theoretical and methodological foundations of sociological analysis. *Sociological Almanac, 4*, 214–220. Retrieved from https://cyberleninka.ru/article/n/sotsialnaya-identichnost-teoretiko-metodologicheskie-osnovaniya-sotsiologicheskogo-analiza.

Barabási, A. L. (2018). *Network science*. Cambridge, UK: Cambridge University Press.

Borgatti, S. P., Mehra, A., Brass, D. J., & Labianca, G. (2009).Network analysis in the social sciences. *Science, 323*(5916), 892–895. https://doi.org/10.1126/science.1165821.

Bourdieu, P. (1989). Social space and symbolic power. *Sociological Theory, 7*(1), 14–25.

Boyd, D., & Ellison, N. (2007). Social network sites: Definition, history, and scholarship. *Journal of Computer-Mediated Communication, 13*(1), 210–230.

Busemann, H., & Schimroszik, N. (2018). *German court rules Facebook use of personal data illegal*. Retrieved from https://www.reuters.com/article/us-germany-facebook/german-court-rules-facebook-use-of-personal-data-illegal-idUSKBN1FW1FI.

Buzan, B., Wæver, O., Waever, O., & de Wilde, J. (1998). *Security: A new framework for analysis*. Boulder, CO: Lynne Rienner Publishers.

Carrington, P. J., Scott, J., & Wasserman, S. (2005). *Models and methods in social network analysis*. Cambridge, UK: Cambridge University Press.

Chitra Devi, J., & Poovammal, E. (2016). An analysis of overlapping community detection algorithms in social networks. *Procedia Computer Science, 89*, 349–358. https://doi.org/10.1016/j.procs.2016.06.082.

Cosmides, L., Tooby, J., & Kurzban, R. (2003). Perceptions of race. *Trends in Cognitive Sciences, 7*(4), 173–179.

Croitoru, A., Crooks, A., Radzikowski, J., & Stefanidis, A. (2017). Geovisualization of social media. In D. Richardson, N. Castree, M. F. Goodchild, A. Kobayashi, W. Liu, & R. A. Marston (Eds.), *The international encyclopedia of geography* (pp. 1–17). London, UK: Wiley. https://doi.org/10.1002/9781118786352.wbieg0605.

Deitrick, W., & Hu, W. (2013). Mutually enhancing community detection and sentiment analysis on Twitter networks. *Journal of Data Analysis and Information Processing, 1,* 19–29.

Easley, D., & Kleinberg, J. (2012). *Networks, crowds, and markets.* Cambridge, UK: Cambridge University Press.

Erikson, E. H. (1968). *Identity: Youth and crisis.* New York, NY: Norton.

European Parliament and the Council of the European Union. (2016). *EU Regulation 2016/679 of April 27, 2016, GDPR.* Luxembourg: Publications Office of the European Union. Retrieved from https://publications.europa.eu/en/publication-detail/-/publication/3e485e15-11bd-11e6-ba9a-01aa75ed71a1.

Facebook for developers: Facebook Platform Policy. (2018). Retrieved from https://developers.facebook.com/policy.

Facebook Help Centre. (2018). *Photos: Tagging—Approve or remove tags.* Retrieved from https://www.facebook.com/help/267689476916031/?helpref=hc_fnav.

Facebook Newsroom. (2018). *Company info.* Retrieved from https://newsroom.fb.com/company-info.

Freud, S. (1949). *The ego and the id.* London, UK: The Hogarth Press.

Granovetter, M. S. (1973). The strength of weak ties. *The American Journal of Sociology, 78*(6), 1360–1380. https://doi.org/10.1086/225469.

Granovetter, M. S. (1985). Economic action and social structure: The problem of embeddedness. *The American Journal of Sociology, 91*(3), 481–510.

Hama, H. (2017). State security, societal security, and human security. *Jadavpur Journal of International Relations, 21*(1), 1–19. https://doi.org/10.1177/0973598417706591.

Huntington, S. P. (2004). *Who are we? The challenges to America's national identity.* New York, NY: Simon and Schuster.

Jain, V., Metta, V., & Roy, P. (2018). Comparative study of sentiment analysis on trending issues on social media. *International Journal of Advanced in Management, Technology and Engineering Sciences, 8*(III), 1739–1750.

Kaplan, A. M., & Haenlein, M. (2010). Users of the world, unite! The challenges and opportunities of social media. *Business Horizon, 53*(1), 59–68. https://doi.org/10.1016/j.bushor.2009.09.003.

Khotilin, M. I., & Blagov, A. V. (2016). Visualization and cluster analysis of social networks. *CEUR Workshop Proceedings, 1638,* 843–850.

Kilduff, M., & Tsai, W. (2003). *Social networks and organisations*. Thousand Oaks, CA: Sage.

Laclau, E., & Mouffe, C. (1985). *Hegemony and socialist strategy: Towards a radical democratic politics*. London and New York: Verso.

Ma, D. (2012). *Visualization of social media data: Mapping changing social network*. Retrieved from https://webapps.itc.utwente.nl/librarywww/papers_2012/msc/gfm/dingma.pdf.

McPherson, M. (1983). Ecology of affiliation. *American Sociological Review, 48*, 519–532.

Melucci, A. (1989). The process of collective identity. In H. Johnston (Ed.), *Social movements and culture* (pp. 41–63). London, UK: Routledge.

Moreno, J. L. (1956). *Sociometry and the Science of man*. New York, NY: Beacon House.

Moreno, J. D. (2014). *Impromptu man* (1st ed.). New York, NY: Bellevue Literary Press.

Mouffe, C. (2005). *On the political*. London, UK: Routledge.

Nedoseka, E., & Sharova, E. (2016). *Problems of regional identity of the population of the Kola North*. Murmansk: MASU.

Obar, J. A., & Wildman, S. (2015). Social media definition and the governance challenge: An introduction to the special issue. *Telecommunications Policy, 39*(9), 745–750. https://doi.org/10.1016/j.telpol.2015.07.014.

Pawlak, Z. (1982). Rough sets. *International Journal of Computer and Information Sciences, 11*, 341–354.

Petrovsky, A. B. (2003). *Spaces of sets and multisets*. Moscow: Editorial URSS.

Russian Federation. (2011). *Federal law about amendments to the Federal law "on personal data" N 261-FZ 2011*. Retrieved from http://www.consultant.ru/document/cons_doc_LAW_117437.

Russian Federation. (2016). *Federal law "on personal data" N 152-FZ 2016*. Retrieved from http://www.consultant.ru/document/cons_doc_LAW_61801.

Scott, J. P. (1979). *Corporations, classes and capitalism*. London, UK: Hutchinson.

Scott, J. (1986). *Capitalist property and financial power: A comparative study of Britain, the United States, and Japan*. Brighton, Sussex: Wheatsheaf Books.

Shalak, V. (2018). *Scai4Twi—Internet content analysis system*. Retrieved from http://www.vaal.ru/show.php?id=210.

Shyak, K. (2018, April 16). Face Recognition on Facebook: What is it? Why Bother? *Lawless.Tech*. Retrieved from https://lawless.tech/face-recognition-on-facebook-what-is-it-why-bother/.

Slobodyanik, N. (2007). Construction of identity in political discourse: To the question of the role of social antagonism. *Political Linguistics, 2*(22), 60–67.

Smirnov, I., Sivak, E., & Kozmina, Y. (2016). In search of lost profiles: The reliability of VKontakte data and its importance in educational research. *Voprosy Obrazovaniya [Educational Studies Moscow], 4*, 106–122.

Tajfel, H., & Turner, J. C. (1986). *The social identity theory of intergroup behaviour*. Chicago, IL: Nelson-Hall.

Takaffoli, M., Sangi, F., Fagnan, J., & Zäıane, O. R. (2011). Community evolution mining in dynamic social networks. *Procedia—Social and Behavioral Sciences, 22*, 49–58. https://doi.org/10.1016/j.sbspro.2011.07.055.

Tan, W., Blake, M. B., Saleh, I., & Dustdar, S. (2013). Social-network-sourced big data analytics. *IEEE Internet Computing, 5*, 62–69.

Tiku, N. (2018). *Facebook is steering users away from privacy protections*. Retrieved from https://www.wired.com/story/facebook-is-steering-users-away-from-privacy-protections/.

Torfing, J. (1999). *New theories of discourse: Laclau, Mouffe and Žižek*. Oxford, UK: Blackwell.

United Nations Development Programme (UNDP). (1994). *Human development report*. New York, NY: Oxford University Press.

Vkontakte. (2017). *License and user agreement*. Retrieved from https://vk.com/licence.

Watts, D. J., & Strogatz, S. H. (1998). Collective dynamics of 'small-world' networks. *Nature, 393*, 440–442. https://doi.org/10.1038/30918.

Wæver, O., Buzan, B., Kelstrup, M., Lemaitre, P., & Carlton, D. (1993). *Identity, migration and the new security agenda in Europe*. London, UK: Pinter.

Wellman, B., Salaff, J., Dimitrova, D., Garton, L., Gulia, M., & Haythornthwaite, C. (1996). Computer networks as social networks: Collaborative work, telework, and virtual community. *Annual Review of Sociology, 22*, 213–238. https://doi.org/10.1146/annurev.soc.22.1.213.

Yang, B., Liu, D., & Liu, J. (2010). Discovering communities from social networks: Methodologies and applications. In B. Furht (Ed.), *Handbook of social network technologies and applications* (pp. 331–346). Boston, MA: Springer. https://doi.org/10.1007/978-1-4419-7142-5_16.

Yuan, T., Cheng, J., Zhang, X., Liu, Q., & Lu, H. (2015). How friends affect user behaviors? An exploration of social relation analysis for recommendation. *Knowledge-Based Systems, 88*, 70–84.

Zadeh, L. A., Abbasov, A. M., & Shahbazova, S. N. (2015). Fuzzy-based techniques in human-like processing of social network data. *International Journal of Uncertainty Fuzziness and Knowledge-Based Systems, 23*(1), 1–14. https://doi.org/10.1142/s0218488515400012.

Zamyatina, N. Y., &Yashunsky, A. D. (2018). Virtual geography of virtual population. *Monitoring of Public Opinion: Economic and Social Changes, 1*, 117–137. https://doi.org/10.14515/monitoring.2018.1.07.

Zhang, J., Fang, Z., Chen, W., & Tang, J. (2015). Diffusion of following links in microblogging network. *IEEE Transactions on Knowledge and Data Engineering, 27*(8), 2093–2106.

CHAPTER 11

The Interconnection Between Digitalisation and Human Security in the Lives of Sámi with Disabilities

Laura Olsén-Ljetoff and Liisa Hokkanen

11.1 Introduction

This chapter focuses on the impact of digitalisation as well as information and communication technologies (ICT) on the everyday lives of Sámi with disabilities living in Finland. The aim is to examine, through a critical

This study was enabled by collaboration between two projects, namely "Enablement Besides Constraint: Human Security and a Cyber Multi-Disciplinary Framework in the European High North (ECoHuCy)", which is financed by NordForsk (no. 81030), and "Strengthening Participation and Preventing Marginalization of Social Work Processes in Disability Services", which is financed by the European Social Fund (no. S20752). In addition to these projects, we would like to thank the interviewees for their significant contribution to the study.

L. Olsén-Ljetoff (✉) · L. Hokkanen
University of Lapland, Rovaniemi, Finland
e-mail: lolsen@ulapland.fi

L. Hokkanen
e-mail: liisa.hokkanen@ulapland.fi

M. Salminen et al. (eds.),
Digitalisation and Human Security, New Security Challenges,
https://doi.org/10.1007/978-3-030-48070-7_11

intersectional lens, how digitalisation and ICT affect the human security of Sámi with disabilities. Sámi persons with disabilities live as a minority within a minority and, therefore, they face some particular challenges due to their social position (Hokkanen, 2017; Olsén, Heinämäki, & Harkoma, 2018).

Strategies and action plans at both the international and national levels emphasise the importance of digital inclusion and the accessibility of digital services and ICT for everyone, with special attention being paid to persons with disabilities (e.g. European Commission, 2010, p. 5; United Nations General Assembly, 2007; Vesanen-Nikitin & Åkermarck, 2017). Studies on the subject indicate that digitalisation and ICT could provide empowering tools for persons with disabilities, provided that such tools were affordable and equally accessible to everyone regardless of disability as well as societal and environmental factors (Darcy, Maxwell, & Green, 2016; Kivistö, 2017; Macdonald & Clayton, 2013; Raja, 2016). Despite a good number of academic, governmental and non-governmental organisation (NGO) publications having dealt with these issues, indigenous persons with disabilities are still not widely discussed. In the present chapter, this issue is addressed by employing a case study approach to examine the particular situation of Sámi with disabilities.

In 2017, two studies involving Sámi persons with disabilities were conducted in Finland (Hokkanen, 2017; Olsén et al., 2018). As an outcome of the study carried out by Olsén et al. (2018), recommendations were presented with regards to actions and policies concerning Sámi with disabilities. In addition to these two studies, Sámi with disabilities living in Finland have been discussed in a few other papers (e.g. Faurie, 2011; Lehtola & Ruotsala, 2017; Magga, 1997, 1999). In Sweden, the situation is no better (e.g. Edin-Liljegren & Flykt, 2017; Uttjek, 2016), whereas in Norway more attempts have been made to consider the issue (e.g. Fedreheim, Melbøe, & Opsal, 2017; Gjertsen, Melbøe, Fedreheim, & Fylling, 2017; Huuva, Montefusco, & Olsen, 2014; Melbøe et al., 2016a, 2016b). Policy recommendations regarding Sámi persons with disabilities (Bengström, 2018) have also been formulated as an outcome of a Nordic project coordinated by the Nordic Welfare Centre on the initiative of the Norwegian Directorate for Children, Youth and Family Affairs. Issues related to digitalisation are touched upon briefly, for example, in the context of rehabilitation and assistive devices, although in general, the impacts of digitalisation and ICT on the everyday security of Sámi with disabilities are not addressed in depth in the research.

The concept of human security provides a framework for our research. In the *Human Development Report* (HDR) published by the United

Nations Development Programme (UNDP) in 1994, economic security, health security, personal security, political security, food security, environmental security and community security are described as the seven dimensions of human security (pp. 24–25). Although digital security is not explicitly mentioned as one of the dimensions, the interconnection between digitalisation and human development—and, in many cases, human security—is a recurring theme in the HDRs. However, only a few scientific publications have focused on digitalisation and human security with regards to people with disabilities (e.g. Alhaboby, Haider, Barnes, & Short, 2016).

A few articles discuss the interfaces of digitalisation, digital security, cybersecurity and human security, but they do not focus on persons with disabilities (Salminen, 2018, 2019; Salminen & Hossain, 2018; Zojer, 2019). Salminen (2019) considers digitalisation to be a phenomenon that penetrates many dimensions of human security and, therefore, integrates the human security perspective into the cybersecurity approach. Zojer (2019) focuses on analysing the interconnections between digitalisation and the seven dimensions of human security. We contribute to the discussion by analysing the significance of digitalisation to human security in the context of a minority group within indigenous people, that is, Sámi with disabilities living as a minority in Finland. The research material consists of 14 interviews conducted with Sámi with disabilities. To respect our interviewees' different positioning when compared to non-disabled Sámi or to Finnish persons with disabilities, we utilise an intersectional approach in our study.

The first part of this chapter introduces the theoretical framework employed in our study and also discusses the concept of human security. After that, the methodology, research material and ethical choices are presented. The second part of the chapter consists of the analysis, which is divided into two sections addressing (1) the intersectionality between digitalisation and personal security and (2) the intersectionality between community security aspects and digitalisation in the lives of Sámi with disabilities. At the end, the conclusions are made based on the discussions and analysis presented in this chapter.

11.2 Sámi with Disabilities Through Intersectionality

The Sámi are recognised as indigenous people within the Constitution of Finland (The Constitution of Finland, 1999, Ch. 2, S. 17, para. 3). The Sámi are one people who live across four states. However, there are different linguistic groups within the Sámi and, further, the traditional livelihoods and culture of the Sámi may vary according to their living circumstances and traditions. In Finland, there are three different Sámi linguistic groups: Skolt Sámi, Inari Sámi and Northern Sámi. In 2015, in Finland, 2.4% of the Sámi spoke Inari Sámi as their mother tongue, 3.1% spoke Skolt Sámi and 17.3% spoke Northern Sámi, while the rest spoke either Finnish or some other language (Lehtola & Ruotsala, 2017, p. 5). People who have learned one of the Sámi languages later in life are not included in the statistics and, therefore, the total number of speakers of the Sámi languages is actually higher than reported (Lehtola & Ruotsala, 2017, p. 5). As the given percentages indicate, Inari and Skolt Sámi speakers are minority groups within the Sámi in Finland. Internationally, there is no commonly agreed definition of those groups, which can be considered as minorities. However, under international law, the term "minority" often refers to national or ethnic, religious and linguistic minorities. When it comes to persons with disabilities, they are mentioned as, for example, persons, who can experience discrimination on the grounds of disability (United Nations, 2010, pp. 1–2).

A person who is a Skolt or Inari Sámi belongs to a linguistic minority within the Sámi, who are, in turn, ethnic and linguistic minorities and indigenous people in Finland.[1] In addition, persons with disabilities are a numerical minority in Finnish society (e.g. Nurmi-Koikkalainen et al., 2017, pp. 34–39). Hence, in this study, we view persons with disabilities or impairments as a minority group. This means that if an Inari or Skolt Sámi person is living with a special need or an impairment, they belong to at least three different minority groups (also Heinämäki, 2019). Belonging to several minorities, as well as having the associated intersecting axes of identity, affect people's perceptions of the impact of digitalisation, for example, on their personal and community security.

The possible threats to personal security listed in the HDR (1994) consist of several forms of threats ranging from "threats from other groups of people (ethnic tensions)" to "threats to self (suicide, drugs use)" (p. 30). A person who is a Sámi and who lives with a special need or an impairment can face, for example, different types of multiple discrimination in their daily life (Heinämäki, 2019; Hokkanen, 2017; Olsén et al., 2018), which have an impact on person's personal security. For

[1] See also Article 27 of the International Covenant on Civil and Political Rights (United Nations General Assembly, 1976).

instance, they may not have access to essential services in their own language or they may be discriminated against on the grounds of their impairment (e.g. Heinämäki, 2019; Olsén et al., 2018).

Belonging to a community can foster a sense of security and play a significant role in life, for example, by providing a cultural identity (UNDP, 1994, p. 31). However, community can also create insecurity in the lives of individuals, for example, through certain traditional practices (UNDP, 1994, p. 31). Sámi with disabilities can face discrimination among the Sámi community, which can, according to the contemporary definition of community security, be viewed as a threat to community security (also UNDP, 2009, pp. 13–14). In addition, community security should, in the context of the Sámi living in Finland, be viewed in light of Sámi people's position within the dominant population and their social structures (Amatulli & Klein, 2018, p. 152). As provided by the Constitution of Finland, the Sámi have a right as indigenous people to maintain and develop their own language and culture (Constitution of Finland, 1999, Ch. 2, S. 17, para. 3), which is also an external component of community security (Amatulli & Klein, 2018, p. 153).

Intersectionality has been identified as a valuable approach when examining issues concerning persons who identify themselves as belonging to more than one minority. In this chapter, we apply intersectionality to analyse the impacts of digitalisation on the personal and community security of a group of persons who are in a minority position in relation to both the dominant society and the Sámi community.

In the socio-legal discussion, the roots of the concept of intersectionality are often traced back to Kimberlé Crenshaw's (1989) article, in which she analyses intersectionality in the context of the discrimination faced by black women (p. 149; also Olsén et al., 2018, pp. 23–24). The concept is often used to concretise how different axes of identity—in the present chapter, ethnicity and disability—position a person in society and to render it possible to define the links between certain axes and power constellations (Van Schuylenbergh, Motmans, & Coene, 2018, p. 123). Intersectionality allows us to perceive the social position of individuals by linking it to different axes or categories (Nguyen & Mitchell, 2014, p. 328). Further, it allows us to view how the positions of these axes are visible in people's lives (Phoenix & Pattynama, 2006, p. 187). In some cases, disability has been dismissed as an intersecting category because it is considered to be only a biological or medical problem (Erevelles, 2011, p. 103; Nguyen & Mitchell, 2014, p. 329). However, intersectional orientation occurs frequently in societal disability studies in which the intersectional approach is employed to study disability as connected to ethnicity, gender, age, class, sexuality or policy (e.g. Hansen, Wilton, & Newbold, 2017).

Constructing the social model of disability (Oliver, 1990) represented a shift in focus for the disability rights movement—both in the context of policy making and in academia—in terms of understanding and conceptualising disability. Rather than being perceived as a feature attributed to an individual, disability came to be regarded as a feature of society: issues that used to be "about *them*" became issues "about *us*". The social model of disability has been developed further in relational models in which intersectionality exists in the basic orientation, although it may not be explicitly conceptualised as such. There is an ongoing debate concerning the "right" way to understand disability and the related phenomena among different strands of critical disability studies and different medical, social, relational, political and human rights models (e.g. Katsui, 2005; Sherry, 2016; Tøssebro, 2004; Vandekinderen & Roets, 2016; Vehmas & Watson, 2016). The social dimensions of phenomena are no longer denied and, recently, policy changes have highlighted the need to also consider these aspects (Oliver & Barnes, 2010). We focus on disability as a societal phenomenon interwoven with other aspects of life, such as Sáminess. Together, those aspects construct the multiple faces of disability at the individual and societal levels. We examine this by adopting the intersectional approach. The social position of Sámi persons with disabilities is rooted in both their *Sáminess* and their *disability* (Hokkanen, 2017; Olsén et al., 2018). In this chapter, we examine how digitalisation is interconnected with human security in the everyday lives of people who live at the intersection of Sáminess and disability.

11.3 Understanding Human Security

The idea behind human security first began to emerge in the social contract theories of the seventeenth century (Hossain, 2013, p. 497), while in the United Nations (UN) context, the concept dates back to the 1940s (Szpak, 2017, p. 81). Following the end of the Cold War, security paradigms evolved from the classical state-centric approach towards a more complex understanding of security (Hossain, 2013; Szpak, 2017, pp. 80–81). In the 1993 HDR, the new concept of human security was introduced, which represented a call for the redefinition of the concept of security with the objective of rendering it more people-centric (UNDP, 1993; also Szpak, 2017, p. 81). Instead of armaments, human development was stressed as a means of achieving human security (UNDP, 1993, p. 2). In 1994, one of the most commonly used definitions of the concept of human security—"freedom from fear and freedom from want"—was formulated and the seven above-mentioned categories of human security were introduced (UNDP, 1994, pp. 24–25).

In 2012, the UN General Assembly adopted a resolution that defined human security as: "[...] an approach to assist Member States in identifying and addressing widespread and cross-cutting challenges to the survival, livelihood and dignity of their people" (Article 3). In line with the definition set out in the 1994 HDR, the notion of human security is considered to include freedom from fear and freedom from want, as well as the right to live with dignity. Everybody, particularly vulnerable people, should have "[...] an equal opportunity to enjoy all their rights and fully develop their human potential" (UN General Assembly, 2012, Article 3, a).

Both indigenous peoples and persons with disabilities are considered to be vulnerable groups, the members of which can be marginalised in societies, and thus special attention is paid to them (e.g. UNDP, 2016, p. 14). Based on global recognition of the continued discrimination faced by persons with disabilities and the general ignorance of their human rights, a need for a special convention was identified by the UN and the Convention on the Rights of Persons with Disabilities and its Optional Protocol (UNCRPD) was adopted in 2006 and entered into force in 2008 (UN General Assembly, 2007). In the Convention's preamble, in paragraph (p), the situation of indigenous persons with disabilities is mentioned due to the risk of them facing multiple or aggravated forms of discrimination. Moreover, in the UN Declaration on Rights of Indigenous Peoples (UNDRIP), the need to pay particular attention to the rights and special needs of persons with disabilities is stressed (UN, 2007, Article 21, para. 2, Article 22).

Human security is strongly interconnected with both human development and human rights (e.g. Zojer, 2019). The 2016 HDR highlights human security as a "precondition for human development" (p. 88), while many development goals, for example, those pursued by the UN's development agencies, are linked to dimensions of human security, as defined in the 1994 HDR (Hossain, Zojer, Greaves,

Roncero, & Sheehan, 2017, p. 57; UN, 2000, 2015; Zojer, 2019). In addition, human rights conventions endorse the human security principles in the context of international law and often impose obligations on states to implement them in national legislation (Hossain et al., 2017, p. 57). Human development is said to be "anchored in human rights" (UNDP, 2016, p. 25), and human rights conventions also promote the core principles of human security (e.g. Hossain et al., 2017, p. 58), which plays a key role in making human development possible (e.g. UNDP, 1994, p. 23).

The notion of human security holds that threats to the security of people extend beyond the need for freedom from physical violence (being free from fear) (e.g. Hossain et al., 2017, p. 54). Instead, the scope of potential threats is seen to include everything that threatens individuals' and communities' survival or their possibilities to fulfil their basic needs, to define and express their identity, or to belong to a certain community (being free from want) (e.g. Hossain et al., 2017, p. 54; Zojer, 2019). In addition to looking at *negative security*, which commonly refers to security from threats, the notion of human security enables us to consider *positive security*, which refers to security as enabling (Hoogensen Gjørv, 2012).

The risk of the concept of human security being used as a means to impose Western values on non-Western cultures has been noted in the academic literature (e.g. Hoogensen Gjørv, & Stuvøy, 2006, p. 216; Hoogensen Gjørv, Bazely, Goloviznina, & Tanentzap, 2014; Hossain et al., 2017, p. 56). In his article entitled *The Security of the Sámi People,* Sheehan (2016) emphasises that there is no commonly agreed upon understanding of the meaning of security. He further notes that the concept is socially constructed and shaped by the particularities of each social context, and he stresses the need to view the Sámi, for example, as a heterogeneous group of people living in diverse circumstances and facing a variety of challenges (Sheehan, 2016, pp. 139–141). Being aware of this, we conducted focused interviews by not giving the interviewees any strict definition of human security beforehand, but rather by discussing the meanings of security and digitalisation in their lives in an open, semi-structured format.

When analysing the impact of digitalisation on the everyday security of Sámi with disabilities, issues related to a person's possibilities to fulfil their basic needs play a significant role. Examples of these possibilities include having adequate financial resources, having access to assistive devices and rehabilitation, education, work and activities, as well as having the possibility to create and maintain social relations and to participate in community and societal life. The impact of digitalisation on a person's everyday life through social media, computer games, different devices and software for e-Health is growing. Additionally, issues related to a person's survival are discussed because digitalisation plays a key role in getting help in an emergency when living in sparsely populated areas.

The concept of human security has been criticised for vagueness, given its flexibility and its ability to cover a wide variety of issues (Paris, 2001, p. 87, 91; Zojer, 2019, p. 301). However, as Zojer (2019) points out, there has been discussion about whether or not there exists a need to create a strict definition of the concept, since it is intended as a tool for identifying threats and challenges faced by people and not as an academic perspective for studying security (pp. 301–302). As an intersectional lens places an emphasis on the individual's and the minority communities' experiences and understanding of security, it reveals the highly multifaceted and fragmented nature of threats as well as the enabling features of security.

In this chapter, human security and, in particular, personal and community security provide a theoretical framework in addition to serving as analytical concepts that will be supplemented by the perceptions of Sámi with disabilities when it comes to the impacts of digitalisation on their lives.

11.4 Digitalisation as a Cross-Cutting Theme in Human Security

Digitalisation and ICT play an ever-growing role in societies worldwide, and various online and offline modes are now blended in people's daily lives (Granholm, 2016). Digitalisation penetrates all seven dimensions of human security (Zojer, 2019). For this reason, the vulnerability of individuals in ever-changing digital environments as well as the significance of the digital divide have been addressed in the research literature (e.g. Raja, 2016; Salminen, 2019).

Digitalisation, particularly in relation to the digital divide, has been discussed to some extent in a number of HDRs over several years (e.g. UNDP, 2001). In the HDRs, the term "digital revolution" begins to appear more frequently from the 2015 report onwards. The positive aspects of the digital revolution are highlighted and the terms "enabling" and "empowering" appear often. However, challenges also emerge. For instance, in the 2015 HDR, the role of digitalisation in transforming working life is thoroughly discussed (UNDP, 2015). In the 2016 HDR, issues related to digitalisation and the digital revolution are again highlighted. It becomes clear that digitalisation has penetrated more and more areas of people's lives, and it has been integrated into an increasing number of social structures. The 2016 HDR continues to embrace the enabling potential of the digital revolution, although it also points out the relevant challenges and, according to our interpretation, the threats to human development and to freedom from want (p. 51). Even though human development and human security should not be understood as being identical (Gómez & Gasper, 2013, p. 3), it is possible to see a strong connection between the two concepts.

In the case of persons with disabilities, accessibility is a focal theme with regards to human rights. In Article 9 of the UNCRPD, the responsibility for ensuring access "to information and communications, including information and communications technologies and systems" for persons with disabilities is highlighted. To achieve this objective, barriers and obstacles to accessibility must be identified and eliminated (Article 9). Further, measures must be taken "to promote other appropriate forms of assistance and support to persons with disabilities to ensure their access to information" (Article 9, f.), "to promote access for persons with disabilities to new information and communications technologies and systems, including the Internet" (Article 9, g) and "to promote the design, development, production and distribution of accessible information and communications technologies and systems at an early stage, so that these technologies and systems become accessible at minimum cost" (Article 9, h) (see also Olsén et al., 2018, pp. 125–132). In addition, internet accessibility is considered to be connected to the realisation of the requirements of both Article 21, concerning freedom of expression and opinion, and Article 29, concerning participation in political and public life (Gustafsson, 2016, p. 12; Olsén et al., 2018, p. 130), which also secures a person's freedom from want.

Finland ratified the UNCRPD in 2016, and issues concerning people with disabilities have been increasingly gaining attention in Finland's human rights policies since then (Olsén et al., 2018, pp. 83–90). Aside from the context of human rights, the accessibility of communications services is discussed in international strategies (e.g. European

Commission, 2010) as well as in national action programmes concerning the accessibility of digital transport and communication services, particularly from the perspective of the opportunities and challenges it represents to persons with disabilities (e.g. Vesanen-Nikitin & Åkermarck, 2017). The lack of services available in sign language and in the Sámi languages, the slow development of new services and the impracticality and inaccessibility of existing services are all mentioned as challenges (Vesanen-Nikitin & Åkermarck, 2017, pp. 47–54). According to recent research, accessibility is not just about the language of services, but also about culturally adequate services characterised by cultural competence and sensitivity (Heinämäki, 2019; Hokkanen, 2017, pp. 96–98). The Finnish Ministry of Transport and Communications published a report in which the barriers or obstacles to the use of transaction and information search services for visually and hearing-impaired persons, for persons with learning difficulties and for elderly people were investigated (Åkermarck, 2014). Many of the aforementioned and other identified issues are connected to human security, although they are neither examined through the framework of human security nor discussed in light of the social position based on the intersection of different minority or identity axes.

11.5 Methodology, Research Material and Ethical Choices

As we are investigating the situation of one indigenous people with internal minority groups, a participatory methodology has been applied whenever possible, while always remaining aware of the high risk that the interviewees could be identified within the community. The original idea for this study stemmed from the research project conducted in 2017 by Olsén, Heinämäki and Harkoma, which focused on human rights and multiple discrimination of Sámi persons with disabilities and sexual and gender minorities (Olsén et al., 2018). The project was carried out by using semi-structured interviews. The interviewee who inspired us to write this chapter, as well as other stakeholders relevant to the topic, was consulted before the decision to write it was made. In addition, in 2017, Hokkanen conducted a study on *Experiences of inclusion and welfare services among Finnish Sámi with disabilities* and, in the narratively oriented, themed interviews conducted during the research process, the linkage between certain human security dimensions and digitalisation emerged.

For the purposes of the present chapter, ten interviews—later "narrative interviews"—conducted as part of Hokkanen's (2017) previous research were used for the analysis. With regards to this particular topic, we conducted four semi-structured interviews—later "focused interviews"—with Sámi persons with disabilities, of which each of us conducted two. In addition, the corresponding author of this chapter approached the topic by consulting an expert who works on a daily basis with matters concerning the minority group in question. The focused interviews were conducted during the winter and spring of 2018.

The interviewees were all recruited via the snowball sampling method using the authors' existing networks, an approach that had worked well in Hokkanen's (2017) earlier research project. Since Sámi persons with disabilities, as a target group, are not easily reachable, the snowball sampling method seemed the most appropriate way to find interviewees (Hokkanen, 2017, p. 30). Although the snowball sampling method has its advantages, including speed and the ability to uncover the silent issues, a risk exists that later informants will be drawn from the same network as earlier informants, meaning that the pictures of phenomena may be narrowed (Heckathorn, 1997). We thus made every effort to avoid the risk of one-sided or selective information by attending to as many different subgroups as was feasible.

All the interviews were recorded and transcribed. We consider combining the narrative and focused interviews in an intersectional manner to represent a fruitful approach for understanding the content and the context of phenomena, which is also ethically essential. The narrative interviews allowed us a wider perspective, while the focused interviews enabled a deeper inspection of the phenomena. Further, the narrative interviews function as a kind of societal reliability test for the focused interviews, while the focused interviews offer us a glimpse of the future possibilities of digitalisation within the framework of human security, if the identified obstacles are removed.

Informed consent to participate in the research was obtained from all the interviewees, and the researchers took appropriate steps to maintain the anonymity of the interviewees throughout the participatory research process. Some interviewees, on their own initiative, engaged in conversations with people about having participated in the study and about the topics they found to be important. All the research material was handled according to the requirements of the European Union's General Data Protection Regulation (European Parliament, 2016).

The analysis process was based on intersectionality. During the first phase of the analysis, both researchers engaged in the close reading of, and listening to, their own material, and they made their own interpretations based on that material. During the second phase, the initial findings were jointly reflected upon in generative discussions, albeit without the sharing of authentic interview material. In this way, new interpretations emerged from the findings, thereby leading to reconstructed understandings of the phenomena. Intersectional discussions and writing played an essential role in the analysis process, during which we shared our earlier understandings and together reconstructed a new synthesis. In addition, we conducted face-to-face and phone discussions with all the persons interviewed for the focused interviews, as well as with other experts, so as to reflect on and test our findings in different situations.

Although all the persons interviewed for the focused interviews reviewed the chapter or, at least, the outcomes of the research prior to submission, the interpretations presented here were made by the authors and so cannot be seen as representing "the truth", or even a generalisation, of the views of Sámi persons with disabilities. Due to the small number of Sámi with disabilities as well as the sensitivity of the topic discussed, verbatim quotations are not used in the text, while details concerning the interviewees' gender, age, place of residence, disability, special needs or linguistic background have been omitted. All the interviewees have attained the age of majority (i.e. 18 years of age in Finland) and live either by themselves or with their families. All the persons interviewed are Sámi and they are living with physical, cognitive or sensory impairments or special needs. Nearly all the interviewees have lived, or still live, in the Sámi homeland.

The ethical aspects of research conducted among the Sámi have recently been discussed by scholars doing research on Sámi issues (e.g. Drugge, 2016), although at the time of writing this chapter, the development of ethical guidelines was still ongoing in Finland (e.g. Nygård & Saus, 2016). However, as both the authors come from outside the Sámi community, although they both have previous experience of conducting research among the Sámi and among persons with disabilities, ethical questions relevant to indigenous and disability studies were taken into careful consideration.

11.6 Concretising the Connections Between Digitalisation and Human Security

In the following two sections, we consider both the negative (threats) and the positive (enabling) aspects of personal and community security when analysing how digitalisation can be seen as an enabling means for strengthening the everyday security of the interviewees.

11.6.1 Intersectionality Between Digitalisation and Personal Security

In this section, we examine the devices, software, computer programmes and applications through which digitalisation manifests in the everyday life of Sámi with disabilities. Additionally, we analyse the ways in which digitalisation is connected to the economic situation, education, work, services and rehabilitation of Sámi with disabilities, which we view as aspects of personal security.

All the interviewees use ICT in some form or other. The appearance of digitalisation in their life is sometimes unrecognised and invisible, while other times it is prominent and significant. They mention devices such as smartphones, computers, smartwatches and global positioning systems (GPS) as features of their daily life, and they describe using the internet, social media and a diversity of software, computer programmes and applications in everyday life.

The narrative interviews indicate that some of the interviewees are experienced with regards to digitalisation, most are not particularly familiar with it and a few consider digitalisation to be unattractive or more valuable for the next generation (also Macdonald & Clayton, 2013; Salminen, 2019, pp. 195–196). Due to its growing significance in daily life, more effort is required to make digitalisation attractive and suitable for everyone as well as to embed it more smoothly in everyday matters.

During the focused interviews, digitalisation emerges as a more integral part of the interviewees' lives. They are well versed in the available devices, the internet and digitalisation, and they are willing to learn new things. They use apps of different kinds related to disabilities or impairments as well as apps that are commonly used in everyday life in Finnish society. Further, they spend a lot of time in digital environments. However, some of them face difficulties in terms of using the apps encountered in everyday life and, therefore, in gaining access to services, which puts them in an unequal position when compared to people who do not face the same challenges related to their disabilities. Phones are mentioned as belonging to the basic infrastructure of safety, and the interviewees consider, for example, the emergency application *112 app* to be helpful in this regard. Digitalisation is also needed to render movement safer in all environments all year round. Safety issues are extremely important when living in sparsely populated areas where the distances are long, only a few people are around and the environmental conditions pose particular challenges during wintertime. When the freedom and safety of movement are restricted, a person's choice of residence is likely to be influenced by that restriction.

In Finland, as elsewhere, persons with disabilities live in poverty and experience financial troubles to a greater extent than other Finns. According to a recent evaluation, about one-fifth of people with disabilities have had to cut consumption significantly in everyday life, compared to just one in ten Finns (Nurmi-Koikkalainen et al., 2017, pp. 41–42). Our interviewees manifest resilience with regards to financial problems, although they mention a lack of money as a reason for not being able to purchase apps or devices that would help them to become more equal to non-disabled and non-Sámi people. A lack of money is also mentioned in the context of housing, travelling, rehabilitation, services, hobbies and relationships, all of which constitute core elements of personal security and all of which are also linked to community security.

In terms of their financial situation, entry into education and the labour market has been shown to be more limited for people with a disability than for those without a disability (Nurmi-Koikkalainen et al., 2017). Our interviewees report having experienced difficulties stemming from educators' limited capability for inclusive pedagogy and their lack of awareness of the opportunities that digital solutions can offer to people with disabilities. Due to the extreme rarity of digital educational solutions in the Sámi languages, Sámi with disabilities are forced to assimilate by using Finnish when inclusive solutions are required. This leaves Sámi with disabilities in a situation in which they experience multiple discrimination on the grounds of their disabilities and of belonging to a linguistic and ethnic minority (also Heinämäki, 2019). Moreover, the situation can be viewed in the context of community security, as the community in which the interviewees live is not able to guarantee equal opportunities for Sámi with disabilities.

Digitalisation is also present in unnoticeable ways in the lives of the interviewees via various tests, medical treatments, therapies and devices. The interviewees do not tend to associate these with digitalisation, but rather with security and welfare. They are solutions that have saved lives and make lives easier and, therefore, they are deeply needed. However, many of the interviewees are uncertain as to the usability and reliability of the solutions as well as unsettled due to their versatile knowledge of the nature of the digitalised world. A number of interviewees are concerned about risks such as the possibility of their assistive devices or information on certain databases being hacked, or of falling victim to identity theft (also Salminen, 2018, p. 199). Personal security and digitalisation hence seem to have a complex connection to welfare services, health care and devices.

To clarify the interconnections between the aspects discussed above, one interviewee reports having invented their own ways of coping with the challenges of everyday life related to their disabilities through self-innovated digitally based solutions. The interviewee notes that some of the software developed to assist persons with disabilities is expensive, might only function on particular—and often too expensive—devices and are only available in Finnish or English. Native speakers of Sámi languages do not have access to such assistive software in their mother tongue. The possibility of expressing one's opinion freely is seen as a basic element of freedom from want. Although digital assistive devices and software serve as a means of enablement in people's lives and also strengthen their sense of personal security, accessibility remains a challenge and the possibility of negative impacts cannot be excluded. The linkages between the different dimensions of human security are obvious (also Zojer, 2019), as a person's economic situation plays a significant role in, and sets limits to the realisation of possibilities.[2]

11.6.2 Intersectionality Between Community Security Aspects and Digitalisation

In this section, we examine both the negative and the positive aspects of digitalisation as well as their connection to community security, particularly the cultural aspects of it seen in the lives of Sámi with disabilities.

[2]See also the definition of economic security contained within the 1994 HDR (pp. 25–26).

The conditions analysed in the previous section partly overlap with the cultural aspects, although the focus here will be on social relations, community, language, the significance of nature and traditional livelihoods and the identity of the interviewees.

Communities in the European High North are facing new challenges concerning their integrity (Zojer, 2019, p. 315). Many people are moving out of the region, for example, to access education (Heleniak & Bogoyavlenskiy, 2014). However, digitalisation can help communities to cope with the challenges stemming from this development (Zojer, 2019). The significance of digitalisation to social networks and relationships is emphasised by most of the interviewees. Digitalisation and social media provide ways of staying in touch with family, relatives, friends, Sámi communities and various peer communities for persons with disabilities. The meaning of family in the widest sense is essential in Sámi culture (e.g. Hokkanen, 2017, pp. 67–68; Uttjek, 2016, pp. 34–36), and when living outside the Sámi homeland region, maintaining contact with family members and the Sámi community is vitally important. This also holds true for family members living outside the homeland. The possibility of following the lives of relatives and Sámi communities via social media appears meaningful even without active participation in the communication.

Digitalisation enables Sámi people to use their languages and to maintain contact with their culture via the internet (also Zojer, 2019), which is particularly important for Sámi with disabilities, for whom travelling long distances can prove challenging. As Heinämäki (2019) notes, migration to the large cities is likely to continue in Finland and this transition will challenge the availability of social and health services and education delivered in Sámi languages and in a culturally sensitive way (see also Sámi Parliament, 2017). According to our findings, digitalisation provides some solutions to this issue, albeit depending on investments in easy-to-use digital services and solutions as well as on an individual's digital skills, awareness and possibilities (also Salminen, 2018).

Sámi people's relationship with nature is presented as a core element of their culture (Hokkanen, 2017, p. 70; Melbøe et al., 2016a, pp. 57–58; Uttjek, 2016, pp. 38–41). In our material, it emerges as an intimate or practical relationship constituting the basis of the Sámi way of life. Digitalisation provides a wide range of tools for traditional livelihoods, including saving and sharing traditional knowledge and expertise, communicating over long distances and with hard to reach locations, and utilising software concerning spatial information. An increasing amount of information on nature and livelihoods is available on the internet, but for example some traditional ecological knowledge (TEK) and skills in Sámi communities are only passed down through generations via practice and oral histories and communities are willing to keep some knowledge to themselves, too (e.g. Helander-Renvall & Markkula, 2017, pp. 105–116; Porsanger & Guttorm, 2011, p. 18). Although the development of ICT and digitalisation enables Sámi persons with disabilities to follow and learn about traditional livelihoods in new ways, challenges remain. For example, owing to the nature of TEK its digitation is problematic, and digitation could have unpredictable impacts on the culture surrounding traditional Sámi livelihoods.

In digital environments, people can decide, how much and to whom they want to disclose about their disability, ethnicity or other life circumstances. In many situations in the offline world, persons with disabilities feel that they are met through their disability, which overshadows both their individual personality (Olsén et al., 2018, pp. 137–139) and their Sáminess. Disability can emerge as a defining factor in their everyday life. Our interviewees report not sharing all the details of their disability openly and to everyone on the internet. This finding differs from the finding of Furr, Carrairo and McArthur (2016) concerning the disclosure strategies of persons with physical disabilities: The persons studied participated on social media on a regular basis. Furr, Carrairo and McArthur identified three strategic approaches to disability disclosure on social media: *open*, *secured* and *limited*. The open approach means full disclosure regarding a disability to disabled and non-disabled audiences on social media. The secured approach refers to open disclosure on social media exclusively for disabled people, but extremely selective disclosure to non-disabled audiences. The limited approach is adopted when a person almost never shares details about their disability to any audience on social media. Most of our interviewees do not share their disability openly in public arenas (i.e. they adopt the limited strategy of disability disclosure on the internet and social media). Their disability disclosure is selective and limited to a few people and, generally, they do not focus on disability on social media. Regarding their Sáminess, the interviewees are more open and so employ all three disclosure strategies.

The significance of communities and social networks established around computer games emerges in the interviews (also Hokkanen, 2017, pp. 72–73). Such communities highlight a person's abilities rather than their disabilities. Computer games provide meaningful activity and pleasure and, on some occasions, also rehabilitation, mental wellbeing and/or escape from reality. However, games can have negative effects on personal security, too—one being addiction. The interviewees are aware of this and have found ways to manage the risk. They also mention bullying within gaming communities, which they perceive as normal. In our view, the "normalisation" of bullying and harsh speech reflects everyday discrimination based on disability (also Burch, 2018). Although computer games provide one interviewee with an important network and a way to pass the time, they still desire more relationships and meaningful action in the offline world.

The decision to hide or cover up disability or ethnicity in the online world is linked to the values of communities and societies in general as well as to the interviewees' previous experiences of discrimination and their fear that the same kinds of experiences might also characterise online relations (on hate speech on the internet, see Burch, 2018). Most of the interviewees reflect on how the digital environment provides a space in which they can decide how to present themselves to different audiences. They are not willing to be restricted by either the notion of disability or that of Sáminess. Rather, they want to present themselves and to be perceived as unique individuals with specific interests. This may be seen as reflecting their feeling of community security as well as how they feel the Sámi, as indigenous people, are positioned in relation to the dominant population in Finland. At the same time, it might reflect their feelings regarding the position of people with disabilities in both Sámi communities and Finnish society.

The interviewees use social media to maintain, develop and vitalise Sámi language, culture and identity. Sámi language networks on social media provide a platform for using the languages in everyday life and for keeping up with developments of the languages. For many of our interviewees, internet-related platforms and social media make it possible to practise and test the language at their own pace and with a selected audience, which is highly important to some persons with disabilities, for whom learning and using written and/or spoken language is demanding. If a person becomes disabled in later life, revitalising and relearning their language, culture and identity through social media can form a significant part of their recovery process. It is noteworthy that in addition to the written Sámi languages, social media and certain software provide access to spoken languages, thereby rendering them more widely available and more usable.

The impacts of digitalisation and ICT on the personal and community security of Sámi persons with disabilities, as discussed in this chapter, do not provide a comprehensive account of the subject. However, our research offers insights into the ways in which an intersectional examination of solitary facts can begin to reveal a world in which everyday human security is not always secure.

11.7 Conclusion

In this chapter, we have analysed—based on both narrative and focused interview materials—how digitalisation and the development of ICT affect the personal and community security of Sámi persons with disabilities. When analysing these issues, it is crucial to be aware of whose understanding of security creates the basis for the analysis. In different cultures and among different minority groups, the meaning of security varies, and no universal definition of security exists (e.g. Sheehan, 2016, pp. 139–141). Additionally, the diversity of disabilities as well as understandings of disability (Katsui, 2005) challenge the different dimensions of security.

Intersectionality provides a targeted approach to our study by allowing us to focus on a minority or marginalised group within a minority, that is, Sámi with disabilities. This means that our interviewees are differently positioned when compared to non-disabled Sámi or to Finnish people with disabilities (also Olsén et al., 2018). Intersectionality thus allows for a more comprehensive and in-depth understanding of the internal heterogeneity of both Sáminess and disabilities.

In our analysis, we first address both the positive and negative impacts of digitalisation on the personal security of Sámi with disabilities. Digitalisation has afforded new possibilities for getting help in an emergency, and it has also transformed medical treatment, rehabilitation, welfare services, therapy and assistive devices. Furthermore, as more knowledge accumulates in this area, digitalisation could be used to enhance the inclusiveness of education systems and labour markets, which should further improve the personal security of Sámi with disabilities. A person's opportunities to benefit from digitalisation are also strongly linked to person's economic situation, as the latest digital solutions are generally expensive and the income of persons with disabilities is generally lower compared to non-disabled persons. Moreover, a lack of financial resources is frequently linked to a person's education and labour market situation.

Community security is the focus of the second section of our analysis, in which we discuss the impacts of digitalisation on social relations, Sámi communities, traditional livelihoods, languages and identity. Social relations, communities as well as Sámi languages and culture are at the core of community security for Sámi people. They are also relevant to personal security because they can increase a person's sense of personal security and provide a buffer against adversities. Digitalisation has provided new forms and forums for the elements required to maintain a culture, including learning and using the Sámi languages, thereby strengthening both community and individuals. Then again, the online and offline worlds are blended, meaning that discriminatory practices from the offline world can enter online communities and communication, and vice versa. Such discriminatory practices make it challenging for an individual to form an identity comprising both Sáminess and disability without being restricted by either aspect. The impacts of digitalisation are often considered to be positive for both individuals and communities, although there are also challenges, for example, in relation to traditional Sámi livelihoods.

Our study shows that, with regards to mainstream digital development, it is important to take into account the special needs of different minority groups, so that such development can equally benefit all of society. The interconnections between personal and community security and digitalisation are evident, and digitalisation is integrated in a complex manner into people's everyday lives, which has been demonstrated in this chapter in the particular context of Sámi with disabilities.

References

Åkermarck, M. (2014). *Paina vihreää nappia! Asiointi- ja tiedonhakupalveluiden käytön esteitä erityisryhmien näkökulmasta* [Press the green button! User obstacles met by special groups in transactions and information searches]. Helsinki: Publications of the Ministry of Transport and Communications 23.

Alhaboby, Z. A., Haider, M. A. K., Barnes, J., & Short, E. (2016). 'The language is disgusting and they refer to my disability': The cyberharassment of disabled people. *Disability & Society, 31*(8), 1138–1143. https://doi.org/10.1080/09687599.2016.1235313.

Amatulli, G., & Klein, J. (2018). Community security of indigenous peoples in the Barents Region. In K. Hussein & D. Cambou (Eds.), *Society, environment and human security in the Arctic Barents region* (pp. 151–167). London and New York: Routledge.

Bengström, J. (2018). *Personer med funktionsnedsattning och samisk-bakgrund—policyrekommendationer* [People with disabilities and Sámi background—policy recommendations]. Nordic Welfare Centre. Retrieved from https://issuu.com/nordicwelfare/docs/policyrekommendationer?e=30952623/58393450.

Burch, L. (2018). 'You are a parasite on the productive classes': Online disablist hate speech in austere times. *Disability & Society, 33*(3), 392–415. https://doi.org/10.1080/09687599.2017.1411250.

Crenshaw, K. (1989). Demarginalizing the intersection of race and sex: A black feminist critique of antidiscrimination doctrine, feminist theory and antiracist politics. *University of Chicago Legal Forum, 1*, 139–167.

Darcy, S., Maxwell, H., & Green, J. (2016). Disability citizenship and independence through mobile technology? A study exploring adoption and use of a mobile technology platform. *Disability & Society, 31*(4), 497–519. https://doi.org/10.1080/09687599.2016.1179172.

Drugge, A. L. (Ed.). (2016). *Ethics in indigenous research: Past experiences—Future challenges.* Umeå: Vaartoe Centre for Sami Research, Umeå University.

Edin-Liljegren, A., & Flykt, K. (2017). *Förutsättningar och villkor för samer med funktionnedsättning i de samiska förvaltningskommunerna i Sverige* [Circumstances and conditions for Sámi people with disabilities in the Sámi administrative municipalities of Sweden]. Stockholm: Nordic Welfare Centre.

Erevelles, N. (2011). *Disability and difference in global contexts: Enabling a transformative body politic.* New York, NY: Palgrave Macmillan.

European Commission. (2010). *European disability strategy 2010–2020: A renewed commitment to barrier-free Europe (Communication from the commission to the European parliament, the council, the european economic and social committee and the committee of the regions).* Brussels 15.11.2010. COM (2010) 636 final. Retrieved from https://eur-lex.europa.eu/LexUriServ/LexUriServ.do?uri=COM:2010:0636:FIN:en:PDF.

European Parliament. (2016). *Regulation (EU) 2016/679 of the European parliament and of the council of 27 April 2016 on the protection of natural persons with regard to the processing of personal data and on the free movement of such data, and repealing directive 95/46/EC (General Data Protection Regulation).* Retrieved from https://publications.europa.eu/en/publication-detail/-/publication/3e485e15-11bd-11e6-ba9a-01aa75ed71a1/language-en.

Faurie, M. (2011). *Saamelainen Kelan asiakkaana. Saamelaisten oikeudet ja asema alkuperäiskansana Kelan toimeenpanemassa sosiaaliturvassa* [A Sámi as a customer in Kela. The rights and status of the Sámi as an indigenous people in the social security implemented by Kela]. Social Insurance Institution of Finland. Sosiaali- ja terveysturvan selosteita 75. Kelan tutkimusosasto. Helsinki: Juvenes Print—Tampereen yliopistopaino Oy.

Fedreheim, G. E., Melbøe, L., & Opsal, K. A. (2017). *Demokratisk deltakelse blant mennesker i en dobbel minoritetssituasjon som same og funksjonshemmet* [Political participation among people in a doubleminority situation as Sámi experiencing disability]. *Norsk statsvitenskapelig tidsskrift, 33*(3–4), 235–253.

Furr, J. B., Carrairo, A., & McArthur, J. A. (2016). Strategic approaches to disability disclosure on social media. *Disability & Society, 31*(9), 1353–1368. https://doi.org/10.1080/09687599.2016.1256272.

Gjertsen, H., Melbøe, L., Fedreheim, G. E., & Fylling, I. (2017). *Kartlegging av levekårene til personer med utviklingshemming i samiske områder* [Study of the living conditions among persons with intellectual disabilities in the Sámi administrative municipalities in Norway]. Stockholm: Nordic Welfare Centre.

Gómez, O., & Gasper, D. (2013). *Human security. A thematic guidance note for regional and national human development report teams.* United Nations Development Programme, Human Development Report Office. Retrieved from http://hdr.undp.org/sites/default/files/human_security_guidance_note_r-nhdrs.pdf.

Granholm, C. (2016). Blended lives: ICT talk among vulnerable people in Finland. *Nordic Journal of Youth Research, 24*(2), 85–101. https://doi.org/10.1177/1103308815613188.

Gustafsson, H. (2016). *YK:n vammaissopimus käyttöön!—käsikirja* [Implementation of FN's Convention on the Rights of Persons with Disabilities—Handbook]. Invalidiliitto—The Finnish Association of People with Physical Disabilities. Retrieved from https://www.invalidiliitto.fi/sites/default/files/2017-03/YK_Toolkit_2016_low.pdf.

Hansen, S., Wilton, R. D., & Newbold, B. (2017). 'There is always this feeling of otherness': Exploring the lived experiences of visually impaired immigrant women in Canada. *Disability & Society, 32*(8), 1121–1141. https://doi.org/10.1080/09687599.2017.1343128.

Heckathorn, D. D. (1997). Respondent-driven sampling: A new approach to the study of hidden populations. *Social Problems, 44*(2), 174–199. https://doi.org/10.2307/3096941.

Heinämäki, L. (2019). Indigenous persons with disabilities: The quest for legal recognition of intersectionality and the prohibition of multiple discrimination using the example of the Sámi people in Finland. In G. Alfredsson, T. Koivurova, D. Cambou & J. Klein (Eds.), *The yearbook of polar law* (vol. 10) (pp. 41–74). Leiden: Koninklijke Brill Nv.

Helander-Renvall, E., & Markkula, I. (2017). On transfer of Sámi traditional knowledge: Scientification, traditionalization, secrecy, and equality. In A. Xanthaki, S. Valkonen, L. Heinämäki, & P. Nuorgam (Eds.), *Indigenous peoples' cultural heritage, rights, debates, challenges* (pp. 104–129). Leiden: Brill Nijhoff.

Heleniak, T., & Bogoyavlenskiy, D. (2014). Arctic populations and migration. In L. J. Nymand & G. Fondahl (Eds.), *Arctic human development report: Regional processes and global linkages* (pp. 53–104). Denmark: Nordic Council of Ministers.

Hokkanen, L. (2017). *Experience of inclusion and welfare services among Finnish Sámi with disabilities.* Stockholm: Nordic Welfare Centre.

Hoogensen, G., & Stuvøy, K. (2006). Gender, resistance and human security. *Security Dialogue, 37*(2), 207–228. https://doi.org/10.1177/0967010606066436.

Hoogensen Gjørv, G. (2012). Security by any other name: Negative security, positive security, and a multi-actor security approach. *Review of International Studies, 38,* 835–859. https://doi.org/10.1017/S0260210511000751.

Hoogensen Gjørv, G., Bazely, D. R., Goloviznina, M., & Tanentzap, A. J. (2014). *Environmental and human security in the Arctic.* New York, NY: Routledge.

Hossain, K. (2013). Securing the rights: A human security perspective in the context of Arctic indigenous peoples. In G. Alfredsson & T. Koivurova (Eds.), *The yearbook of polar law* (vol. 5) (pp. 493–522). Leiden: Brill Nijhoff.

Hossain, K., Zojer, G., Greaves, W., Roncero, M. J., & Sheehan, M. (2017). Constructing Arctic security: An inter-disciplinary approach to understanding security in the Barents region. *Polar Record, 53*(1), 52–66. https://doi.org/10.1017/S0032247416000693.

Huuva, K., Montefusco, M., & Olsen, T. (2014). *Kunskapsläge: Personer med funktionsnedsättning och samisk bakgrund* [Investigating the information: People with disabilities and Sámi background]. Stockholm: Nordic Welfare Centre.

Katsui, H. (2005). *Towards equality: Creation of the disability movement in Central Asia.* Department of Social Policy Research Reports 1/2005. Helsinki University.

Kivistö, M. (2017). *Vammaisten nuorten teknologisten toimijuuksien rakentuminen digitalisoituvassa yhteiskunnassa. Katsaus laadulliseen tutkimukseen* [Construction of the technological agency of young people with disabilities in an increasingly digitalized society]. *Nuorisotutkimus, 35*(4), 35–49.

Lehtola, R., & Ruotsala, P. (2017). *Saamenkielisten palveluiden nykytilakartoitus. Saamelaisten lasten, nuorten ja perheiden palvelut* [Survey on the current situation regarding services in Sámi languages. Services for Sámi children, young people and families]. The Sámi Parliament, Posken saamelaisyksikkö. Retrieved from http://www.sosiaalikollega.fi/uutiset/hankkeet/lape/saam_nykytila_lanupe_110117.

Macdonald, S. J., & Clayton, J. (2013). Back to the future, disability and the digital divide. *Disability & Society, 28*(5), 702–718. https://doi.org/10.1080/09687599.2012.732538.

Magga, R. (1997). Vammaispalvelut [Disability services]. In A. Välimaa & E. Laiti-Hedemäki (Eds.), *Birge ja eale dearvan. Saamelaisten sosiaali—ja terveyspalveluprojektin loppuraportti* [Take care and farewell. The Final Report of the Sami Social and Health Services Project] (pp. 34–36). Sosiaali—ja terveysministeriön monisteita. Helsinki: Ministry of Social Affairs and Health.

Magga, R. (1999). *Geas diehu, gos doarjaga. Saamelaisten vammaisten, vanhusten ja pitkäaikaissairaiden tiedotus- ja tukiprojektin loppuraportti* [Who gives information, where to find support. The Final report on the information and support project for the Sámi people with disabilities or long-term illnesses and Sámi elderly]. Inari: The Sámi Parliament, Kynnys—The treshold Association.

Melbøe, L., Hansen, K. L., Johnsen, B. E., Fredreheim, G. E., Dinesen, T., Minde, G. T., & Rustad, M. (2016b). Ethical and methodological issues in research with Sami experiencing disability. *International Journal of Circumpolar Health, 75*(1), 1–7. https://doi.org/10.3402/ijch.v75.31656.

Melbøe, L., Johnsen, B. E., Fedreheim, G. E., & Hansen, K. L. (2016a). *Situasjonen til samer med funksjonsnedsettelse* [Situation of people with disabilities and Sámi background]. Stockholm: Nordic Welfare Centre.

Nguyen, X. T., & Mitchell, C. (2014). Inclusion in Vietnam: An intersectionality perspective on girls with disabilities and education. *Childhood, 21*(3), 324–338. https://doi.org/10.1177/0907568214524459.

Nurmi-Koikkalainen, P., Ahola, S., Giiler, M., Halme, N., Koskinen, S., Luoma, M. L., ... Väyrynen, R. (2017). *Tietoa ja tietotarpeita vammaisuudesta: Analyysia THL:n tietotuotannosta* [Disability-related information and information needs—An analysis of data production activities of the National Institute for Health and Welfare]. Discussion Paper 38. Helsinki: National Institute for Health and Welfare.

Nygård, R. H., & Saus, M. (2016). Emphasizing indigenous communities in social work research ethics. *International Social Work, 59*(5), 666–678. https://doi.org/10.1177/0020872816646815.

Oliver, M. (1990). *The politics of disablement*. Basingstoke: Macmillan.

Oliver, M., & Barnes, C. (2010). Disability studies, disabled people and the struggle for inclusion. *British Journal of Sociology of Education, 31*(5), 547–560. https://doi.org/10.1080/01425692.2010.500088.

Olsén, L., Heinämäki, L., & Harkoma, A. (2018). *Human rights and multiple discrimination of minorities within minorities: Sámi persons with disabilities and sexual and gender minorities*. Juridica Lapponica 44. Finnish and Sámi edition: Olsén, L., Heinämäki, L., & Harkoma, A. (2017). *Vähemmistöjen sisäisten vähemmistöjen ihmisoikeudet ja moniperustainen syrjintä: saamelaiset vammaiset henkilöt ja seksuaali- ja sukupuolivähemmistöt; Unnitloguid siskkáldas unnitloguid olmmošvuoigatvuođat ja máŋggadáfot vealaheapmi: sápmelaš lámisolbmot ja seksuála- ja sohkabealleunnitlogut*. Rovaniemi: University of Lapland.

Paris, R. (2001). Human security: Paradigm shift or not air? *International Security, 26*(2), 87–102.

Phoenix, A., & Pattynama, P. (2006). Intersectionality. *European Journal of Women's Studies, 13*(3), 187–192. https://doi.org/10.1177/1350506806065751.

Porsanger, J., & Guttorm, G. (2011). Árbediehtu-fágasuorggi huksen. In J. Porsanger & G. Guttorm (Eds.), *Working with traditional knowledge: Communities, institutions, information systems, law and ethics* (pp. 13–57). DIEĐUT 1/2011. Alta: Bjørkmanns Trykkeri AS.

Raja, D. S. (2016). *Digital dividends: Bridging the disability divide through digital technologies.* Background paper, World Development Report. World Bank Group. Retrieved from http://pubdocs.worldbank.org/en/123481461249337484/WDR16-BP-Bridging-the-Disability-Divide-through-Digital-Technology-RAJA.pdf.

Salminen, M. (2018). Digital security in the Barents region. In K. Hossain & D. Cambou (Eds.), *Society, environment and human security in the Arctic Barents region* (pp. 187–204). Routledge Explorations in Environmental Studies. Oxford and New York: Routledge.

Salminen, M. (2019). Refocusing and redefining cybersecurity: Individual security in the digitalising European High North. In G. Alfredsson, T. Koivurova, D. Cambou & J. Klein (Eds.), *The yearbook of polar law* (vol. 10) (pp. 321–356). Leiden: Koninklijke Brill Nv.

Salminen, M., & Hossain, K. (2018). Digitalisation and human security dimensions in cybersecurity: An appraisal for the European High North. *Polar Record, 54*(2), 108–118. https://doi.org/10.1017/S0032247418000268.

Sámi Parliament. (2017). *Sámi parliament statement to the Ministry of Justice, 30 March 2017.* Dnro 184/D.a.2/2017.

Sheehan, M. (2016). The security of the Sámi people. In K. Hossain & A. Petrétei (Eds.), *Understanding the many faces of human security: Perspectives of northern indigenous peoples.* (pp. 139–154). Leiden: Brill; Boston: Nijhoff.

Sherry, M. (2016). A sociology of impairment. *Disability & Society, 31*(6), 729–744. https://doi.org/10.1080/09687599.2016.1203290.

Szpak, A. (2017). Human security of the indigenous peoples in the Arctic. The Sami case. *International Studies, 20*(1), 75–96. https://doi.org/10.1515/ipcj-2017-0018.

The Constitution of Finland. (731/1999) (1999). Retrieved from https://publicofficialsfinancialdisclosure.worldbank.org/sites/fdl/files/assets/law-library-files/Finland_Constitution_2000_en.pdf#targetText=The%20constitution%20of%20Finland%20is,and%20promote%20justice%20in%20society.&targetText=Section%204%20%2D%20The%20Territory%20of,territory%20of%20Finland%20is%20indivisible.

Tøssebro, J. (2004). Introduction to the special issue: Understanding disability. *Scandinavian Journal of Disability Research, 6*(1), 3–7. https://doi.org/10.1080/15017410409512635.

United Nations. (2000). *Millennium development goals*. Retrieved from https://www.un.org/millenniumgoals/.

United Nations. (2007). *United nations declaration on rights of indigenous peoples*. A/RES/61/295, Resolution adopted by the General Assembly, 61/295 (13.9.2007). Retrieved from https://www.un.org/development/desa/indigenouspeoples/wp-content/uploads/sites/19/2018/11/UNDRIP_E_web.pdf.

United Nations. (2010). *Minority rights: International standards and guidance for implementation*. United Nations. Human Rights Office of the High Commissioner. Retrieved from https://www.ohchr.org/Documents/Publications/MinorityRights_en.pdf.

United Nations. (2015). *Sustainable development goals*. Retrieved from https://sustainabledevelopment.un.org/?menu=1300.

United Nations Development Programme. (1993). *Human development report 1993*. New York and Oxford: Oxford University Press.

United Nations Development Programme. (1994). *Human development report 1994*. Oxford: Oxford University Press.

United Nations Development Programme. (2001). *Human development report 2001: Making new technologies work for human development*. New York: Oxford University Press.

United Nations Development Programme. (2009). *Community security and social cohesion. Towards a UNDP approach*. Retrieved from http://www.undp.org/content/dam/thailand/docs/CommSecandSocialCohesion.pdf.

United Nations Development Programme. (2015). *Human development report 2015: Work For human development*. New York: Author.

United Nations Development Programme. (2016). *Human development report 2016: Human development for everyone*. New York: Author.

United Nations General Assembly. (1976). *International covenant on civil and political rights, A/RES/2200A(XXI)*. Adopted and opened for signature, ratification and accession by General Assembly resolution 2200A (XXI) of 16 December 1966, entry into force 23 March 1976, in accordance with Article 49. Retrieved from https://www.ohchr.org/en/professionalinterest/pages/ccpr.aspx.

United Nations General Assembly. (2007). *Convention on the rights of persons with disabilities and its optional protocol*. A/RES/61/106. Resolution adopted by the General Assembly, 61/106, (24.1.2007). Retrieved from https://www.un.org/disabilities/documents/convention/convoptprot-e.pdf.

United Nations General Assembly. (2012). *A/RES/66/290, Resolution adopted by general assembly on 10 September 2012, 66/290.* Follow-up to paragraph 143 on human security of the 2005 world summit outcome. Retrieved from https://www.unocha.org/sites/dms/HSU/Publications%20and%20Products/GA%20Resolutions%20and%20Debate%20Summaries/GA%20Resolutions.pdf.

Uttjek, M. (2016). *Levnadsförhållanden bland samer med funktionsnedsättning i Sverige* [Living conditions among the Sámi people with disabilities in Sweden]. Stockholm: Nordic Welfare Centre.

Van Schuylenbergh, J., Motmans, J., & Coene, G. (2018). Transgender and non-binary persons and sexual-risk: A critical review of 10 years of research from feminist intersectional perspective. *Critical Social Policy, 38*(1), 121–142. https://doi.org/10.1177/0261018317732478.

Vandekinderen, C., & Roets, G. (2016). The post(hu)man always rings twice: Theorising the difference of impairment in the lives of people with 'mental health problems'. *Disability & Society, 31*(1), 33–46. https://doi.org/10.1080/09687599.2015.1119037.

Vehmas, S., & Watson, N. (2016). Exploring normality in disability studies. *Disability & Society, 31*(1), 1–16. https://doi.org/10.1080/09687599.2015.1120657.

Vesanen-Nikitin, I., & Åkermarck, M. (2017). *Making digital transport and communication services accessible. Action Programme 2017–2021.* Publication 15. Helsinki: Ministry of Transport and Communication of Finland.

Zojer, G. (2019). The interconnectedness of digitalisation and human security in the European High North: Cybersecurity conceptualised through the human security lens. In G. Alfredsson, T. Koivurova, D. Cambou & J. Klein (Eds.), *The yearbook of polar law,* (vol. 10) (pp. 297–320). Leiden: Koninklijke Brill Nv.

CHAPTER 12

Emerging Pathogeneses and Satellite Telemetry: Containing Contagion in the European High North

Alexandra L. Carleton

12.1 Introduction

In 2014, an interesting piece of research concerning the presence of *Toxoplasma gondii* in a New Zealand fur seal (*Arctocephalus forsteri*) was presented at the Australian Society for Parasitology (Donahoe, Rose, & Slapeta, 2014). *T. gondii* is a pathogen with a fondness for feral felines, and it is prevalent on the Australian mainland. The discovery suggests the ability of this pathogen to find a naïve host and, given that the coastal range of *A. forsteri* includes sub-Antarctic Macquarie Island (Taylor, 1992, p. 115), the real possibility of pathogen translocation to endemic sub-polar and even polar regions. Pathogen transmission in animals of the Arctic and sub-Arctic is a human security issue in the sense that our knowledge of zoonoses (diseases capable of being transmitted from animals to humans) is still in its infancy. The reliance on the part of many endemic Arctic communities on similarly endemic animals, such as reindeer, for their livelihoods means that the prevalence of disease directly impacts such livelihoods.

A. L. Carleton (✉)
University of Sydney, Sydney, Australia

M. Salminen et al. (eds.),
Digitalisation and Human Security, New Security Challenges,
https://doi.org/10.1007/978-3-030-48070-7_12

Beginning with the scientific evidence regarding the emergence of pathogens (disease-causing bacteria, viruses, fungi, protozoa, parasites) and the associated outbreaks or incidents (where a group or an individual, respectively, is affected by the same pathogen or a pathogen not usually seen in the area) in relation to Arctic fauna, this chapter examines the reporting of pathogenic outbreaks in various parts of the Arctic over time. It discusses the impact that emerging pathogeneses have on local livelihoods and the potential for human development as well as on the prospects for disease mapping. Finally, it examines the use of a digitalised incident reporting mechanism and the role of a decentralised collaborative network.

Digital banks of outbreak and surveillance data have previously been employed in a variety of situations. Digital frameworks promote human security by facilitating the timely reporting of incidents of disease, both actual and potential (that is, whether confirmed by a veterinary professional or otherwise). Timeliness is essential in terms of both curbing epidemics and monitoring outbreak situations, and the sooner identification can occur, the sooner research can be conducted and appropriate preventative measures can be put in place to prevent further spread. It remains to be seen, however, whether a system can be utilised for the extensive Arctic region that allows for both access to and use of health surveillance. This is a complicated task because current trading interests in the region render those aspects that may hinder such opportunities undesirable counter-narratives. With the increasing understanding of the transmission of disease, the migration of disease and the mutability of many diseases, including the susceptibility of human populations, this issue cannot be labelled as undesirable because the consequences of missing a pathogen of high human virulence are too great. It is important, in the context of the law and governance of the Arctic, to place health and livelihood—and, therefore, emerging pathogens—at the centre of the debate.

Although there exist several definitions of the Arctic as a latitudinal and geomorphological place, for the purposes of this chapter, the European High North is taken to encompass both those Arctic and sub-Arctic latitudes that represent lowered biodiversity, extreme lows in temperature, high levels of exposure due to low tree coverage and a high level of pathogen naïveté with regards to the ecosystem. The European High North, as a less discriminate description of the terrain and water reserves that lie north of the 66 degree latitude, may be part of the

Arctic and the sub-Arctic. This chapter focuses on the European High North, although as this is an emergent field of professional activity and as the ecosystem of the High North is shared beyond Europe, scientific evidence from the United States is also relevant.

The governance of the Arctic must consider both the spatial endemism and that which gives rise to threats to its future resilience. Climate change affects biodiverse regions in ways yet to be fully understood. Our understanding of the effects of a changing climate on bio-uniformity has some catching up to do. In terms of the development of codes for increased human activity, tourism, consideration of the socio-economic development of indigenous peoples, greater inclusion and opportunities for development, the health, not only of the human population but also the animal population is of vital importance. In particular, given the increasing awareness of zoonotic (animal-to-human transmissibility) and anthrozoonotic (human-to-animal transmissibility) diseases, humanity can no longer consider itself to stand apart from the ecosystems in which it moves and seeks to inhabit.

12.2 Current Scientific Output on Emerging Pathogens in the Arctic and Sub-Arctic: An Emerging Picture

In recent years, a number of pathogens have emerged as possible causes of disease and death in incidents involving Arctic fauna. Reports of incidents involving Arctic endemic species that have emerged over the last twenty years point to the endemic susceptibility of such species to specific disease-causing pathogens. In fact, issues concerning emerging pathogens are so important that they have their own place on the Arctic Health information portal.[1]

Several emerging pathogens have been shown to affect Arctic artiodactyla (caribou, musk ox, etc.), including Brucella spp., *Coxiella burnetii*, Chlamydophila spp. (Afema, Beckmen, Arthur, Huntington, & Mazet, 2017, p. 311) and *Erysipelothrix rhusiopathiae* (Forde et al., 2016). Focusing on artiodactyla in relation to understanding emerging pathogeneses provides evidence of High North infected taxon/surviving

[1] https://arctichealth.nlm.nih.gov/climate_change/70/climate_change_and_human_health/73/emerging_pathogens.

at higher latitudes, discounting domestic dogs, cats and equids. While domestic animals are carriers of disease and so ought to be included when seeking to understand pathogen transmission across high latitudes more widely, for the purposes of understanding emerging pathogens at higher latitudes, it is the research that examines wild herds that is most important. This is because both the evidence it may provide concerning pathogens moving into more remote terrain, due to the traditionally restricted access to wild animals of the north such as polar bears (Hueffer, O'Hara, & Follmann, 2011, p. 6), and its potential to illustrate the impact that altering temperatures, landscapes and landscape use have on endemic species. Furthermore, the migratory preference of some of these species, their latitudinal confinement and their genetic homogeneity are all unique characteristics that may contribute to a disease prevalence in the European High North that is unlikely to have been seen or addressed before.

Scientific research examining climate change and its effects on pathogen prevalence has emerged more clearly since the early twenty-first century, following the compiling of earlier research results (Harvell et al., 2002; Tilman et al., 2001). It has been demonstrated that pathogens with high virulence (due to certain virulence factors), such as influenza, Campylobacter (Townsend, Frett, McGarvey, & Taff, 2018), herpesvirus and Erysipelothrix, are capable of moving both across species and across environments. The ranges of host and pathogen are determined by the types of habitat, temperature and moisture, where reindeer congregate during the summer months around bogs that also provide fertile ground for mosquitoes, as demonstrated by Laaksonen et al. (2010, p. 7). Burek, Gulland, and O'Hara (2008) ascribe the emergence of infectious disease to four factors, namely altered transmission rates (through the altered frequency of interaction of or increased density of possible hosts), increased survival, carrier–host and vector range expansion, and altered host susceptibility due to lowered bodily resilience (pp. S127–S128). In other words, there exists a need to understand the virulence factors, the transmission routes and the migratory patterns of susceptible species.

Virulence factors—those factors associated with a pathogen that render it a disease-causing agent—are understood for a large number of known pathogens. Parasites, which feed on and often mature inside their hosts, are currently experiencing an expanded range and altered life cycle in the northern latitudes. *Setaria tundra*, which is responsible for outbreaks of disease in reindeer and moose in Finnish Lapland, is a filarioid nematode whose range has expanded due to increasing temperatures in the region (Laaksonen et al., 2010, p. 10). Another nematode, *Trichinella nativa*, is the most common of all the Trichinella species at high latitudes, and it has shown resilience to freezing temperatures across the Arctic, with particular freeze tolerance being exhibited towards carnivore hosts such as foxes (clinical study carried out by Davidson, Handeland, & Kapel, 2008, as cited in Hueffer et al., 2011, p. 5).

The emergence of bacteria and protozoa responsible for outbreaks has also been recognised. *Vibrio parahaemolyticus* was found to be responsible for outbreaks in oysters and northern sea otters in 2004 and 2005, respectively, indicating an extension of range to Prince William Sound, Alaska (although absolute clarity on whether or not this is a new disease in the region is lacking; Burek et al., 2008, p. S128). There was an outbreak of pasteurellosis in Norway in 2006 (Ytrehus, Bretten, Bergsjø, & Isaksen, 2008). The transmission of pathogens in the Pasteurellaceae family, such as *Pasteurella multocida* and *Mannheimia haemolytica*, is thought to be due to the high density of agricultural animal hosts such as sheep, the proximity to wild animals such as the bighorn sheep (*Ovis canadensis*) (Burek et al., p. S129) as well as extraordinary weather events such as increasing temperatures (Ytrehus et al., 2008, p. 219). Other potential invading bacterium include *Mycoplasma bovis* (*M. bovis*), an infectious and contagious disease and notifiable to the World Organisation for Animal Health (OIE) that causes tuberculosis in cattle and severe outbreaks among bison, and *Mycoplasma ovipneumoniae* (Handeland et al., 2014).

Among the viridae, phocine distemper virus (morbillivirus) has been responsible for a number of outbreaks. It was thought that the northern expansion of sub-Arctic marine mammals such as the grey seal may have brought new morbillivirus to naïve populations of marine mammals such as arctic seals, likely due to increasing water temperature (Burek et al., 2008, p. S129). In both 1988 and 2002, this occurred through outbreaks of phocine distemper virus off the Danish coast (Härkönen et al., 2006), which killed 60% of North Sea harbour seals the first time and 47% the second time (Hammond, Pomeroy, Hall, & Smith, 2005, p. 2563). The disease was found to have spread from the Skagerrak/ Kattegat areas of Denmark and the Wadden Sea to the coast of Scotland (Hammond et al., 2005, p. 2563). Yet, harbour seals are sedentary (Härkönen et al., 2006, p. 115). It was hypothesised that the source may have been the harp seal (*Phoca groenlandica*), which moved during the stated periods to northern European waters, thereby coming into contact with the harbour seal populations. However, this only occurred in 2002 (Hammond et al., 2005, p. 2563). Härkönen et al. (2006) hypothesised that grey seals acted as subclinical infected carriers of the virus between the Arctic and North Sea seal populations as a migratory species, but this remains unproven. The morbillivirus has also been found in polar bear populations (Follmann, Garner, Evermann, & McKeirnan, 1996), as has the canine distemper virus (antibodies to this virus were found in polar bears by Tryland et al., 2005), which is a disease of domestic dogs. In 2014, a pinniped (New Zealand seal) found moribund in Australia was determined to be infected with *Toxoplasma gondii*, typically a pathogen of feral felines (Donahoe et al., 2014). Thus, naïve species are not only susceptible to virulent pathogens (due to having no resilience), as, particularly when it involves transmission to either avian or pinniped species that migrate to Antarctica or the Arctic, these new pathogens are also capable of invading naïve ecosystems.

12.3 The Need for Vigilance: The Reasons Behind Emerging Pathogeneses in the European High North

It is believed that these emerging pathogeneses stem from "unprecedented ecological change, largely driven by climate change" and that, as a result, infectious pathogens are increasingly prevalent at higher latitudes (Forde et al., 2016, p. 1759). Studies concerning emerging pathogens (Afema et al., 2017; Forde et al., 2016) in the Arctic and Boreal ungulates in particular demonstrate a desire on the part of the scientific community to understand the effects of environmental and landscape changes on the resilience of herd species. This is important if we are to stand any chance of understanding and halting the rise of new endemics or pandemics in the northern latitudes.

There is evidence that alterations in climate are responsible for pathogen range expansion, transmission and survival, which is perhaps made all the more possible by the worsening body condition of those animals not suited to a warmer ecosystem (Burek et al., 2008). Temperature can affect the rate of transmission, the development of parasite vectors and the movement of humans and animals northward, with consequent new hosts, new carriers and new environmental exposure (Altizer, Ostfeld, Johnson, Kutz, & Harvell, 2013; Bradley, Kutz, Jenkins, & O'Hara,

2005; Harvell et al., 2002). Warmer temperatures affect the spread, viability and transmission rate of pathogens (Hueffer et al., 2011). Warmer temperatures also encourage the development and sustained viability of both cestodes and nematodes (parasitic infections of the gastrointestinal system), encouraging them to move northwards and to undergo development in order to survive (Jenkins et al., 2013). Arthropods are also predicted to have expanded ranges into northern areas, where transmission through direct and indirect contact with humans and domestic animals will occur more frequently (Hueffer et al., 2011, p. 2), which may cause destructive sequelae.

It is not only the vectors that carry disease that are moving northward but also potential carrier/host animals, as warming temperatures (Short, Caminade, & Thomas, 2017) can increase the risk of pathogen transmission and thus increase the chances of disease occurrence. The northward migration of pathogens and their vectors as well as the increased transmission to new hosts have been demonstrated in tick-borne encephalitis (TBE) carried by *Ixodid* ticks (Tokarevich et al., 2011), in the expansion of the range of Avian Cholera in the Arctic Common Eider and the Northern Common Eider in the Canadian Arctic (Iverson, Forbes, Simard, Soos, & Gilchrist, 2016) and in the emergence of Lyme disease in Canada as a result of the expansion of the geographic range of *Borrelia burgdorferi* (Ogden, Lindsay, Morshed, Sockett, & Artsob, 2009). In particular, parasites and their increased transmission have been linked to warmer weather. The ectoparasite *Trichodectes canis* has appeared in Alaska (Woldstad, 2010, as cited in Hueffer, 2011), while *Ixodes*-related diseases have been found in parts of Canada and sub-Arctic Russia (Short et al., 2017). Moreover, the nematodes *Umingmakstrongylus pallikuukensis* and *Setaria tundra* are emerging as disease problems in musk oxen and reindeer (*Rangifer tarandus*), respectively (Hueffer et al., 2011; Kutz, Hoberg, Polley, & Jenkins, 2005).

Parasites may adjust their life cycles to suit the environment. *O. Gruehneri*, a major abomasal nematode of caribous and reindeer, has reduced its infestation during the hottest part of the Arctic summer, although it now has two annual transmission peaks rather than one (Altizer et al., 2013, p. 2). Parasites have peak thermal conditions and their survival depends on the "ability of the parasite to thermoregulate, adapt (including shifts in phenology), and shift geographic ranges" (Lafferty & Mordecai, 2016, p. 3). Rising temperatures will expose the Arctic and its fauna to increased infestation by parasites. As noted by Lafferty and Mordecai

(2016), "[p]otential poleward shifts in suitability for tropical diseases predicted from the thermal performance curve can cause alarm in countries at high latitudes" (p. 4).

Migratory species and the patterns of many of the Arctic artiodactyla (caribou, reindeer, bison), birds and cetacean species are under-studied not only with regards to host susceptibility to new pathogens but also to the potential of carrying emergent pathogens along migratory routes and depositing them in new environments. Arctic foxes (*Vulpes lagopus*) are a known carrier of the rabies virus and they have wide ranges (only being territorial during breeding season), which adds to the dispersion possibility of the pathogen (Hueffer et al., 2011, p. 6). The rabies virus can survive in a frozen carcass (Hueffer et al., 2011, p. 6); however, this may also mean that with warmer temperatures and the resultant decomposition of rabid carcasses, the virus may no longer be as resilient.

Although the global spread of disease may be limited by the inability of pathogens to adapt to changes in temperature more generally (Lafferty & Mordecai, 2016), the increased susceptibility of the Arctic to emerging pathogenesis remains high due to both its increased attractiveness as its temperatures change and it being a region that has no former resilience. While general predictions are challenging (Lafferty & Mordecai, 2016), modelling has demonstrated that pathogens increase with temperatures (Harvell et al., 2002, p. 2160). Further, climate change can extend the range of pathogens (Short et al., 2017) to newly susceptible hosts, including humans. Earlier research suggested that the geographical range of pathogens moves towards the poles (Harvell et al., 2002, p. 2161) as global temperatures increase. Together with the predicted "movement of people from endemic areas to nonendemic regions of the world" (Short et al., 2017, p. 3), the exposure of people, animals and environments alike to invasive pathogens increases.

Keeping in mind Hueffer et al.'s (2011, p. 4) warning against extrapolating data to new latitudes, this knowledge is relevant to the European High North as those species adapted to living in the extreme and naïve Arctic environment may bear similar adaptations and, therefore, fragilities to the environment and to previously unmet pathogens. Fauna adapted to an Arctic environment have a lower threshold of tolerance or resilience to pathogens from more southern latitudes (Geist, 1985, p. 355).

12.4 The European High North and Emerging Pathogeneses

Studies of emerging pathogeneses in the European High North have been conducted without a concomitant concern for national or regional borders. This is important because pathogens do not respect national or regional borders. Table 12.1 lists the prior studies that were carried out within the European High North. As can be seen from the table, the studies on morbillivirus (including PDV) and Pasteurella show similar results to those found in the American High North.

Laaksonen et al. (2010) identified a connection between the temperature gradient and the prevalence of the filarioid nematode *Setaria tundra* in reindeer and moose populations in northern Finland.

Clinically associated with peritonitis in reindeer, *S. tundra* requires a mosquito vector for transmission and, due to the development of greater areas of frost in sub-Arctic ecosystems, reindeer are forced to stay in denser-forested, mosquito-infested areas (Laaksonen et al., 2010). In Laaksonen et al.'s (2010) study, a mean summer temperature exceeding 14 degrees Celsius was found to be positively correlated with the emergence of the filarioid nematode in three populations, namely at Kevo, Kuusamo and Sodankylä. Climate data were retrieved from the Finnish Meteorological Institute stations at Kevo, Kuusamo and Sodankylä for the period 1962–2004, which covers all the years since outbreaks of *S. tundra* have been documented.

Over the five years between 1987 and 1992, Follmann et al. (1996) found morbillivirus antibodies in high levels in polar bears (*Ursus maritimus*) along the Bering, Chukchi and East Siberian seas. Some 191 blood samples collected from adults, sub-adults and cubs were tested for serum antibodies. The prevalence was found to be under 50% for all the groups, with the adults having the highest prevalence (43.9%), although this percentage indicates neither titre levels nor seasonality. Härkönen et al. (2006) examined two PDV epidemics (in 1988 and 2002) that resulted in the deaths of more than 23,000 and 30,000 harbour seals (*Phoca vitulina*), respectively. The PDV was found in areas of Denmark, the Wadden Sea and the coast of Scotland among discrete harbour seal populations. The exact route of transmission is, as yet, unknown. Between 1990 and 1998, during testing of the seroprevalence of antibodies in the High North of Norway, polar bears were found to have been exposed to canine distemper virus and calicivirus (Tryland et al., 2005). In 2006,

Table 12.1 Examples of studies from the Northern latitudes of Europe

Study	*Year(s) of research/ Publication*	*Results of study*	*Region*
Laaksonen et al.	1964–2004/2010	Known outbreaks of *Setaria tundra* in reindeer (1973 and 2003) and moose (1989) indicated the outbreaks to be associated with temperatures above 14°C	FinlandFinnish Lapland—Kevo, Kuusamo and Sodankylä
Follmann et al.	1987–1992/1996	High morbillivirus antibody titres found in polar bears (*Ursus maritimus*)	Russia—Bering, Chukchi and East Siberian seas
Härkönen et al.	1988/2002	Phocine distemper virus (PDV) epidemics resulted in the deaths of more than 23,000 harbour seals (*Phoca vitulina*) in 1988 and 30,000 in 2002	Outbreak started in Anholt, Denmark
Tryland et al.	1990–1998/1996	Polar bears have been exposed to canine distemper virus (CDV) and calicivirus	Svalbard, Norway
Ytrehus et al.	2006/2008	Outbreak of pasteurellosis detected in musk oxen	Norway
Davidson et al.	2008/2008	Parasite Parasitic nematodes can modify their host-specific encystment so as to better survive (findings of a clinical trial, not found in the field)	Norway

Source Author's own creation

musk oxen in Norway suffered an outbreak of Pasteurellosis (Ytrehus et al., 2008). Due to the few examples as well as the varied and sporadic data collection, no particular disease, animal or area has been subjected to the consistent, measurable indexing of pathogens. Indeed, three of the above-mentioned examples are responses to an outbreak in wildlife, and only two specifically tested for possible past infection (Follmann et al., 1996; Tryland et al., 2005). As such, the evidence is anecdotally interesting, although not quantitatively sufficient to draw even general conclusions. The findings do, however, form part of a narrative regarding emerging disease in European High North populations that do have sufficient weight to be worthy of some attention.

Not only do such results reflect the potential for pathogenic spread across similar latitudes and ecosystems once resilience has been established, they also indicate that a regional approach may be in the best interests of all concerned. Given the likelihood of expanding trade in the European High North, human consumption of relevant products and the livelihoods attached to them, this raises important human security questions.

12.5 Human Security: Ecosystem Resilience, Trade, Livelihood and Living

Zoonoses and anthrozoonoses are of increasing importance (Hossain, 2016, p. 416). Human livelihoods, involving both those communities that live directly on the fauna of the Arctic and those that trade with it, are likely to be affected. Previous non-exposure in remote communities increases the risk to lives and livelihoods (Jenkins et al., 2013) because such communities are scientifically and immunologically naïve. That is, they have not been exposed to a particular pathogen and so have not built resilience to it. For example, "water-borne protozoans such as Giardia, Cryptosporidium, and Toxoplasma may be emerging causes of human disease in a warming North" (Jenkins et al., 2013, p. 35). The same is true of animals. Where pathogens find residence within new hosts without necessarily killing them, those hosts are classified as carriers. In such cases, the morbidity (or infection status) does not necessarily result in disease-induced mortality (or death). Pathogens that find new hosts and favourable environmental conditions may survive in carrier species or the soil for a certain period, particularly if they are saprophytic. Taking this point further, as species are hunted and traded across borders, pathogens move with them, making health a trans-border problem and food security an issue (Jenkins et al., 2013).

Regardless of the reasons, whether due to climate alterations, landscape use changes resulting from tourism and industry, or increased road infrastructure creating greater access and higher mobility, pathogen incursion into the high northern latitudes impact the health of humans as well as the health and survival of various herd species, with the latter often being a source of sustenance for indigenous peoples (Hueffer et al., 2011, p. 2). The Sámi of the European High North, for example, rely on traditional activities (such as reindeer herding) to provide them with the resources necessary for survival (Hossain, 2016, p. 420; Ross, Johansson, Ingman, & Gyllensten, 2006, p. 554). Mammals such as reindeer play a huge role in Arctic communities, as food that forms part of subsistence living, tradable products and cultural items. As such, alterations will affect trade, food security and the livelihood/identity nexus (Hueffer et al., 2011, p. 18). The presence of new diseases hinders the ability to trade and thus live off the products derived from these animals. Consequently, pathogens constrain both indigenous and local livelihoods where they depend upon the health of local species.

Food security is not simply a human concern. Indeed, evidence suggests that the resilience of sub-Arctic bird populations in Alaska is determined from the bottom-up by the availability of food supplies and the flexibility of both foraging behaviour and diet (Goyert, Garton, & Poe, 2018, p. 987). Populations of animals that use the same pastures or feed on the same salt licks may also transmit pathogens across species, which represents untested ground as wild species translocate and agricultural practice moves northward, thereby creating a new dynamic. This was the case with musk oxen and sheep in Norway, who through feeding on the same salt lick may have passed the cause of pneumonia between them (Handeland et al., 2014). The intensification of husbandry as human consumption demands grow also increases the interspecies transmission of disease. Further, with increasing understanding and developing technology in terms of assessing the disease status of endemic populations of animals, the impacts on human security via trade will have an impact as well. For example, reindeer husbandry is undergoing intensification in Sweden (Kautto et al., 2012, p. 64). Due to the endemic presence of pestivirus and alphaherpesvirus in those populations, there is a chance for transmission to occur to domestic animals that graze in the same pastures, which would "cause significant economic losses" (Kautto et al., 2012, p. 65).

Humans are another migratory species that may be both the sufferer of a new illness and the carrier. "Predicted changes include ... increased pathogen transmission through direct and indirect contact with humans and their domesticated animals" (Hueffer et al., 2011, p. 18) and increased contact with human waste (Hueffer et al., 2011). Changes to water use and water-borne transmission, fomite transfer on boots and tools, transportation via food and animals, and increases in mining and tourism (Jenkins et al., 2013) are all involved in pathogenic spread. Pathogens that can be linked to increased tourism or industry may become a reason for constraining human development, as the health of wildlife has been directly linked to landscape disturbance in the American Arctic (Kutz et al., 2015, p. 2). The resilience of the environment is thus lowered with increasing pathogenic transmission by migratory species, the influx of humans and the associated international trade (Liu et al., 2013, p. 1). In turn, the transmission of disease to new hosts impacts food security, livelihood, subsistence and culture (Hueffer et al., 2011; Kutz et al., 2015, p. 4) and, therefore, human security.

Resistance to pathogenic outbreak is host-specific, and it differs among populations (Barbosa & Palacios, 2009, p. 1110). Thus, even when there is no disease occurrence in one location due to some new activity, the same will not necessarily be true in another location. Disease emergence, prevalence and spread are all location- (Lafferty & Mordecai, 2016) and biodiversity-dependent (that is, the lower the number of different species across the ecosystem, the lower the resilience to pathogenicity). This can be seen from the research results mentioned above.

Pathogen and disease can also be a result of, and cause, ecological niche changes (Harvell et al., 2002). Biota shifts brought about by animals translocating due to food scarcity, for example, creates new interspecies interactions (Altizer et al., 2013, p. 516), as has occurred between polar and grizzly bears in northern Alaska and eastern Siberia (Burek et al., 2008). Ecosystem "simplification" through land conversion from natural habitat to productive agricultural land (Tilman et al., 2001, pp. 282–283) can also bring about alterations in migration that can increase host–pathogen susceptibility and the environmental build-up of pathogens (Altizer et al., 2013, p. 516), particularly when the development of a pathogen depends upon the host harbouring a stage of its development. Furthermore, such life cycles and the transmissibility of causative agents "including those causing disease in humans, agricultural systems, and free-living animals and plants—are inextricably tied to climate" (Altizer et al., 2013, p. 1). Healthy wildlife populations are critical for ecosystem integrity and human physical (food security), social and economic health throughout the Arctic (Kutz et al., 2015, p. 4).

12.6 Field Surveillance in the High North: The First Step in Digital Recording

In 2018, veterinary medical professionals were acknowledged as being instrumental in the development of knowledge and the conducting of investigations concerning pathogenic interactions. For this reason, following the health surveillance meeting of the U.S. Programme on Health Surveillance in the Arctic (U.S. Embassy, 2018) veterinarian experience was sought in relation to examining the productivity of surveillance efforts (Meidenbauer, 2017). As veterinarians are able to pathologically examine and diagnose disease in animals (Ytrehus et al., 2008, p. 217), their involvement in surveillance is crucial. The number of investigation and/or research units in the Circumpolar North, including those formed as part of governmental programmes, demonstrates the importance of keeping an eye on Arctic health.[2] The need for investigative teams stems from the fact that wildlife disease surveillance requires a network that is both "expansive … [and] often do[es] not exist in remote and developing areas" (Iverson et al., 2016, p. 12). Community-based research groups using local knowledge have thus been advocated (Iverson et al., 2016).

Veterinarian and field surveillance efforts have previously looked at emerging pathogeneses in the Arctic when there is a mass/unusual mortality event (MME/UME) (Iverson et al., 2016; Kutz et al., 2005). Such events attract attention and scientific investigative interest, although much of the research in response to major events is sporadic. Regular interventions are difficult and dependent on funding (Meidenbauer,

[2]For example, the Committee for Arctic Medical Research, the Arctic Council's Arctic Monitoring and Assessment Programme, the Sustainable Development Working Group, the Arctic Contaminants Action Programme, the Conservation of Arctic Flora and Fauna, Emergency Prevention, Preparedness and Response, and the Protection of the Arctic Marine Environment (Ruscio, Brubaker, Glasser, Hueston, & Hennessy, 2015).

2017).[3] Both MME and UME have been linked to changing temperatures, altered seasonal weather conditions and human development (Kutz et al., 2015). Yet, major pathogens of increasing virulence, such as *Pasteurella multocida* and *Erysipelothrix rhusiopathiae* (Uhart, Gallo, & Quintana, 2018), which are capable of transmission between people, wild animals such as musk oxen and a variety of birds (Kutz et al., 2015; Uhart et al., 2018) and domestic animals such as swine, may be responsible for more incidents than even those recorded. Further, the MME among musk oxen (*Ovibos moschatus*) may indicate an inability to adapt to the rapidly changing environment and increasing comorbidity (Afema et al., 2017).

Field surveillance can clarify the parameters of a changing northern latitude climate by examining the nuances of ecosystem changes over time. The Alaskan seabird monitoring conducted for 40 years by Goyert et al. (2018) resulted in a thorough understanding of the differences of various avian species in terms of responding to environmental alterations at sub-Polar latitudes. The results of that research have given us two sentinel species for northern latitude environment shifts: the black-legged kittiwake (*Rissa tridactyla*), which has struggled to acclimatise, making it a sentinel for environmental decline, and the thick-billed murres (*Uria lomvia*), which occupies the same latitudinal sphere as the black-legged kittiwake but has adaptive foraging behaviour that, contrariwise, makes it a sentinel for the resilience of a species in the face of change (Goyert et al., 2018). Murres and kittiwakes have both increased in the Arctic, although they have declined in southern Alaska (Goyert et al., 2018, p. 976). Hence, while species may decline in traditional environments at sub-polar latitudes, they may increase at polar latitudes.

New research, better funding and the adequate implementation of information are all key to furthering natural resource governance in the polar regions. Finally, while the work of field veterinarians is vital to the initial investigation and identification of outbreaks, the use of local partners, especially the Arctic's indigenous peoples (Iverson et al., 2016), is also useful in terms of maintaining such a surveillance system. Field researchers may be able to identify singular pattern-based pathogeneses and their pathogens based on experience, although confirmed laboratory

[3]Recognised as major resource needs within the Arctic Investigations Programme Strategic Plan 2016–2020. https://www.cdc.gov/ncezid/dpei/aip/AIPStrategicPlan-508.pdf.

testing away from the site would be required for credible identification. Veterinarians can perform point-based on-site diagnostics with sufficient gross description and morphological interpretation to be immediately useful. In fact, it is the work of field veterinarians in particular that provides high-value information to surveillance in the case of specific and immediate diagnostics. The challenge remains building a digital network that is dynamic enough to receive new intelligence concerning outbreaks or novel incidents.

For the European High North, without entering the political game of collaboration, each national surveillance system could have a series of triggers that notify other members, most likely when there is a concern for airborne transmission for a pathogen or migratory herds seeking new pasture across borders, for example.

12.7 Disease Mapping: Evidence of Its Beneficial Use for Scientific Collaboration

The pooling of data so as to create spatial maps is embedded within the operations of global organisations dedicated to monitoring animal and human health, such as the World Animal Health Information System (WAHIS) used by the OIE. It has been recognised that countries that use such systems are more likely to predict the next outbreak and to be better prepared for its effects (Mörner, Obendorf, Artois, & Woodford, 2002, p. 67). Satellite telemetry, radio tracking as well as microsatellite and isotope analysis are all part of the mapping toolkit when it comes to mapping disease transmission routes, likely spread and host susceptibility. Mapping has the ability to disseminate data on emerging outbreaks and to learn from past outbreaks. Jenkins et al. (2013) identified the enhanced surveillance associated with increased and emerging pathogeneses as being key to improving the chances of mitigating the risks associated with zoonotic disease.

Animal disease surveillance programmes in the Arctic, on the whole, have only been around since 1996, following the formulation of, for example, the Arctic Council's Arctic Monitoring and Assessment Programme; the Sustainable Development Working Group; the Arctic Contaminants Action Programme; the Conservation of Arctic Flora and Fauna; the Emergency Prevention, Preparedness and Response; and the Protection of the Arctic Marine Environment (Ruscio et al.,

2015). However, institutions for health monitoring in the region have been around since 1957, with the presence of the Committee for Arctic Medical Research providing additional longevity to the field study. A "One Health" approach integrating animal surveillance into human health surveillance could take advantage of this. One Health, as an advanced and integrated approach, provides advance notice of likely or possible public and human health dangers. Such an approach requires innovation and collaboration, and it will provide the critical time necessary to achieve a buffer, thereby increasing the chances of a resilient response, particularly for peoples attached to their environment (Ruscio et al., 2015). Building a "common linked system" for integrated animal and human health and "[i]ntegrating animal data into existing, well-established human health surveillance systems would reduce the resources needed while still providing the advanced capabilities that are available for human health surveillance" (Meidenbauer, 2017, p. e147).

Disease mapping may also serve as a tool for human security by helping to build remote community resilience to pathogenic outbreaks. Through being able to map disease outbreaks and build predictive models based on ecosystem determinants and temperature fluctuations, with the accompanying knowledge of pathogens and hosts (Altizer et al., 2013), we can build community and ecosystem resilience. In endemic regions in particular, given the sporadic nature of the research and the vast improvement that has occurred in terms of identifying pathogens, current information outlining the spatial and temporal variation of pathogens is crucial to understanding which pathogens may be exogenous so as to curb outbreaks and aid in decision-making (Barbosa & Palacios, 2009, p. 1111). This will require a combination of science (Ruscio et al., 2015, p. 1) and technology. Both the Arctic as a region and the American Arctic currently have programmes running in this regard, from which lessons can be taken and used in relation to the European High North.

Disease mapping has previously been used with regards to rabies outbreaks (Smith, Lucey, Waller, Childs, & Real, 2002), with the partial northern migratory behaviour of the American crow, which is the carrier of a number of pathogens, being uncovered (Townsend et al., 2018). Further, such mapping has allowed for the forecasting of the rate of development of an Arctic nematode lungworm (Kutz et al., 2005). In the American Arctic, disease mapping is undertaken by national surveillance units. The U.S. Programme on Health Surveillance in the

Arctic, the International Circumpolar Surveillance system and the Arctic Investigations Programme of the Centre for Disease Control and Prevention (CDC) are all examples of such systems. The Alaska One Health Group "maintains a current web accessible One Health Map, which provides a visual aide to help track emerging and trending events. The maps include events screened from news reports, selected posts from the Local Environmental Observer (LEO) Network and additions made by group members" (Ruscio et al., 2015, p. 3). Similarly, a number of monitoring systems already exist, such as the Canadian Cooperative Wildlife Health Centre (CCWHC), the Circumpolar Arctic Rangifer Monitoring and Assessment Network (CARMA), the Northern Contaminants programme, the Nunavik Research Centre, the Makivik Corporation, the Eastern Canadian Arctic Seabird Ecology and Health Programme, and the Alaska Department of Fish and Game Wildlife Health Surveillance Programme (Kutz et al., 2005, pp. 5–6). Other institutional structures can also be employed in relation to surveillance, including regional cooperatives, abattoir checks and wildlife autopsies (Mörner et al., 2002, p. 71). The use of indigenous peoples as networks of primary responders has been encouraged (Altizer et al., 2013, p. 518), although this form of reporting may lack credibility as it is not clinically based. A suggestion here is that any initial observations should be clinically described or verified by veterinarians who have the ability to clinically diagnose and assess pathogeneses.

The primary cause for concern relates to the collection of the data used to create the spatial maps. Information and data regarding emerging pathogens and the effects of pathogeneses on both the ecosystem and the host are in short supply. It is only in the last ten years that there has been an awareness of the gaps in the data concerning polar regions (Barbosa & Palacios, 2009). All information on free-ranging animal populations in endemic regions, particularly those with large migratory routes and those with increased interactions with agriculture, needs to be widely reported, shared (Mörner et al., 2002, p. 72) and disseminated rather than necessarily having to pass the laborious, time-consuming and academic requirements of peer-reviewed journals. Veterinarians in particular have the professional capacity to report without peer critique on every case. The surveillance of wildlife disease in the Arctic needs to be more consistent and to be conducted for broader reasons than simply economic (for the verification of trade status) (Mörner et al., 2002, pp. 71–72) or reputational reasons.

12.8 Creating a Digital Framework for Reporting Pathogen Incidents

A United States survey found three critical gaps in animal health/disease surveillance have been identified, namely the lack of "integrated human and animal health surveillance, *real-time animal health data collection*, and companion animal surveillance" (Meidenbauer, 2017, p. e147, emphasis added). Such gaps significantly impact human security with regards to the timeliness of catching potential disease outbreaks, establishing what level of risk (virulence) is present for human transmission and cordoning off relevant areas. A slow human response to outbreaks has been criticised in the past, for example, in the case of the 2009 H1N1 influenza virus pandemic, where the European Union recognised that for prevention plans to be effective, more than simply cooperation between national authorities is required. Indeed, flexibility and adaptability, speed of action commensurable to the risk, current and relevant information, and the consistent and "closer involvement of health professionals at every stage in preparing and applying the necessary strategies" are all desired (Watson, 2011, p. 1).

Difficulties associated with creating a digital framework include locating dynamic (that is, actively changing) local and regional data; the lack of an easy reporting mechanism (particularly the time delay when there are digital outages) and, therefore, accurate current knowledge, particularly regarding zoonoses; and the failure to recognise the relevance of animal monitoring to human medicine (an example here would be tick-related disease, given its increasing prevalence in the European High North, as discussed above) (Meidenbauer, 2017). Kutz et al. (2005, pp. 11–12) suggest the formation of an Arctic Wildlife Health Observation Network to pool, expand and coordinate the effective capabilities of a number of different monitoring bodies. Research and field investigations into the pathogenesis of mortality events have previously been sporadic, although field investigations may be improving due to better diagnostic ability. Whether the growing number of UME/MME reports is due to a greater focus on the issue or because such events are indeed on the increase cannot be determined with certainty.

Creating a digital framework that is dynamic, where new incidences can be immediately reported, coded, viewed by others and included as part of established data sets, would render the continual monitoring of the health of Northern Europe and the Arctic more timely. Three types or levels of monitoring could be applied. Combining them would generate the best picture in terms of the surveillance of disease: (i) the use of pathogen biomarkers such as phycocyanin (Singh, Ranjan, & Verma, 2015); (ii) the use of animal-environmental associated biomarkers such as the level of an isotope in feathers and the individual tagging of migratory species to determine the level of spread (Townsend et al., 2018); and (iii) the use of biosphere or ecosystem variables that measure the net primary productivity (NPP) of a place, such as the temperature, pasture cover, ice depth and rainfall (Crabtree et al., 2009), the land surface temperature (LST), changes in vegetation (NDVI) or evapotranspiration (ETa) (Singh et al., 2015) as well as changes in the established reference levels of any of these (Singh, 1989, p. 989). Each of these approaches has been tentatively associated with tracking certain aspect of disease outbreak (Singh et al., 2015; Townsend et al., 2018, p. 971). The use of microsatellites, satellite telemetry, radio tracking devices and thermal imaging combined with known thermal performance curves (Lafferty & Mordecai, 2016) can also be employed.

Digitising pathogen outbreaks can contribute to the maintenance of a dynamic spatial map of disease and its possible causes. This can be achieved by utilising the knowledge and diagnostic abilities of multiple people, and it may lead to the better management of objectives and outcomes, not just in relation to the European High North but to the entire Arctic. As such, the region should be considered as a whole.

A database project is not intended to replace the need for academic journal reporting on clinical trials or longer systematic reviews. Instead, it is intended to collectivise the current ad hoc clinical and field veterinarian experience with regards to uncovering potential new disease occurrences. It can collect data/evidence that does not have to wait for systemic review or for a diligent researcher to run through an academic critique. In pragmatic terms, it would require field veterinarians to have access to a platform to report clinical observations and diagnoses, together with likely pathogeneses, to a central system. This would form the basis of the knowledge necessary to create predictive spatiotemporal models of disease outbreak as well as real-time updates on disease outbreaks that fuel mapping systems and predictive forecasters. When creating such a system for surveillance, there are also certain extra-scientific considerations, which are addressed below.

12.8.1 Regional Endemism

Northern Europe must be considered as a region and as part of the Arctic/sub-Arctic. Species distribution models (SDMs) of pathogens consider the whole geography rather than just hot-spots and easily accessible areas (Liu, Rohr, & Li, 2013, p. 2), thereby highlighting transmission stressors over the region. Liu et al.'s (2013) study of the spread of the deadly *Batrachochytrium dendrobatidis*, for example, considered fundamental niche (FN) and pressure predictors (PP). For *B. dendrobatidis*, such consideration showed that the introduced hosts, species richness and international trade all had effects on the distribution of the pathogen (Liu et al., 2013).

However, consideration of the Arctic as a whole represents a better approach for a number of reasons: (i) because the Arctic shares similar attributes across borders; (ii) because pathogens that are able to infiltrate one part of the Arctic are likely to have increased resilience in another part, particularly when temperature is a factor for virulence; and (iii) because an approach applied in one area may offer lessons for another.

12.8.2 Trade

Systems of surveillance, as aggregates of large volumes of information regarding the health statuses of countries, will play a tremendous role in both international trade discussions and the use of new technologies in the future. The WAHIS provides the current status of major diseases and aims, in conjunction with the OIE, to provide mapped and real-time information on disease occurrence, thereby placing organisations such as the OIE in a primary power position when it comes to validating/legitimising and invalidating/delegitimising trade arrangements. An infection-free status ought to only be declared when evidence is available based on sampling and diagnostic procedure, including for wild animals (Mörner et al., 2002, p. 67), which have greater capacity to roam and carry pathogens across contiguous land zones.

12.8.3 Privacy

Global systems may give rise to questions regarding privacy, technology ownership and data use. In particular, given the competing and sometimes divergent interests of industries, investors, governments and individuals, concerns over the "funding, data acquisition, data confidentiality, and identification of desired stakeholders" render the pooling and supply "of animal health data a challenge" (Meidenbauer, 2017, p. e147). The primary ethical consideration associated with the collecting and disseminating of animal health data, concerns the likely collision with international privacy standards. Without something akin to a global "public benefit" exception to privacy, where health and disease are the major concerns, such surveillance systems are unlikely in the short-term to gain plurilateral diplomatic support. Despite the advantages offered by such a system, the surveillance concept still ought to be debated and negotiated heavily, as any breach in the territorial (literal or cyber) sanctity of nations may be said to infringe on the concept of sovereignty. For this reason, the European Union (2016) has sought to ensure that personal control of data is well regulated under the General Data Protection Regulation (GDPR).

12.8.4 Language

Digital reporting, recording and dissemination need to consider the issue of language: a surveillance system that uses local indigenous peoples such as the Sámi, Inuit, Yupik, Sakha and Nenets requires the capability of language plurality. Similarly, there are five nations that are contiguous to the Arctic, the so-called Arctic Five (Russia, Denmark, Norway, Canada and the USA) plus Iceland which may be considered a sixth, most of which have different national languages, plus the additional members that form the Arctic Eight, which add an additional three languages (Sweden, Iceland and Finland).

12.9 Conclusion: Endemic Vulnerability

When seeking a better approach to the European High North and to human security, an understanding of the pathogens that have gained entry into the Arctic and managed to demonstrate some viability must be achieved. Even in relation to the practice of veterinary medicine in areas where pathogens and virulence are known, human security has been threatened anew by, for example, genetic mutations, temperature shifts and alterations in land use. Human security in the European High North will not be assured until all matters affecting human livelihoods are considered.

Arctic endemism poses a unique challenge with regards to the deliverance of systems of disease surveillance and management, not only because higher latitudes pose obstacles to the researcher but also because MME/UME may have greater consequences for naïve fauna in the region. With no or little prior exposure, the high host susceptibility to opportunistic pathogens means a greater chance of contracting disease.

Research into pathogen transmission and rates is still in its infancy. It may require an intergovernmental mission to redress the current scattered approach of independent researchers and laboratories and to create an Arctic Surveillance Network or Arctic Laboratory for Research into Infectious Disease.

A digital framework for reporting disease occurrences and likely pathogeneses in Arctic and Boreal regions represents a vital tool for information sharing and for developing long-term possibilities for data analysis and health management. Such a tool is arguably more valuable than filing reports with either specific academic journals or government archives. The digital world is rapidly changing the way we view and engage with our world. It has the capability to encapsulate data on both the micro and macro scales, which should enhance efforts to protect our environment. As noted by Crabtree et al. (2009), such an ecosystem monitoring and analysis system offers the benefit of aiding in decision-making concerning control programmes and conservation planning.

The endemic vulnerability of the high latitudes makes understanding disease transmission all the more important. At the same time, however, increased surveillance and keener interest runs the risk of forcing these quieter latitudes into the mainstream and, with every new pair of eyes wishing to behold the white stillness of the High North, another step is taken towards making this last wilderness less of an endemic area and more the domain of humans. In increasing our understanding of the pathogeneses of the Arctic, we also run the risk of introducing anthrozoonoses into the region before we fully appreciate the impact that they will have.

References

Afema, J. A., Beckmen, K. B., Arthur, S. M., Huntington, K. B., & Mazet, J. A. K. (2017). Disease complexity in a declining Alaskan muskox (*Ovibos Moschatus*) population. *Journal of Wildlife Disease, 53*(2), 311–329.

Alaska Veterinary Pathology Services, P.O. Box 773072, Eagle River, Alaska 99577 USA Alaska Veterinary Pathology Services, P.O. Box 773072, Eagle River, Alaska 99577 USA. Altizer, S., Ostfeld, R. S., Johnson, P. T. J., Kutz, S., & Harvell, C. D. (2013). Climate change and infectious diseases: From evidence to a predictive framework. *Science, 341*(6145), 514–519. https://doi.org/10.1126/science.1239401.

And Todd M. O and Todd M. O, Barbosa, A., & Palacios, M. J. (2009). Health of Antarctic birds: A review of their parasites, pathogens and diseases. *Polar Biology, 32*, 1095–1115. https://doi.org/10.1007/s00300-009-0640-3.

Bradley, M. J., Kutz, S. J., Jenkins, E., & O'Hara, T. M. (2005). The potential impact of climate change on infectious diseases of Arctic fauna. *International Journal of Circumpolar Health, 64*(5), 468–477.

Burek, K. A., Gulland, F., & O'Hara, T. M. (2008). Effects of climate change on Arctic marine mammal health. *Ecological Applications, 18*(2), S126–S134.

Crabtree, R., Potter, C., Mullen, R., Sheldon, J., Huang, S., Harmsen, J., … Jean, C. (2009). A modeling and spatio-temporal analysis framework for monitoring environmental change using NPP as an ecosystem indicator. *Remote Sensing of Environment, 113*(7), 1486–1496.

Davidson, R. K., Handeland, K., & Kapel, C. M. (2008). High tolerance to repeated cycles of freezing and thawing in different trichinella nativa isolates. *Parasitology Research, 103*(5), 1005–1010. https://doi.org/10.1007/s00436-008-1079-0.

Donahoe, S., Rose, K., & Slapeta, J. (2014). *Toxoplasmosis associated with a non-archetypal type II-like Toxoplasma gondii strain in a New Zealand fur seal (Arctocephalus forsteri) from New South Wales, Australia*. Australian Society for Parasitology, 50th Anniversary Conference, Canberra, 2014.

Ecological Applications Ecological applications effects of climate change on Arctice effects of climate change on Arctic, Follmann, E. H., Garner, G. W., Evermann, J. F., & McKeirnan, A. J. (1996). Serological evidence of morbillivirus infection in polar bears (*Ursus maritimus*) from Alaska and Russia. *The Veterinary Record, 138*, 615–618. https://doi.org/10.1136/vr.138.25.615.

Forde, T. L., Orsel, K., Zadoks, R. N., Biek, R., Adams, L. G., Checkley, S. L., … Kutz, S. J. (2016). Bacterial genomics reveal the complex epidemiology of an emerging pathogen in Arctic and Boreal ungulates. *Frontiers in Microbiology, 7*(7), 1759. https://doi.org/10.3389/fmicb.2016.01759.

Geist, V. (1985). On Pleistocene bighorn sheep: Some problems of adaptation, and relevance to today's American megafauna. *Wildlife Society Bulletin (1973–2006), 13*(3), 351–359. Retrieved from http://www.jstor.org/stable/3782506.

Frances M. D. Gulland, Frances M. D. Gulland, Goyert, H. F., Garton, E. O., & Poe, A. J. (2018). Effects of climate change and environmental variability on the carrying capacity of Alaskan seabird populations. *The Auk: Ornithological Advances, 135(4)*, 975–991. https://doi.org/10.1642/AUK-18-37.1.

Hammond, J. A., Pomeroy, P. P., Hall, A. J., & Smith, V. J. (2005). Identification and real-time PCR quantification of Phocine distemper virus from two colonies of Scottish grey seals in 2002. *Journal of General Virology, 86*, 2563–2567. https://doi.org/10.1099/vir.0.80962-0.

Handeland, K., Tengs, T., Kokotovic, B., Vikøren, T., Ayling, R. D., Bergsjø, B., ... Bretten, T. (2014). Mycoplasma ovipneumoniae—A primary cause of severe pneumonia epizootics in the Norwegian muskox (Ovibos moschatus) Population. *PLoS ONE, 9*(9), e106116. https://doi.org/10.1371/journal.pone.0106116.

Härkönen, T., Dietz, R., Reijnders, P., Teilmann, J., Harding, K., Hall, A., ... & Thompson, P. (2006). The 1988 and 2002 phocine distemper virus epidemics in European harbour seals. *Diseases of Aquatic Organisms, 68*(2), 115–130.

Harvell, C. D., Mitchell, C. E., Ward, J. R., Altizer, S., Dobson, A. P., Ostfeld, R. S. & Samuel, M. D. (2002). Climate warming and disease risks for terrestrial and marine biota. *Science, 296*(5576), 2158–2162. https://doi.org/10.1126/science.1063699.

Hossain, K. (2016). Securitizing the Arctic Indigenous peoples: A community security perspective with special reference to the Sámi of the European high north. *Polar Science, 10*(3), 415–424.

Hueffer, K., O'Hara, T. M., & Follmann, E. H. (2011). Adaptation of mammalian host-pathogen interactions in a changing arctic environment. *Acta Veterinaria Scandinavica, 53*, 17. https://doi.org/10.1186/1751-0147-53-17.

Iverson, S. A., Forbes, M. R., Simard, M., Soos, C., & Gilchrist, H. G. (2016). Avian cholera emergence in Arctic-nesting northern common eiders: Using community-based, participatory surveillance to delineate disease outbreak patterns and predict transmission risk. *Ecology and Society, 21*(4), 12. https://doi.org/10.5751/ES-08873-210412.

Jenkins, E. J., Castrodale, L. J., Rosemond, S. J., Dixon, B. R., Elmore, S. A., Gesy, K. M., ... & Thompson, R. C. (2013). Tradition and transition: Parasitic zoonoses of people and animals in Alaska, northern Canada, and Greenland. *Advances in Parasitology, 82*, 33–204.

Kautto, A. H., Alenius, S., Mossing, T., Becher, P., Belák, S., & Larska, M. (2012). Pestivirus and alphaherpesvirus infections in Swedish reindeer (*Rangifer tarandus tarandus* L.). *Veterinary Microbiology, 156*, 64–71.

Kutz, S., Bollinger, T., Branigan, M., Checkley, S., Davison, T., Dumond, M., ... Orsel, K. (2015). Erysipelothrix rhusiopathiae associated with recent widespread muskox mortalities in the Canadian Arctic. *Canadian Veterinary Journal, 56*(6), 560–563.

Kutz, S. J., Checkley, S., Simard, M., Soos, C., Black, S., & Duignan, P. (2013). *The need for a sustainable Arctic wildlife health observation network*. AOS 2013 White Paper. University of Calgary. Retrieved from https://vet.ucalgary.ca/files/vet/kutz-et-al-arctic-wildlife-health-observation-network-wp.pdf.

Kutz, S. J., Hoberg, E. P., Polley, L., & Jenkins, E. J. (2005). Global warming is changing the dynamics of Arctic host-parasite systems. *Proceedings of the Royal Society B: Biological Sciences, 272*, 2571–2576.

Laaksonen, S., Pusenius, J., Kumpula, J., Venäläinen, A., Kortet, R., Oksanen, A., & Hoberg, E. (2010). Climate change promotes the emergence of serious disease outbreaks of filarioid nematodes. *Ecohealth, 7*(1), 7–13. https://doi.org/10.1007/s10393-010-0308-z.

Lafferty, K. D., & Mordecai, E. A. (2016). The rise and fall of infectious disease in a warmer world. *F1000Research, 5*. https://doi.org/10.12688/f1000research.8766.1.

Liu, X., Rohr, J. R., & Li, Y. (2013). Climate, vegetation, introduced hosts and trade shape a global wildlife pandemic. *Proceedings of the Royal Society B: Biological Sciences, 280*, 20122506. https://doi.org/10.1098/rspb.2012.2506.

Marine Mammal Health Marine Mammal Health, Meidenbauer, K. L. (2017). Animal surveillance: Use of animal health data to improve global disease surveillance. *Online Journal of Public Health Information, 9*(1), e147. https://doi.org/10.5210/ojphi.v9i1.7737.

Mörner, T., Obendorf, D. L., Artois, M., & Woodford, M. H. (2002). Surveillance and monitoring of wildlife diseases. *Revue scientifique et technique* [International Office of Epizootics], *21*(1), 67–76.

Ogden, N. H., Lindsay, L. R., Morshed, M., Sockett, P. N., & Artsob, H. (2009). The emergence of Lyme disease in Canada. *Canadian Medical Association Journal, 180*(12), 1221–1222. https://doi.org/10.1503/cmaj.080148.

Regulation (EU) 2016/679 (General Data Protection Regulation) Official Journal L119, 04.05.2016.

Ross, A. B., Johansson, A., Ingman, M., & Gyllensten, U. (2006). Lifestyle, genetics, and disease in Sami. *Croatian Medical Journal, 47*(4), 553–565.

Ruscio, B. A., Brubaker, M., Glasser, J., Hueston, W., & Hennessy, T. W. (2015). One health—A strategy for resilience in a changing Arctic. *International Journal of Circumpolar Health, 74*, 27913. https://doi.org/10.3402/ijch.v74.27913.

Short, E. E., Caminade, C., & Thomas, B. N. (2017). Climate change contribution to the emergence or re-emergence of parasitic diseases, change and infectious diseases. *Infectious Diseases Journal (Auckl), 10*. https://doi.org/10.1177/1178633617732296.

Singh, A. (1989). Digital change detection techniques using remotely-sensed data. *International Journal of Remote Sensing, 10*(6), 989–1003. https://doi.org/10.1080/01431168908903939.

Singh, R., Ranjan, K., & Verma, H. (2015). Satellite imaging and surveillance of infectious diseases. *Journal of Tropical Diseases and Public Health*, S1-004. https://doi.org/10.4172/2329-891x.s1-004.

Smith, D. L., Lucey, B., Waller, L. A., Childs, J. E., & Real, L. A. (2002). Predicting the spatial dynamics of rabies epidemics on heterogeneous landscapes. *Proceedings of the National Academy of Sciences of the United States of America, 99*, 3668–3672.

Taylor, R. H. (1992). New Zealand fur seals at the antipodes Islands. *Journal of the Royal Society of New Zealand, 22*(2), 107–122. https://doi.org/10.1080/03036758.1992.10420809.

Tilman, D., Fargione, J., Wolff, B., D'Antonio, C., Dobson, A., Howarth, R., ... Swackhamer, D. (2001). Forecasting agriculturally driven global environmental change. *Science, 292*(5515), 281–284. https://doi.org/10.1126/science.1057544.

Tokarevich, N. K., Tronin, A. A., Blinova, O. V., Buzinov, R. V., Boltenkov, V. P., Yurasova, E. D., & Nurse, J. (2011). The impact of climate change on the expansion of *Ixodes persulcatus* habitat and the incidence of tick-borne encephalitis in the north of European Russia. *Global Health Action, 4*, 8448. https://doi.org/10.3402/gha.v4i0.8448.

Townsend, A. K., Frett, B., McGarvey, A., & Taff, C. C. (2018). Where do winter crows go? Characterizing partial migration of American Crows with satellite telemetry, stable isotopes, and molecular markers. *The Auk: Ornithological Advances, 135*(4), 964–974. https://doi.org/10.1642/AUK-18-23.1.

Tryland, M., Neuvonen, E., Huovilainen, A., Tapiovaara, H., Osterhaus, A., Wiig, O., & Derocher, A. E. (2005). Serologic survey for selected virus infections in polar bears at Svalbard. *Journal of Wildlife Diseases, 41*, 310–316. https://doi.org/10.7589/0090-3558-41.2.310.

Uhart, M. M., Gallo, L., & Quintana, F. (2018). Review of diseases (pathogen isolation, direct recovery and antibodies) in albatrosses and large petrels worldwide. *Bird Conservation International, 28*(2), 169–196.

U.S. Embassy. 2018. "Canadians to attend U.S. program on health surveillance in the Arctic", 15 May, U.S. Embassy & Consulates in Canada. https://ca.usembassy.gov/canadians-to-attend-u-s-program-on-health-surveillance-in-the-arctic/. Accessed 18 October 2019.

Watson, R. (2011). European parliament criticizes H1N1 pandemic response. *BMJ, 342*, d1639.

Woldstad, T. M. (2010). *Trichodectes canis, an invasive ectoparasite of Alaskan wolves: Detection methods, current distribution, and ecological correlates of spread* (Unpublished master's dissertation). University of Alaska Fairbanks, Fairbanks, Alaska.

Ytrehus, B., Bretten, T., Bergsjø, B., & Isaksen, K. (2008). Fatal pneumonia epizootic in musk ox (*Ovibos moschatus*) in a period of extraordinary weather conditions. *EcoHealth, 5*(2), 213–223. https://doi.org/10.1007/s10393-008-0166-0.

PART V

Conclusions

CHAPTER 13

Moving the Human Being Into the Focus of Cybersecurity

Gerald Zojer

13.1 Introduction

The global trend of rapid digitalisation is also affecting societies in the northernmost parts of Finland, Norway and Sweden—the European High North. While there exist many similarities in terms of how digitalisation impacts societies on a global scale, the peculiarities of a given region result in distinct sets of opportunities and challenges concerning digitalisation facing the local population. Digitalisation is advanced in the Nordic countries, which are considered to be among the global leaders in this regard (European Commission, 2019). Yet, the characteristics of this relatively peripheral region, such as its sparse population, the long distances to urban centres and the harsh climate characterised by long and dark winters, render the European High North distinct. Moreover, it is also the homeland of the Sámi, an indigenous people that includes several small linguistic groups. At the same time, these regional specificities have led to the area being a testing ground for the advancement of

G. Zojer (✉)
Northern Institute for Environmental and Minority Law,
Artic Centre, University of Lapland, Rovaniemi, Finland
e-mail: gerald.zojer@ulapland.fi

M. Salminen et al. (eds.),
Digitalisation and Human Security, New Security Challenges,
https://doi.org/10.1007/978-3-030-48070-7_13

new digital solutions, for example, digital public services (telemedicine, online education, e-governance, etc.) and smart technologies.[1] National and regional digitalisation and development policies, as well as business interests, have all served to advance the digital transformation of the European High North. The motifs behind these trends are the promise of increasing (cost) efficiency, the promotion of (new) business opportunities and the potential for providing (public) services in a digital format.

Due to ever-advancing digitalisation, the functioning of society is becoming increasingly dependent on the availability of sufficient information and communication technologies (ICT) infrastructure. To protect the societal benefits that can be derived from digitalisation, information security and cybersecurity frameworks have been established. Further, to protect the infrastructure deemed critical for the functioning of entire societies, states have endorsed cybersecurity strategies, which essentially shift some of the responsibility for the provision of information security to the national level (Salminen, 2019). However, these strategies are (a) state-centric and (b) focused on protecting information and information infrastructure, which means that they fail to acknowledge the region-specific impacts of digitalisation and they fall short in terms of acknowledging how people experience the increasing use of digital technologies in everyday life. Consequently, these policy frameworks fail to address many of the impacts of digital technologies that are perceived as posing a challenge to human wellbeing (Salminen, 2018; Salminen & Hossain, 2018; Zojer, 2019).

13.2 Digital Infrastructure

The countries of the European High North have well-developed ICT infrastructure, while the population of the region has relatively good ICT skills. In 2019, Finland, Norway and Sweden were among the top-ranked countries included in the European Commission's Digital Economic and Society Index, and they are also considered to be among the global leaders with regards to digitalisation (European Commission, 2019). Yet, the development of the ICT infrastructure in the European

[1] For example, in the Muonio municipality in Finland, there is a smart road test section where different kinds of sensors are tested so as to improve road safety in areas with abundant wildlife or a cold climate. Additionally, autonomous vehicles are tested there in winter road conditions (Finnish Transport Infrastructure Agency, 2017).

High North still lags behind somewhat when compared to the more southern parts of the relevant countries. While the infrastructure in the north is well developed along the main roads, the coverage in more remote areas remains less developed. Equal access to the internet is, however, increasingly important because public services are being digitised and physical services dismantled (see Kirchner, Chapter 6; Kivivirta, Viinamäki, & Selkälä, Chapter 9). This trend is even more significant due to the fact that emergency services are set to become increasingly digitised (Yle, 2019). Moreover, access to ICT and the internet is important if an individual is to be socially included: communication with relatives and friends, access to information, as well as political participation are increasingly reliant on the ability to go online. As Kirchner (Chapter 6) argues, access to the internet is not only essential for avoiding human insecurity, as access to sufficiently fast mobile internet is also a concern from the perspective of international human rights law. Indeed, such access is practically a prerequisite for the effective enjoyment of numerous human rights, for example, the right to receive information and the right to exercise freedom of expression. This fact has been acknowledged by the United Nations Human Rights Council (2016), which has identified the intentional disruption of access to, as well as the prevention of the dissemination of, information from the internet to be a violation of human rights. After all, due to the advanced degree of digitalisation, the distinction between online and offline has diminished (Kirchner, Chapter 6; Shishaev, Fedorov, & Datyev, Chapter 10).

Today, many communities in the European High North receive a broadband connection through mobile networks such as 4G, which has a coverage of over 94% in all the countries of the European High North (European Commission, 2019). Yet, an increasing number of submarine cables are being laid to connect different Arctic regions. The Arctic Council (2017) considers fibre-optic-based infrastructure to be the primary means of connecting continents and Arctic countries (Arctic Economic Council, 2016). When linked to them, northern communities may benefit from the construction of new submarine telecommunications cables, such as the Arctic Connect Project, which aims to connect China and Finland. However, the infrastructure in the north is exposed to the cold climatic conditions typical of the region, which pose specific risks, such as damage caused by the movement of ice. As making repairs to Arctic subsea cables might not be feasible during the winter, reliable backup infrastructure needs to be put in place, which adds to the infrastructure cost. Nonetheless, the absence of an international authority covering submarine-cable-related issues hampers the advancement and sustainability of such projects (Saunavaara, Chapter 8; Shvets, Chapter 7).

13.3 Society and Environment

ICT can clearly contribute to the safeguarding of human rights and, therefore, to human security. However, digitalisation does create new insecurities and/or challenges for people. Not every individual sees the value of digitised services or applications, plans to adopt them or perceives advantages associated with using them. Further, the accessibility of digital services is limited by their availability in local languages, keeping in mind the fact that in the European High North a number of minority languages are spoken[2]; by the need for software and hardware designs that are suitable for individuals with special needs; by the requirement for certain computer skills on the part of individuals; and by the need for the individual to know how to access and use such services. This state of affairs creates new forms of exclusion and challenges for individuals who cannot access digital services, which in turn poses a threat to equality within a society (Gulbrandsen & Sheehan, Chapter 5; Olsén-Ljetoff & Hokkanen, Chapter 11). ICT and/or social media can prove enabling when used as a tool for maintaining minority languages or cultural practices, as they allow, for example, individuals to connect with members of the same language or cultural group, even if the group members have moved away or are separated by long distances. Moreover, digital technologies, such as video cameras, personal computers and the internet, can be used to document, share, store or disseminate local and/or traditional knowledge (Hirshberg & Petrov, 2014; Hossain, 2019; Olsén-Ljetoff & Hokkanen, Chapter 11; Pettersen, 2011). While communities have generally adapted to both cultural influences and technological advancements, some locals perceive, for example, the products of the global entertainment industries to challenge the continuity of local cultures (Aarsæther, Riabova, & Bærenholdt, 2004; Hossain, 2019;

[2] Minority language groups are especially likely to be marginalised in the online world (Dymet, 2019). For example, the Skolt Sámi language is only spoken by around 300 people and so is considered to be "severely endangered" by the United Nations Educational, Scientific and Cultural Organization (Moseley, 2010).

Zojer & Hossain, 2017, p. 45). After all, the utilisation of socio-technic systems shapes the behaviour and choices of people, and it has the potential to disrupt traditional culture and identity (Dutton & Graham, 2014) because technology is socially constructed and embeds culture and politics (e.g. Bijker, Hughes, & Pinch, 2012; Latour, 2004; MacKenzie & Wajcman, 1999; Winner, 1980). The use of social media can prove more problematic in the European High North than in other parts of the relevant countries. Indeed, due to the low population density and the fact that most people know each other in small villages, social media can easily expose individuals, their values and their interests or (political) views, thus posing a threat to privacy (see also Shishaev et al., Chapter 10). ICT can be beneficially used for environmental monitoring, thereby contributing to human health and/or food security (Carleton, Chapter 12). However, both the production and use of ICT have a negative environmental impact. Such technologies contribute, for example, to 4% of the total global greenhouse gas emissions, while their energy consumption is growing by 9% per year (Efoui-Hess, 2019). As the average temperature increase in the Arctic is already twice as high as the global average (ACIA, 2005; Anisimov et al., 2007), the European High North is significantly affected by this trend. This is true even if the impacts of ICT infrastructure are neither direct nor visible and are, therefore, considered to be minimal, for instance, in the case of submarine cables (Saunavaara, Chapter 8; Shvets, Chapter 7).

13.4 A Human-Centred Cybersecurity Approach

ICT offer many opportunities for societal development, although it must be acknowledged that their growing use has led to increased dependency on them. With the aim of protecting and advancing the societal benefits of digitalisation, the countries of the European High North have endorsed a number of information security and cybersecurity policies intended to safeguard the information infrastructure considered critical for the functioning of society. While these strategies do address the importance of information infrastructure, they fall short in terms of addressing the challenges that people face in their digitalising daily life. Moreover, the strategies are state-centric and fail to address region-specific peculiarities (Salminen, 2019; Salminen & Hossain, 2018; Salminen, Zojer, & Hossain, Chapter 2; Zojer, 2019). Yet, considering the fact that digital technologies are promoted as tools that are beneficial for society, the overarching aim of a national cybersecurity strategy should be the advancement of human well-being. Utilising the language of security means that digitalisation is deemed relevant for sound societal development, while the securitisation of an issue by a state can justify the allocation of high levels of resources to it (Buzan, Wæver, & Wilde, 1998, p. 23; cf. Hoogensen Gjørv, 2012, p. 846). Therefore, this volume argues that cybersecurity ought to (a) focus on both the enabling and constraining effects of digitalisation in relation to people's everyday life and (b) be better contextualised so that it also takes regional characteristics into account. After all, the interests, fears and needs of both individuals and communities with regards to digitalisation differ on the basis of, for example, their socio-economic and/or cultural groupings.

Attempts to address the shortcomings of the current understanding of cybersecurity can learn from the development of the security approach in the field of international relations. During the 1980s and 1990s, the security discourse widened. On the one hand, new sectors of security were included, such as environmental and economic security, while on the other hand, the focus of security shifted from national (military) security to individuals and sub-state communities as the referent objects of security (Buzan et al., 1998; Heininen, 2013; Hoogensen Gjørv, Bazely, Goloviznina, & Tanentzap, 2014; McSweeney, 1999). The adoption of the human security concept by the United Nations Development Programme in its 1994 Human Development Report accelerated the move towards a human-centred security approach that is concerned with safeguarding the wellbeing of both individuals and communities.

However, there is no universally accepted definition of human security. In fact, scholars and governmental bodies have used the concept in many different ways. Some interpretations of the concept are rather narrow, focusing on the protection of human beings from physical harm and positioning states as securitising actors, that is, as the actors that define security threats and the measures required to respond to them. It could be argued that such a narrow human security focus is present, to some extent at least, in national cybersecurity strategies, as such strategies refer to certain key areas of human security. According to such a reading, mainstream cybersecurity is concerned with, for example, economic prosperity and economic security or with personal security in terms of privacy. For instance, Filimonova and Portugal-Ramirez (Chapter 3) argue that the Russian information security strategy considers economic, community and political security aspects. Yet, this does not mean that such an approach is human-centric, as it does not allow either individuals or communities to identify challenges, threats or opportunities by themselves, nor does it position the individual or community as the referent object of security.

A more common understanding of human security includes "the protection of the vital core of all human lives in ways that enhance human freedoms and human fulfilment", including "processes that build on people's strengths and aspirations. It means creating political, social, environmental, economic, military and cultural systems that together give people the building blocks of survival, livelihood and dignity" (Commission on Human Security, 2003, p. 4). When applied in the context of digitalisation, Collins (Chapter 4), for example, defines the degree of security according to the degree to which ICT enhance or constrain the agency of human beings. Giving people agency allows them to identify what they perceive to be threats. This may include actions by the state or the international system, especially the international economic system. Such agency enables people to vocalise their interests, needs and fears and, therefore, what needs to be protected. It empowers them to address issues that originate from state actions that are detrimental to their wellbeing. It hence renders the human being a securitising actor as well as a referent object of security (Collins, Chapter 4; Hoogensen Gjørv, 2012; Hossain, Zojer, Greaves, Roncero, & Sheehan, 2017; Salminen, Zojer, & Hossain, Chapter 2). A broad understanding of human security conceptualises culture, identity and human progress as things that need to be protected, and it thus positions a multi-disciplinary approach to human wellbeing at the very core of security. While a human security approach also provides the political flexibility necessary for advancing human rights, human rights provide the theoretical and legal rigidity required for advancing human security.

13.5 Conclusion

Mainstream cybersecurity approaches aim to safeguard both information and information infrastructure so that societies can benefit from the opportunities offered by digitalisation. As several chapters in this volume have underlined, ICT can promote human security in many ways. However, the relevant strategies fall short in terms of addressing the challenges and insecurities that are created by digitalisation. Considering the significant impacts of digital technologies on societal development, policies and strategies must be more sensitive to the needs and challenges that people experience in their digitalising everyday life. Moreover, due to region-specific characteristics and cultural practices, policies in this field need to be more sensitive to local peculiarities.

Many issues related to digitalisation are similar on a global level. However, this volume presents numerous examples of how digitalisation affects a particular region—the European High North—given its specific cultural, socio-economic and environmental characteristics—in a distinct way. The chapters in this book highlight how various areas of human security are affected by digitalisation, and they also provide evidence of how mainstream cybersecurity approaches fail to recognise the challenges and insecurities that originate from digitalisation.

This volume argues that a human security approach serves to broaden the scope of cybersecurity. It focuses on the security of both individuals and (sub-state) communities. Such a multi-disciplinary, human-centred cybersecurity approach has the breadth and flexibility necessary to identify and address region-specific desires, needs and challenges, which builds on people's experiences with digitalisation. This approach treats people and communities as both objects and subjects of cybersecurity. It does not question the importance of traditional cybersecurity concerns, but instead widens the questions of security that are related to digitalisation to also include its constraining effects on people. When applied in practice, a human-centred cybersecurity approach can help with the design of meaningful and targeted policies that position human wellbeing at the very core of digitalisation matters.

References

Aarsæther, N., Riabova, L., & Bærenholdt, J. O. (2004). Community viability. In N. Einarsson, J. Nymand Larson, A. Nilsson, & O. R. Young (Eds.), *Arctic human development report* (pp. 139–154). Akureyri Island: Stefansson Arctic Institute.

ACIA. (2005). *Arctic climate impact assessment*. Cambridge and New York, NY: Cambridge University Press.

Anisimov, O. A., Vaughan, D. G., Callaghan, T. V., Furgal, C., Marchant, H., Prowse, T. D., ... Walsh, J. E. (2007). Polar regions (Arctic and Antarctic). In M. L. Parry, O. F. Canziani, J. P. Palutikof, P. J. van der Linden, & C. E. Hanson (Eds.), *Climate change 2007: Impacts, adaptation and vulnerability* (pp. 653–685). Cambridge: Cambridge University Press.

Arctic Economic Council. (2016). *Arctic broadband. Recommendations for an interconnected Arctic.* Telecommunications Infrastructure Working Group, Arctic Economic Council. Retrieved from http://arcticeconomiccouncil.com/wp-content/uploads/2017/01/AEC-Arctic-Broadband.pdf.

Bijker, W. E., Hughes, T. P., & Pinch, T. (Eds.). (2012). *The social construction of technological systems: New directions in the sociology and history of technology* (anniversary ed.). Cambridge, MA: MIT Press.

Buzan, B., Wæver, O., & de Wilde, J. (1998). *Security: A new framework for analysis.* Boulder, CO: Lynne Rienner Publishing.

Commission on Human Security. (2003). *Human security now.* New York, NY: United Nations.

Council, Arctic. (2017). *Telecommunications infrastructure in the Arctic: A circumpolar assessment.* Fairbanks and Tromsø: Arctic Council Secretariat.

Dutton, W. H., & Graham, M. (2014). Introduction. In M. Graham & W. H. Dutton (Eds.), *Society and the internet: How networks of information and communication are changing our lives* (pp. 1–22). Oxford and New York, NY: Oxford University Press.

Dymet, M. (2019). Digital language divide in the European High North: The level of online presence of minority languages from Northern Finland, Norway and Sweden. *The Yearbook of Polar Law Online, 10*(1), 245–274. https://doi.org/10.1163/22116427_010010012.

Efoui-Hess, M. (2019). *Climate crisis: The unsustainable use of online video: The practical case for digital sobriety.* Retrieved from https://theshiftproject.org/wp-content/uploads/2019/07/2019-02.pdf.

European Commission. (2019). *The Digital Economy and Society Index (DESI).* Retrieved from https://ec.europa.eu/digital-single-market/en/desi.

Finnish Transport Infrastructure Agency. (2017, November 7). *The finnish transport agency's intelligent road Aurora has been opened, and testers from all over the world are now welcome to try out new technology in the fields of automated driving and infrastructure.* Välyä. Retrieved from https://vayla.fi/web/en/-/the-finnish-transport-agency-s-intelligent-road-aurora-has-been-opened-and-testers-from-all-over-the-world-are-now-welcome-to-try-out-new-technology-i.

Heininen, L. (2013). 'Politicization' of the environment, and environmental politics and security in the Circumpolar North. In B. S. Zellen (Ed.), *The fast-changing Arctic: Rethinking Arctic security for a warmer world* (pp. 35–55). Calgary, AB: University of Calgary Press.

Hirshberg, D., & Petrov, A. N. (2014). Education and human capital. In J. Nymand Larsen & G. Fondahl (Eds.), *Arctic human development report: Regional processes and global linkages* (pp. 347–396). Copenhagen: Nordic Council of Ministers.

Hoogensen Gjørv, G. (2012). Security by any other name: Negative security, positive security, and a multi-actor security approach. *Review of International Studies, 38*(04), 835–859. https://doi.org/10.1017/S0260210511000751.

Hoogensen Gjørv, G., Bazely, D. R., Goloviznina, M., & Tanentzap, A. J. (Eds.). (2014). *Environmental and human security in the Arctic*. London and New York, NY: Earthscan from Routledge.

Hossain, K. (2019). The evolving information-based society and its influence on traditional culture: Framing community culture and human security of the Sámi in the European High North. *The Yearbook of Polar Law Online, 10*(1), 275–296. https://doi.org/10.1163/22116427_010010013.

Hossain, K., Zojer, G., Greaves, W., Roncero, J. M., & Sheehan, M. (2017). Constructing Arctic security: An inter-disciplinary approach to understanding security in the Barents region. *Polar Record, 53*(1), 52–66. https://doi.org/10.1017/S0032247416000693.

Human Rights Council. (2016). *The promotion, protection and enjoyment of human rights on the internet* (Resolution Adopted by the Human Rights Council No. A/HRC/RES/32/13). Retrieved from https://undocs.org/en/A/HRC/RES/32/13.

Latour, B. (2004). *Politics of nature: How to bring the sciences into democracy*. Cambridge, MA: Harvard University Press.

MacKenzie, D. A., & Wajcman, J. (Eds.). (1999). *The social shaping of technology* (2nd ed.). Buckingham and Philadelphia, PA: Open University Press.

McSweeney, B. (1999). *Security, identity, and interests: A sociology of international relations*. Cambridge and New York, NY: Cambridge University Press.

Moseley, C. (Ed.). (2010). *Atlas of the world's languages in danger* (3rd ed.). Retrieved from http://www.unesco.org/culture/languages-atlas/en/atlasmap.html.

Pettersen, B. (2011). Mind the digital gap: Questions and possible solutions for design of databases and information systems for Sami traditional knowledge. *Dieđut, 1*, 163–192.

Salminen, M. (2018). Digital security in the Barents region. In K. Hossain & D. Cambou (Eds.), *Society, environment and human security in the Arctic Barents region* (pp. 187–204). London and New York, NY: Routledge.

Salminen, M. (2019). Refocusing and redefining cybersecurity: Individual security in the digitalising European High North. *The Yearbook of Polar Law Online, 10*(1), 321–356. https://doi.org/10.1163/22116427_010010015.

Salminen, M., & Hossain, K. (2018). Digitalisation and human security dimensions in cybersecurity: An appraisal for the European High North. *Polar Record, 54*(2), 1–11. https://doi.org/10.1017/S0032247418000268.

Winner, L. (1980). Do artifacts have politics? *Daedalus, 109*(1), 121–136.

Yle. (2019, December 9). Police response times up to an hour slower in rural areas. *Yle*. Retrieved from https://yle.fi/uutiset/osasto/news/police_response_times_up_to_an_hour_slower_in_rural_areas/11108792.

Zojer, G. (2019). The interconnectedness of digitalisation and human security in the European High North: Cybersecurity conceptualised through the human security lens. *The Yearbook of Polar Law, 10*(1), 297–320. https://doi.org/10.1163/22116427_010010014.

Zojer, G., & Hossain, K. (2017). *Rethinking multifaceted human security threats in the Barents region: A multilevel approach to societal security*. Rovaniemi: University of Lapland Printing Centre.

The manufacturer's authorised representative in the EU is Springer Nature Customer Service Centre GmbH, Europaplatz 3, 69115 Heidelberg, Germany. If you have any concerns regarding our products, please contact ProductSafety@springernature.com

Printed and bound by CPI Group (UK) Ltd, Croydon, CR0 4YY
28/09/2026
02224060-0001